Premises and Conclusions

PREMISES AND CONCLUSIONS

Symbolic Logic for Legal Analysis

ROBERT E. RODES, JR.
Notre Dame Law School

HOWARD POSPESEL
University of Miami

Prentice Hall, Upper Saddle River, New Jersey 07458

Library of Congress Cataloging-in Publication Data

Rodes, Robert E.
    Premises and conclusions: symbolic logic for legal analysis/
Robert E. Rodes and Howard Pospesel.
     p. cm.
    Includes index.
    ISBN 0-13-262635-7
    1. Law—Methodology. 2. Logic, Symbolic and mathematical.
3. Symbolism in law. I. Pospesel, Howard II. Title
K213.R59 1997
340'. 11—dc20
                                         96–33572
                                            CIP

Acquisitions editor: Angela Stone
Editorial/production supervision: Alison D. Gnerre
Cover designer: Karen Salzbach
Buyer: Nicholas Sklitsis

© 1997 by Prentice-Hall, Inc.
Simon and Schuster/A Viacom Company
Upper Saddle River, New Jersey 07458

All rights reserved. No part of this book may be reproduced, in any form or by
any means, without permission in writing from the publisher.

Printed in the United States of America
10 9 8 7 6 5 4 3 2 1

ISBN: 0-13-262635-7

Prentice-Hall International (UK) Limited, *London*
Prentice-Hall of Australia Pty. Limited, *Sydney*
Prentice-Hall Canada, *Toronto*
Prentice-Hall Hispanoamericana, S.A., *Mexico*
Prentice-Hall of India Private Limited, *New Delhi*
Prentice-Hall of Japan, Inc., *Tokyo*
Simon & Schuster Asia Pte. Ltd., *Singapore*
Editora Prentice-Hall do Brasil, Ltda., *Rio de Janeiro*

CONTENTS

FOREWORD   vii

PREFACE   ix

1   INTRODUCTION   1
- §1.1   Arguments, 1
- §1.2   Key Terms, 4
- §1.3   Logic and the Law, 8

2   PROPOSITIONAL LOGIC   17
- §2.1   And, 17
- §2.2   If, 27
- §2.3   Not, 47
- §2.4   Or, 59
- §2.5   Iff, 73
- §2.6   Derived Rules, 80
- §2.7   Truth Tables, 96

3   PREDICATE LOGIC   113
- §3.1   Symbolization, 114
- §3.2   Proofs, 124
- §3.3   Propositional Analogues, 139
- §3.4   Interpretations, 152
- §3.5   Relations, 160
- §3.6   Identity, 183
- §3.7   Analysis by Instantiation, 199

4   DISPUTATIONS   207
- §4.1   What Are Disputations? 207
- §4.2   Denials, 209
- §4.3   Pleading, 214
- §4.4   Distinguo, 220
- §4.5   Affirmative Defenses, 230

5 PITFALLS AND PARADOXES 235

§5.1 Necessary and Sufficient Conditions, 235
§5.2 Problematic Universal Statements, 245
§5.3 The Paradox of Material Implication, 259
§5.4 Quantifier Scope, 273
§5.5 Existential Import, 285

6 INTENSIONAL CONTEXTS 297

§6.1 Extensional and Intensional Contexts, 298
§6.2 Recognizing Intensional Contexts, 302
§6.3 Examples and Strategies, 311
§6.4 Intent, Risk, and Inchoate Crimes, 319

7 CONCLUSION 333

§7.1 Analogical and Deductive Legal Reasoning, 333

APPENDICES

1 Deontic Logic, 336
2 Formation Rules, 347
3 Alternative Symbols, 350
4 Solutions to Starred Exercises, 351
5 Rules of Inference, 374

INDEX 377

TABLE OF CASES 385

TABLE OF STATUTES AND RULES 387

FOREWORD

David T. Link
The Joseph A. Matson Dean and Professor of Law
Notre Dame Law School

I have always enjoyed the writings of Professors Howard Pospesel and Robert Rodes. This present collaboration between two fine minds is a most significant contribution to legal analysis and thinking. It will be important to lawyers, judges, legislators, law administrators, legal educators, law students, pre-law students and virtually everyone who has an occasion to analyze and debate the law.

To properly analyze the law law professionals must employ at least two different reasoning styles. On the one hand, it is important for them to be capable of making most of their judgments and decisions based on intuition. Certainly the great law thinkers, whether they acknowledge it or not, make judgments based on what is just, fair, equitable, morally right, workable, achievable, and conforms to one or more other instinctive criteria. It is the usual mode of law thinking. However, because of the stare decisis system, and other standards for uniformity and consistency in interpreting the law, every law professional must prove his or her position logically.

Judging intuitively, but proving logically; when I was a practitioner, I noticed that constantly switching from one mode to the other was a dilemma for most law professionals. This is probably because cognitive and intuitive are at opposite sides of the Jungian circle of personality types, and one's natural reasoning inclinations tend to interfere with the responsibility to use both the intuitive and logical modes. The expression "hard cases make bad law" may be little more than a recognition that in some cases lawyers and judges have failed to logically prove an intuitive decision, while in others, logic was not tested against instinct.

As an educator also, I have seen the intellectual struggle between judging intuitively and proving logically. Some applicants to law school with excellent undergraduate grade-point averages do poorly on the Law School Admissions Test. Virtually all such applicants argue that they do poorly on standardized tests. I suppose they see this as a stress situation which will not affect them in their law studies, but it seems clear to educators that such students actually have a skills deficiency in a very important field. They really don't know much about formal logic. And those students who do get into law school may experience difficulty because of an imbalance of intuitive and logical skills. Some students do well in the first year and taper off in the advanced courses. Others struggle in the basic courses before excelling in the second and third years. We used to think that this was because some students lost interest or energy after their first year, while others finally "hit their stride." I now think that the explanation lies elsewhere—that the first year of law school is traditionally logic based, while intuition is much more of the analytical core of most advanced courses.

Now comes a book that brings the two forms of reasoning some synergy. I have for many years recommended Howard Pospesel's books, along with Layman Allen's logic game, *WFF 'N Proof,* to pre-law students who were struggling with the LSAT, and to first-year law students who were simply struggling. Students who took my recommendations had amazing results. Now there is something better. There is a book which teaches lawyers, law students, and pre-law students about logic, but in the language of legal dilemma. The book which follows makes the application of symbolic logic to law very clear and demonstrates how easy it is to prove logically an instinctively good decision.

PREFACE

Two theses underlie this volume: that much legal reasoning is deductive, and that deductive legal reasoning can be adequately analyzed with the methods of standard symbolic logic if those resources are employed imaginatively.

In Chapters Two and Three we present the elements of standard logic and apply those techniques to legal materials. Beginning with the final section of Chapter Three we expand the emphasis on using symbolic logic in the analysis of legal arguments.

We have tried to make our system neutral as between competing jurisprudential theories. Accordingly, we have had to proceed without committing ourselves to any definite empirical equivalent or any single logical form for terms importing legal obligation. It is for this reason that we have relegated deontic logic to an appendix. Some jurisprudential theories would permit a much more extensive use of it but no such use would be theory-neutral.

Some problems that logicians tend to pass over or leave to more advanced courses have been introduced here because they are too frequently encountered in legal reasoning to be ignored or postponed. Among these are the paradox of material implication and the problem of intensionality. We cannot claim to have solved these problems, but we believe we have developed strategies for avoiding them without running away from legal arguments that threaten to raise them. The book also contains several innovations, including a streamlined technique for evaluating certain predicate arguments (Section 3.3), a method for simplifying the analysis of predicate arguments (Section 3.7), and a scheme for representing denials, pleadings, and affirmative defenses (Chapter Four).

We believe that the book can be used effectively for either a class or a seminar in law school, or a component of an introductory Legal Method course. It can also be recommended to law students for independent study. In an undergraduate course, it will provide a useful preparation for the study of law and for the LSAT. While it covers topics that are not generally found in introductory Logic courses, it is well within the ability of intelligent and interested undergraduates. Instructors, however, who wish to omit any of the more uncommon topics can do so without affecting the continuity of the book. The first three chapters are presupposed by all subsequent sections, but the succeeding chapters may be freely omitted or abridged.

Rudyard Kipling, in one of his *Just So Stories*, has a hedgehog teaching a tortoise to curl up into a ball, and the tortoise in return teaching the hedgehog to swim. In the end, they find they have both turned into armadillos. After a number of years of collaboration between logician and lawyer, we sometimes feel that something of the kind has happened to us. If our colleagues from one discipline or the other find too little of the tortoise or the hedgehog here, and too much of the armadillo, we apologize.

We are grateful to Leonard Carrier, Ed Erwin, Alan Goldman, Susan Haack, Robert Lane, Craig Munns, and Harvey Siegel for reading and commenting on portions of the book, and to Craig Munns for checking the *Solutions Manual*. We wish also to thank our friends and former students for providing examples discussed in the book: Pat Anastasio, Adelsi Arias, Philip Cahill, Catherine Ceder, Mary Ann Clark, Nancy Sue Day, Ruthanne Deakin, Dania Garcia, Alan Goldman, Mitchell Johnson, Jason Jonas, Sarah Marrero, Janice-Marie McDonald, Andrew Moran, Leo De la Peña, Richard Perez, Daisy Rosen, Rochelle E. Rubin, Tom Runyan, Miguel F. Ugarte, and Judith Taylor Walter. Each of the following supplied several examples: James DeMarco, Margarita Kristoff, Rachel Mathason, Denise Oehmig, Supryia Ray, Marlene Rodriguez, Jonathan Sanders, Rey Sio, Richard R. Talda, and Harold Zellner. We are particularly indebted to David Marans for providing seventeen of our examples. We would also like to thank the publisher's reviewers, Scott Brewer, Harvard Law School; Patricia Sayre, Saint Mary's College; and Ron Laymon, Ohio State.

1

Introduction

§1.1 ARGUMENTS

One of us (we will not tell you which) wandered absent-mindedly into a library rest room one day without paying too much attention to the sign on the door, and was suddenly gripped by a neurotic fear that he had walked into the wrong place. The anxiety dissolved when he spotted a urinal. In an instant he made a logical inference that can be given this formal statement:

> **All rest rooms with urinals are men's rooms.**
> **This rest room has a urinal.**
> **Therefore, it is a men's rest room.**

All of us in our everyday lives are constantly drawing logical inferences in this way, and encountering logical arguments advanced by others.

In the legal profession, of course, we have even more occasion than other people to use logical arguments. Both common law and statutory rules are regularly applied through reasoning like the "Rest Room" argument above:

**Contracts without consideration are unenforceable.
This contract is without consideration.
Therefore, it is unenforceable.**

**Everyone who had more than $2,000 of income last year must file a tax return.
Anne had more than $2,000 of income last year.
Therefore, Anne must file a tax return.**

A pleading is a special form of logical argument called an *enthymeme*. Here is Pleading Form 9, which is attached to the Federal Rules of Civil Procedure:

COMPLAINT FOR NEGLIGENCE

1. Allegation of jurisdiction.
2. On June 1, 1936, in a public highway called Boylston Street in Boston, Massachusetts, defendant negligently drove a motor vehicle against plaintiff who was then crossing said highway.
3. As a result plaintiff was thrown down and had his leg broken and was otherwise injured, was prevented from transacting his business, suffered great pain of body and mind, and incurred expenses for medical attention and hospitalization in the sum of one thousand dollars.
Wherefore plaintiff demands judgment against defendant in the sum of . . . dollars and costs.

This is called an "enthymeme" because one of the premises is missing. It is understood, rather than stated. The whole argument made by this pleading is:

**Whoever negligently drives a motor vehicle against another person and causes damages is liable for damages.
The defendant negligently drove a motor vehicle against another person and caused damages.
Therefore (="wherefore"), the defendant is liable for damages.**

The first premise doesn't have to be stated in the pleading because it is a principle of law that everyone understands from first-year Torts. A defendant who claimed that there is no such principle of law would move to dismiss the complaint under Federal Rule 12(b)(6) for "failure to state a claim upon which relief can be granted." In this case, of course, such a motion would be frivolous. Once you supply the missing premise, the argument of Form 9 becomes just like the "Rest Room" argument and the other two arguments set out above.[1]

---

[1]The reasoning in Form 9 also permits a deeper analysis. See exercise 6 on page 175.

Arguments can be assessed in two quite different ways. On the one hand, we may determine the truth or falsity of the premises occurring in an argument. This may be called assessing the argument's *content*. On the other hand, we may determine whether the conclusion follows from the premises. When we evaluate an argument on this score we are assessing its *form*. In general, *logic* is concerned with evaluating the form rather than the content of arguments. When we remove the content from the "Rest Room" argument, this form remains:

All *D* are *E*
f is *D*
Therefore *f* is *E*

The other arguments set out above have the same form. This form is a good one; if an argument exhibits it, its conclusion follows from its premises.

This distinction between form and content may be further illustrated with the help of these examples:

ARGUMENT ONE	ARGUMENT TWO
All Protestants are Lutherans.	All Lutherans are Protestants.
Some Protestants are Methodists.	Some Protestants are Methodists.
Thus, some Lutherans are Methodists.	Thus, some Lutherans are Methodists.

Since these two arguments have false conclusions, each must incorporate some error. The mistake in argument one is obvious: its first premise is false (a mistake of content). The premises of argument two are true. However, it has a different type of defect: its conclusion does not follow from its premises. The argument exhibits the following defective form:

All *D* are *E*
Some *E* are *F*
Thus, some *D* are *F*

What is the form of argument one?

An argument does not establish the truth of its conclusion unless it has *both* correct content *and* a good form. One purpose of education is to prepare people for the job of evaluating the content of the arguments they encounter. One purpose of a logic course (and perhaps the main purpose) is to enhance people's native ability to evaluate the form of the arguments they meet. Another purpose of a logic course is to impress upon the student the importance of identifying arguments and assessing them in both of these ways (in terms of content and form).

But why should we want to assess the arguments that come our way? Because assessing the arguments for or against a proposition or claim is often our best way of arriving at the truth. This is, of course, what a judge does, and lawyers must do it too if they are to persuade judges or predict what judges will decide. It is also what people do in their practical affairs, and lawyers must do it if they are to advise people about those affairs. Also, practical considerations aside, people—and lawyers are people too—like to believe what is true and hate to find they have believed what is false.

§1.2 KEY TERMS

Before proceeding further, it will be helpful to clarify some of the terms that have occurred above and to introduce additional ones. Each term in this list requires comment:

argument
conclusion
premise
statement
validity
invalidity
entailment
logical equivalence
truth
falsity
deduction
induction

Argument is a fundamental logical concept. An argument is a set of statements, one of which (the *conclusion*) supposedly follows from the others (the *premises*). (Including the word 'supposedly'[2] broadens the definition so that it covers bad arguments as well as good ones.) Arguments are the expressions

---

[2]Throughout the book we employ a convention concerning quotation marks that is regularly observed in logic as well as other disciplines. When a word (or other bit of language) is being mentioned, we place it within single quotation marks. Compare the following sentences:
  (A) Elephants are gray.
  (B) 'Elephants' is a noun.
Logicians call this the "use-mention distinction." Note that sentence A *uses* (but does not mention) a three-syllable word, while B *mentions* that word.

in language of pieces of reasoning. The constituents of arguments are statements. A *statement* is a sentence that is true or false. The class of statements is roughly the class of declarative sentences. Some examples:

STATEMENTS	NONSTATEMENTS
Dianne Feinstein is a Democrat.	Do you have the time?
All contracts without consideration are unenforceable.	All rise.
	Look out!
Some Egyptians are blonds.	Objection overruled.[3]

The two main branches of logic are deductive and inductive logic. In deductive logic we are concerned with dividing arguments into two classes: (1) those whose conclusions follow with necessity from their premises; and (2) all other arguments. Arguments in the first class are called *valid*; those in the second class are called *invalid*. Here are two equivalent definitions of 'validity':

A valid argument is one having a form such that *if* all its premises are true, *then* its conclusion must also be true.

A valid argument is one having a form such that it is impossible that its premises are all true and its conclusion false.

Because "validity" is such an important concept in deductive logic there are several equivalent ways of saying that an argument is valid. For example:

The premises of the argument *entail* its conclusion.
The conclusion of the argument *follows with necessity* from its premises.
The conclusion of the argument *follows logically* from its premises.

Note that 'validity' as we have defined it is a matter of form, not content. A valid argument may contain false statements, and an invalid one may be composed exclusively of truths. The examples displayed at the top of page 6 illustrate this. Both statements in the "Mailer" argument are false. In spite of this defect, it exhibits impeccable form. If the premise of the "Mailer" argument *were* true, then the conclusion *would also be* true. Both statements in the

---

[3]This is what is called a *performative utterance*. By saying "Objection overruled," the judge is actually overruling the objection, rather than telling us about something that has already happened. So the sentence is neither true nor false because it does not purport to convey information. Therefore it is not a statement. If you want to learn more about performative utterances, read *How to Do Things with Words* by J.L. Austin (Cambridge, MA: Harvard University Press, 1962).

VALID	INVALID
Norman Mailer is a Greek and he is also a dentist. Therefore, he is a Greek.	Toni Morrison is an American. Therefore, she is an American and she is also a novelist.

"Morrison" argument are true. But in spite of this virtue, its form is poor. The conclusion of the "Morrison" inference simply does not follow from its premise. The "Mailer" and "Morrison" arguments have these forms:

"MAILER"	P and Q. Therefore P.	valid form
"MORRISON"	P. Therefore P and Q.	invalid form

Within the class of invalid arguments one can find all the possible combinations of the truth and falsity of constituent statements. Here are several examples:

TRUE PREMISE AND TRUE CONCLUSION	Some Supreme Court Justices are not women. Hence, some women are not Supreme Court Justices.
FALSE PREMISE AND TRUE CONCLUSION	All Protestants are Lutherans. Hence, all Lutherans are Protestants.
TRUE PREMISE AND FALSE CONCLUSION	Dianne Feinstein is either a Senator or she is a Representative. Hence, she is a Representative.

In the class of valid arguments one can find all combinations except one. Two examples are displayed on the next page. Among valid arguments one combination—namely, "(all) true premises and a false conclusion"—is excluded by the definition of 'validity'. The virtue of validity lies in this feature—that it is *truth preserving*. A valid argument can never take you from truth to falsity.

As we noted above, the relation of *entailment* is linked to the concept of validity. One statement entails a second just when the argument whose premise is the first statement and whose conclusion is the second is valid. We can extend the notion so that it holds between sets of statements. Set one en-

TRUE PREMISE AND TRUE CONCLUSION	Some Lutherans are Democrats. Hence, some Democrats are Lutherans.
FALSE PREMISE AND TRUE CONCLUSION	Dianne Feinstein is a Representative. Hence, she is either a Senator or she is a Representative.

tails set two if every argument having set one as its premise set and a statement from set two as its conclusion is valid.

Logical equivalence is mutual entailment. Two statements (or two sets of statements) are logically equivalent if and only if each entails the other. Logical equivalence is an important concept in legal logic because two statements (or sets of statements) have the same content if and only if they are logically equivalent.

The term 'valid' as we have defined it does not exactly correspond to the term 'valid' of ordinary discourse.[4] (Since the technical term has a precise meaning and the ordinary term does not, their meanings could not correspond completely.) The principal difference is that the ordinary term encompasses both form and content while the technical term is restricted to form alone. When we assess a piece of reasoning as "valid" in nontechnical discourse, we mean to praise both its structure and its substance. The 'valid' of everyday language would not be applied to arguments (such as the "Mailer" and "Morrison" examples) that are defective in either respect.

To avoid confusion, let's agree (while we are engaged in logic) to apply the adjective 'valid' ('invalid') to arguments and not statements, and to apply 'true' ('false') to statements and not arguments. A statement is *true* when it accords with the actual state of affairs, while one that does not accord is *false*.

The distinction between *deduction* and *induction* has been drawn in different ways in different disciplines and at different times. One popular distinction is that "deduction" is reasoning from the general to the particular, and "induction" is reasoning in the opposite direction. Legal scholars have often drawn the distinction in this way. For instance, Sir Carleton Kemp Allen, in his classic *Law in the Making*,[5] begins his discussion of precedent by saying that "the process of judicial decision may be regarded as either *deductive* or *inductive*." On the first theory, the judge's "decision is deduced directly from general to particular—from the general legal rule to the particular circumstances before him." But on the second theory, the judge has to search for the legal rule "in the learning and dialectic which have been applied to particular facts. Thus he is always reasoning inductively."

---

[4] Of course, both meanings are quite different from what we mean when we speak of a valid or invalid legal transaction.
[5] (Oxford, 1927), pp. 107–8.

For a variety of reasons, contemporary legal scholars tend to use terms other than "induction" to describe the process of arriving at legal rules through analysis of precedents, and contemporary logicians decline to distinguish deduction from induction in terms of the concepts of "particular" and "general." The distinction is now customarily drawn by logicians in terms of the kind of connection that links the premises of an inference to its conclusion. In a valid *deduction* the connection is one of *necessity*, while in an *induction* the connection is one of (greater or lesser) *probability*. Here is an example:

The stores at this end of the shopping mall are crowded now.
Thus, the stores at the other end are also crowded.

The premise makes the conclusion probable, but it is possible for the conclusion to be false even though the premise is true (that is, the conclusion does not follow from the premise with necessity). The inference will be classified as an "induction" by logicians.[6]

While logicians and legal theorists offer different definitions for these concepts, in most cases they will agree on whether a given inference is to be counted deductive or inductive. Inferences of both kinds abound in everyday life and in legal contexts. The emphasis in this book will be on the study of deductive legal inferences. The reason is simple: at the present time logicians know much more about deductive logic than they do about inductive logic.

§1.3 LOGIC AND THE LAW

Deductive reasoning has occupied a place in the legal arena since the founding of legal institutions. Here is an example that reaches back about 2,400 years. One of the charges brought against Socrates by Meletus (in the trial that led to the death of Socrates) was that he was an atheist. Here is a portion of the trial "transcript," according to Socrates' follower, Plato (who we know was present at the trial):[7]

---

[6]We note two views about deduction and induction that are gaining acceptance among many logicians. The first idea is that the distinction between deduction and induction is *not* a distinction between types of arguments at all but rather between types of *standards for assessing arguments*. On this view you can describe an argument as "deductively valid" (because it meets the strict standards of deductive logic) or "inductively strong" (because it meets the less stringent standards of inductive logic) or neither (because it fails to satisfy either set of standards), but you cannot correctly describe it as "deductive" or "inductive." This approach was developed by Brian Skyrms in *Choice and Chance* (Belmont, CA: Dickenson, 1966).

The second view gaining acceptance holds that reasoning comes in not two varieties, but three; the third argument type is called "abduction" or "inference to the best explanation." The premises of an abduction describe phenomena in need of explanation, and the conclusion is a hypothesis that purports to provide that explanation. This approach was pioneered by the nineteenth-century American logician Charles Peirce.

[7]"Apology" in *The Last Days of Socrates*, trans. Hugh Tredennick (Baltimore: Penguin Books, 1989), pp. 57–58.

Socrates: *Tell me honestly, Meletus, is that your opinion of me? Do I believe in no god?*
Meletus: *No, none at all, not in the slightest degree. . . .*
Socrates: *Is there anyone who believes in supernatural activities and not in supernatural beings?*
Meletus: *No.*
Socrates: *But if I believe in supernatural activities, it follows inevitably that I also believe in supernatural beings . . . Do we not hold that supernatural beings are either gods or the children of gods? Do you agree or not?*
Meletus: *Certainly.*
Socrates: *Then if I believe in supernatural beings, as you assert, if these supernatural beings are gods in any sense we shall reach the conclusion . . . If on the other hand these supernatural beings are bastard children of the gods by nymphs or other mothers, as they are reputed to be, who in the world would believe in the children of gods and not believe in the gods themselves?*

Socrates' reasoning in this passage is captured in these two connected arguments:

If there are supernatural activities, then there are supernatural beings.
Any supernatural being is either a god or the child of a god.
If there are children of gods, then there are gods.
So, that there are supernatural activities implies that there are gods.

Socrates believes that there are supernatural activities.
Socrates has proved that the existence of supernatural activities implies that there are gods. [See previous argument.]
One who believes a claim and proves that it implies another claim also believes the second claim.
Therefore, Socrates believes that there are gods.

Are these arguments valid or invalid? How can we tell for sure? This text provides the tools for answering these questions. (These two arguments are exercises in Sections 3.2 and 3.5.)

Now for a more recent example of the relevance of logic to law, consider this exchange between Congressmen Latta, McClory, and Mann at a House Judiciary Committee meeting concerning Richard Nixon's involvement in the Watergate scandal:[8]

Clerk: *Mr. McClory moves to postpone for ten days further consideration of whether sufficient grounds exist for the House of Repre-*

---

[8]Meeting of July 26, 1974. This example was given by Daniel Bonevac in *Deduction: Introductory Symbolic Logic* (Palo Alto, CA: Mayfield Publishing Company, 1987), pp. 2–3.

sentatives to exercise constitutional power of impeachment unless by 12 noon, eastern daylight time, on Saturday, July 27, 1974, the president fails to give his unequivocal assurance to produce forthwith all taped conversations subpoenaed by the committee which are to be made available to the district court pursuant to court order in United States v. Mitchell . . .

Mr. Latta: *. . . I just want to call [Mr. McClory's] attention before we vote to the wording of his motion. You moved to postpone for ten days unless the president fails to give his assurance to produce the tapes. So, if he fails tomorrow, we get ten days. If he complies, we do not. The way you have it drafted I would suggest that you correct your motion to say that you get ten days providing the president gives his unequivocal assurance to produce the tapes by tomorrow noon.*

Mr. McClory: *I think the motion is correctly worded; it has been thoughtfully drafted.*

Mr. Latta: *I would suggest you rethink it . . .*

Mr. Mann: *Mr. Chairman, I think it is important that the committee vote on a resolution that properly expressed the intent of the gentleman from Illinois [Mr. McClory], and if he will examine his motion he will find that the words fail to need to be stricken . . .*

Mr. McClory: *If the gentleman will yield, the motion is correctly worded. It provides for a postponement for ten days unless the president fails tomorrow to give his assurance, so there is no postponement for ten days if the president fails to give his assurance, just one day. I think it is correctly drafted. I have had it drafted by counsel, and I was misled originally, too, but it is correctly drafted. There is a ten-day postponement unless the president fails to give assurance. If he fails to give it, there is only a twenty-four-hour or there is only a twenty-three-and-a-half-hour day.*

Misunderstanding runs rampant throughout this exchange.

(1) Congressman Latta maintains that Congressman McClory's motion (to postpone for ten days unless the president fails to assure) has these consequences:

If the president fails to give his assurance, we get a ten-day postponement. If he gives his assurance, we do not get a ten-day postponement.

Is he correct about this?

(2) Mr. Latta further says that Mr. McClory's intent is captured by the following (rather than by the original motion):

The committee will postpone for ten days providing the president gives his assurance.

Is Mr. Latta right to assert that this statement is distinct from the original motion?

(3) Congressman Mann says that Mr. McClory's intent is captured by the following:

> The committee will postpone for ten days further consideration unless the president gives his assurance.

Is Mr. Mann correct?

(4) Finally, Mr. McClory claims that his motion has the following consequence:

> There is no ten-day postponement if the president fails to give his assurance.

Is this in fact entailed? If you feel unsure about the answers to these four questions, the study of legal logic can benefit you.

Logical words like 'unless', 'fails', 'not', and 'providing' play a critical role in legal discourse. Understanding their meaning and the meaning of sentences structured around them is essential for those who draft, read, or apply the law as well as those who advise clients on matters pertaining to the law. One purpose of this text is to elucidate these and similar terms. In an exercise following Section 2.7 you will employ the techniques of symbolic logic to clarify the statements made in this committee meeting and to answer the questions raised above.

The ability to detect, reveal, and eliminate ambiguity and vagueness in statutes and other kinds of legal discourse is an important legal skill. Symbolic logic has a role to play in this activity, especially when the type of ambiguity involved is *amphiboly* (i.e., ambiguity grounded in poor sentence structure). For an example consider the Fifth Amendment to the United States Constitution:

> *No person shall be held to answer for a capital, or otherwise infamous crime, unless on a presentment or indictment of a grand jury, except in cases arising in the land or naval forces, or in the militia, when in actual service in time of war or public danger . . .*

It is not clear whether the expression 'when in actual service in time of war or public danger' attaches just to 'in the militia' or to all of 'in the land or naval forces, or in the militia'. This unclarity makes a big difference, especially to someone "in the land or naval forces" who has been accused of committing a crime during peacetime. In their opinions written for *O'Callahan* v. *Parker,* 395 U.S. 258 (1969), Justices Douglas and Harlan disagreed on the interpretation of the amendment. It is a significant fact about logic that no acceptable formula of logic will preserve the amphiboly present in the amendment. When you symbolize the amendment, you are forced by the rules governing formula construction to opt for one meaning or the other. Learning how to symbolize sentences can enhance your ability to detect amphiboly—both in other people's writing and in your own. (The amphiboly in the Fifth Amendment is the subject of an exercise following Section 5.2.)

Consider as a final example this piece of legislative drafting:

Draft A
No judge shall grant probation to any person who shall have been convicted of robbery, burglary or arson, and who at the time of the perpetration of said crime or any of them or at the time of his arrest was himself armed with a deadly weapon (unless at the time he had a lawful right to carry the same), nor to a defendant who used or attempted to use a deadly weapon upon a human being in connection with the perpetration of the crime for which he was convicted, nor to one who in the perpetration of the crime of which he was convicted wilfully inflicted great bodily injury or torture, nor to any such person who has ever been previously convicted of a felony in this state or previously convicted in any other place of a public offense which would have been a felony if committed in this state.

Now consider John Doe who is to be sentenced for a felony conviction for selling heroin. Doe was unarmed at the time of the sale, but was carrying a gun unlawfully when arrested. He has one previous conviction on a charge that is a misdemeanor in this state. Would Doe be ineligible for probation under the terms of the proposed legislation displayed above? You may find that the opacity of the draft makes the question hard to answer.

Would Doe be ineligible for probation according to the following proposed legislation?

Draft B
A person shall be ineligible for probation if one or more of the following apply:

(1) he has been convicted of robbery, burglary or arson, and, at the time of the perpetration of the crime or at the time of arrest, was armed with a deadly weapon which he had no lawful right to carry;

(2) in perpetrating the crime of which he was convicted he used (or attempted to use) a deadly weapon upon a human being or wilfully inflicted great bodily injury or torture;

(3) he has been previously convicted of a felony in this state;

(4) he has been previously convicted in another place of a public offense which would have been a felony if committed in this state.

Doe does not satisfy the first description because he has not been convicted of robbery, burglary, or arson, and he does not satisfy any of the other descriptions either. So, he is not rendered ineligible for probation under the terms of Draft B. But what about Draft A? As it turns out, the two drafts are logically equivalent; they have just the same content, and therefore Doe is not rendered ineligible under the terms of either draft.

You can see that the ability to recognize the equivalence of, or differences between, two complex statements would be important both to a legislator and to a lawyer who is trying to understand and apply legislation or execute a legal document. Determining whether two statements of policy or two descriptions of fact are equivalent, consistent, or incompatible, or whether one entails the other are matters belonging to the realm of logic. It is the aim of this book to enhance your ability to determine the logical content of blocks of prose. In this way your comprehension of legislation or legal instruments will be enhanced. The value to a student of the law of this increased comprehension is obvious.

The language and techniques of symbolic logic explained in the following chapters will play an important part in developing your logical skills. As an example of how symbolic logic can aid one's logical intuition, note that discovering that Drafts A and B above have *the same symbolization* in logical notation is one way of coming to see that they are equivalent in content. That symbolization, soon to be intelligible, is this:

(x)({[Ax & (Bx v Cx) & −Dx] v Ex v Fx v Gx v Hx} → −Ix)[9]

The logic presented in this book is called "symbolic" logic because it employs symbols such as these in the evaluation of arguments. There are several different systems of symbolic logic in use; the one presented here is known as the "standard" system. This standard system is simpler than the others, but in our exposition we have introduced some methods and strategies not usually found in the presentation of this system. We believe that these refine-

---

[9]   Ax= x has been convicted of robbery, burglary or arson

  Bx= x was armed with a deadly weapon at the time of the perpetration of the crime

  Cx= x was armed with a deadly weapon at the time of his arrest

  Dx= x had a lawful right to carry the deadly weapon that he was carrying

  Ex= in perpetrating the crime of which he was convicted x used or attempted to use a deadly weapon upon a human being

  Fx= in perpetrating the crime of which he was convicted x willfully inflicted great bodily injury or torture

  Gx= x has been previously convicted of a felony in this state

  Hx= x has been previously convicted in another place of a public offense which would have been a felony if committed in this state

  Ix= x is eligible for probation

  (x)= for each x

   v= or

   −= not

$p \rightarrow q$= if p then q

ments make it possible to express most deductive legal arguments in the standard system. The elements of standard symbolic logic are presented in the next two chapters; the refinements are introduced in subsequent chapters.[10]

EXERCISES

*1. Identify the false statements in this set. (Solutions for starred exercises are provided in Appendix Four.)
   (a) Some arguments with true premises and true conclusions are valid, while others are invalid.
   (b) Some arguments with true premises and false conclusions are valid, while others are invalid.
   (c) Some arguments with false premises and true conclusions are valid, while others are invalid.
   (d) Some arguments with false premises and false conclusions are valid, while others are invalid.
   (e) An argument with one premise is valid if and only if the premise entails the conclusion.
   (f) Two statements are logically equivalent if and only if each entails the other.
   (g) Some false statement is logically equivalent to a true statement.

2. Logic is concerned with determining the content of statements. Consider, as an example, this statement made by an Orange County (Florida) sheriff's spokesperson:

   > Unless they're in your house, you're not allowed to shoot anyone unless you're in imminent danger.[11]

   What does this statement mean? Does it have the same content as any of the following statements?
   (a) You are allowed to shoot people if they are in your house and you are in imminent danger.
   (b) You are allowed to shoot people only if they are in your house and you are in imminent danger.

---

[10]Logicians from Aristotle onward have plied their craft with the help of special symbols. But modern symbolic logic was chiefly developed in the work of certain nineteenth-century mathematician-logicians, especially George Boole, Augustus De Morgan, Charles Peirce, and Gottlob Frege. Their work was extended by Bertrand Russell and Alfred North Whitehead in the monumental *Principia Mathematica*, published between 1910 and 1913. See William and Martha Kneale's *The Development of Logic* (Oxford: The Clarendon Press, 1962) for a valuable treatment of these historical developments.

[11]"Robbery Victim May Face Charges," *Miami Herald* (November 12, 1994), p. 5B.

(c) You are allowed to shoot people if they are in your house or you are in imminent danger.

(d) You are allowed to shoot people only if they are in your house or you are in imminent danger.

(e) If a person is not in your house and you are not in imminent danger, you are not allowed to shoot that person.

Symbolic logic provides tools for answering questions like these. (The sheriff's statement is the subject of an exercise following Section 3.5.)

*3. Logic is also concerned with determining what follows from what. How keen is your logical intuition about such matters? This exercise will help you find out. For each of the following pairs of sentences, judge whether the second sentence follows from (is entailed by) the first, and whether the first follows from the second. Answers are provided in Appendix Four. Incidentally, all of these problems can be solved with the techniques presented in the book.

(a) Sue takes the LSAT or the GMAT, but not both.
Sue takes the LSAT if and only if she does not take the GMAT.

(b) If suicide is permissible so is assisted suicide, and vice versa.
If either suicide or assisted suicide is impermissible, then neither is permissible.

(c) Left turns permitted from left lane only.
Only left turns permitted from left lane.

(d) If anyone is found guilty, Laurence Powell will be.
Only Laurence Powell will be found guilty—if indeed anyone is found guilty.

(e) If some people are in fetters, we are all in fetters.
For any two people, if the first is in fetters, so is the second.

(f) If some members of the jury have been bribed, then some of them will vote acquittal.
If every jury member has been bribed, then all of them will vote acquittal.

(g) There is no substance that possesses no attribute.
Every attribute belongs to some substance.

(h) (*lyrics*) "Everybody loves my baby, but my baby don't love nobody but me."
I am my baby.

(i) Whoever degrades someone other than himself or herself degrades Walt Whitman.
Whoever degrades someone other than Walt Whitman degrades Walt Whitman.

(j) Waldo admires all who do not admire themselves.
Waldo admires some self-admirer.

2
Propositional Logic

§2.1 AND

We will call a statement that has no parts which are themselves statements *simple*. Some examples:

Sandra O'Connor is a Supreme Court Justice.

The federal courts have exclusive jurisdiction over patent cases.

Any statement that is not simple we will call *compound*. Some sample compound statements:

The lawyers made their closing arguments *and* the court instructed the jury.

If the defendant pleads guilty, *then* there will be no trial.

Either the defendant is guilty *or* the prosecution witnesses are lying.

The defendant will be found guilty *if and only if* the prosecution proves its case beyond a reasonable doubt.

Each of these four statements consists of two simple statements and an expression that connects the simple statements. These connecting expressions (which are italicized in the above examples) are called, naturally enough, *statement connectives*. The following statement is also compound:

It is not the case that defendant's plea of guilty is the result of force.

It consists of the locution 'it is not the case that' and a simple statement. The locution is called a *statement connective* in spite of the fact that it attaches to one statement instead of connecting two.

There are many statement connectives in English (or in any natural language), but these five are of special importance in logic:

and
if . . . then
it is not the case that
either . . . or
if and only if

This chapter treats these five connectives, the compound statements that can be constructed with their help, and the arguments that involve these compound statements. The study of the forms of arguments whose validity or invalidity rests on the pattern of compound statements composing them is called *the logic of statements* or, more commonly, *propositional logic*. (We will use the terms 'statement' and 'proposition' interchangeably for now. A distinction that will become relevant later is drawn on page 301.)

A statement consisting of two (or more) constituent statements joined by the connective 'and' is called a *conjunction*. The component statements are termed *conjuncts*; they may be simple or compound statements. A sample conjunction:

(S1) Defendant refused to ACCEPT the tender and refused to make the CONVEYANCE.

We facilitate our work by employing symbols. Two kinds of symbols will suffice for the present. (1) *Capital letters* are used to abbreviate simple statements. Where possible, the letter chosen to abbreviate a statement will be the first letter of a prominent word in the statement. In this text the words selected to supply statement abbreviations are printed entirely in capital letters; this will help us keep track of the symbols chosen. An example: the first conjunct in S1 will be symbolized by the letter 'A'. Although the convention we have chosen for indicating letters involves printing *words* in capital letters, do not think that an abbreviating letter represents a word or any linguistic expression shorter than a statement. Simple statements (occurring in one problem) that convey the same information or have the same content will be abbreviated by the same letter. For example, these statements would be symbolized with the same letter:

> Marvin struck Norton.
> Marvin hit Norton.
> Norton was struck by Marvin.
> Marvin committed a battery on Norton.
> Norton was the victim of a battery perpetrated by Marvin.

(2) The statement connective 'and' is abbreviated by the *ampersand* symbol (&). The ampersand is preceded by a capital letter representing the first conjunct and followed by a capital letter representing the second conjunct. S1 is symbolized by the formula F1.

> (F1) **A & C**

This formula is read "A and C" or "A ampersand C."

Several English connective expressions are equivalent (from the standpoint of logic) to the connective 'and'. Some appear (in italics) in the following list of statements:

> OPRAH stays *but* PAULA leaves.
> Oprah stays, *however* Paula leaves.
> Oprah stays, *moreover* Paula leaves.
> Oprah stays *although* Paula leaves.
> Oprah stays *yet* Paula leaves.
> Oprah stays *even though* Paula leaves.

Each of these statements is symbolized 'O & P'. The italicized expressions are not completely synonymous; nevertheless, there is a common factor in their meanings that justifies our treating them in a group. (Did you notice that the preceding sentence is a conjunction?) The meaning of each term is such that a person who assents to a compound statement built around that term is automatically committed to accepting both of the constituent statements.

Note that the information contained in the above statements could also be conveyed in the following way:

Oprah stays. Paula leaves.

But because these are two separate sentences, we will not symbolize them by the single formula 'O & P'. Rather, we will use the two formulas: 'O' and 'P'. In making this choice, we are not following any principle of logic. We are simply adopting the following convention for the sake of uniformity:

One English sentence will be represented by *one* formula.

Of course, if 'O' and 'P' are both true, then 'O & P' is true also, and vice versa.

The above conjunctive statements about Oprah and Paula are easy to pull apart because each conjunct has its own subject and predicate. But a conjunctive statement can have one subject with several predicates, like S2:

(S2) Sam opened his book and began to read.

Or it can have one predicate and multiple subjects, like S3:

(S3) Joe and Mabel are witnesses.

Before these can be symbolized, we have to turn them into English sentences that can be pulled apart:

(S2A) Sam OPENED his book and Sam began to READ.
(S3A) JOE is a witness and MABEL is a witness.

Then we can employ the following formulas:

(F2) O & R
(F3) J & M

But not all statements that look like S2 or S3 can be broken up in this way. For instance, S4 can, but S5 cannot:

(S4) The voters of Indiana elect 50 SENATORS and 100 REPRESENTATIVES to the Legislature.

(S5) The Indiana Legislature consists of 50 senators and 100 representatives.

So S4 can be symbolized with a compound formula, F4:

(F4) S & R

(S = The voters of Indiana elect 50 senators to the Legislature, R = The voters of Indiana elect 100 representatives to the Legislature.) But S5 will have to be symbolized with a single letter. That is because it is not equivalent to the following conjunction:

> The Indiana Legislature consists of 50 senators and the Indiana Legislature consists of 100 representatives.

Each conjunct of this latter conjunction is false and so the conjunction is false. Therefore, it cannot be equivalent to the true S5.

Determining whether a particular statement is compound like S4 or simple like S5 is a matter of carefully analyzing what it means. We hope that this kind of analysis of legal statements will help you get into the habit of considering the precise meanings of words, and so help make you a better lawyer.

Before concluding this discussion of symbolization we should say something about determining the overall structure of arguments. How does one tell which statements in an argument function as premises and which as the conclusion? Of course, there is no substitute for understanding the passage that contains the argument. However, certain key terms are fairly reliable indicators. These expressions follow premises and introduce conclusions:

> therefore, so, hence, thus, consequently, it follows that, . . . proves that, . . . shows that

And these expressions regularly follow conclusions and introduce premises:

> since, because, for

Neither of these lists is exhaustive.

It often helps to identify the conclusion first. As there is only *one* conclusion per argument, the remaining statements (assuming they all belong to the argument) will be premises. The conclusion may appear at the beginning, in the middle, or at the end of an argument.

The logical content of the 'and' connective is captured in the following two observations:

(1) If a conjunction is true, then so is each of its conjuncts.

(2) If two (or more) statements are true, then so is the conjunction formed from them.

If you have doubts about either of these observations, test it with examples. We can base two *rules of inference* on these observations. A rule of inference is a rule that sanctions a move from one (or more) statement(s) to another statement—a rule that allows you to deduce or derive one statement from others. In order to be correct or valid a rule of inference must be *truth preserving*—that is to say, the rule must never permit the derivation of falsity from truth.

These two valid rules of inference are based on the observations made in the preceding paragraph:

The Ampersand Out Rule (&O): From a conjunction derive any conjunct.

The Ampersand In Rule (&I): From two or more statements derive a conjunction formed from them.

The first of these rules is called "Ampersand Out" because it licenses the move *from* a formula containing an ampersand, and the second rule is called "Ampersand In" because it licenses a move *to* a formula that contains an ampersand. Each of these rules provides a certain degree of flexibility. For instance, the Ampersand Out Rule justifies the derivation of F2 from F1 and also justifies the derivation of F3 from F1.

(F1) A & C
(F2) A
(F3) C

The Ampersand In Rule sanctions the derivation of F6 from F4 and F5, but it also sanctions the derivation of F7 from F4 and F5.

(F4) B
(F5) D

(F6) B & D
(F7) D & B

The adoption of these two rules of inference leads to the introduction of the logical technique known as *formal proof*.[1] This is a technique for demonstrating the validity of arguments. A formal proof consists of a numbered list of statements where the statements at the top of the list are the premises of the argument and the statement at the bottom is the conclusion of the argument. Every statement beneath the premises is justified by appeal to some correct rule of inference (such as the two Ampersand Rules). (This explanation of "formal proof" will be tightened as we go along.)

We illustrate the technique by proving the validity of this symbolized argument:

A & C ⊢ C & A

(The *turnstile* symbol, '⊢', is short for 'therefore'; it is placed before the conclusion of a symbolized argument.) The formal proof for this argument:

(1) A & C   A
(2) A   1 &O
(3) C   1 &O
(4) C & A   3,2 &I

The column on the right is the *justification* column. It gives for each statement the basis for its inclusion. We adopt 'A' as an abbreviation for 'Introduced as an assumption'. (For now, all assumptions in a proof will be premises of the argument being validated.) The justification entry on line 2 ('1 &O') is short for 'Derived from line 1 by the Ampersand Out Rule'. The fourth justification entry is short for 'Derived from lines 3 and 2 by the Ampersand In Rule'. Completing the proof demonstrates the validity of the argument. Why? If an argument can be divided up into a series of steps and if each of the steps is correct, then the argument itself is also correct. Showing that all of the steps are correct serves to show that the argument is correct. We know that the steps are correct because they are made in accordance with valid rules of inference. This method of formal proof is the most important device logicians have for demonstrating validity. It has applications in every branch of logic.

---

[1] The type of proof system presented in this book was introduced by Gerhard Gentzen in 1934. Gentzen's approach differed from earlier proof systems chiefly in two ways: he provided an "In" rule and an "Out" rule for each logical connective, and he did not employ axioms. Logicians call his approach "natural deduction." See "Untersuchungen über das logische Schliessen," *Mathematische Zeitschrift*, vol. XXXIX (1934), pp.176–210 and 405–431.

Note that the above proof would remain correct if the statements on lines 2 and 3 were switched. For most valid arguments there will be more than one correct formal proof. Logicians prefer shorter proofs to longer ones and organized proofs to disorganized ones, but for our purposes any proof that is free of error will be fully acceptable.

For a second example of a formal proof we establish the validity of this argument:

E & F, G & H ⊢ E & H

(The premises of this argument are separated by a comma.)

(1)   E & F   A
(2)   G & H   A
(3)   E   1 &O
(4)   H   2 &O
(5)   E & H   3,4 &I

Some conjunctive statements contain more than one 'and' (or equivalent words); statement S8 is an example:

(S8) I have PLANTED, Apollos WATERED, but God gave the INCREASE. (I Cor. 3:6)

S8 is symbolized as follows:

(F8) P & W & I

S8 (F8) is a *double* (or *multiple*) conjunction. There is no limit to the number of conjuncts that a conjunction may contain.

The Ampersand Out and Ampersand In Rules apply to conjunctions containing two or more ampersands. We can illustrate this by constructing a proof for this argument:

P & W & I ⊢ I & W & P

(1)   P & W & I   A
(2)   P   1 &O
(3)   W   1 &O

(4) I   1 &O
(5) I & W & P   4,3,2 &I

So far in this section we have concentrated on valid arguments involving conjunctions, but of course there are invalid arguments as well. For example:

**Jesse Jackson is a DEMOCRAT.
So, Jesse Jackson is a Democrat and he is a CALIFORNIAN.**

In symbols:

D ⊢ D & C

We know this argument to be invalid because we recognize that it has a true premise and a false conclusion. The conclusion is false because one of the things it claims (that Jackson is a California resident) is false. Of course, many invalid arguments have true conclusions. We need another way to demonstrate the invalidity of such arguments. In Section 2.7 we describe such a method.

If you attempt to construct a formal proof for the "Jackson" argument, you will be unable to do so (without making a mistake). But note that failure to complete a proof is *not* proof of invalidity. That is so because there is no foolproof way to show that the failure is due to the invalidity of the argument rather than to the lack of ingenuity of the proof constructor.

EXERCISES

1. Symbolize each statement using the suggested notation.
    (a) (*newspaper column*) "I am BLACK and I am CUBAN."
   *(b) (*I Cor. 4:10*) "We are FOOLS for Christ's sake, but you are WISE in Christ."
    (c) (*children's book*) "And although he was a very SMALL ghost, Georgie had a really BIG idea."
    (d) (*Julius Caesar*) "I CAME, I SAW, I conquered." (Let 'A' abbreviate 'I conquered'.)
    (e) (*folk song*) "Jelly, jelly, jelly, got jelly on my MIND."
   *(f) (*folk song continued*) "Jelly KILLED my father, DROVE my mother stone blind, But I LOVE my jelly roll."

(g) (*Ted Kennedy*) "I am an AMERICAN and a CATHOLIC; I LOVE my country and TREASURE my faith."

2. Indicate for each sentence whether it is simple or a conjunction. If it is a conjunction, identify its conjuncts. Count a sentence as a conjunction if it can be readily restated in conjunctive form.
  (a) (*tax instructions*) "Line 5c is the sum of lines 5a and 5b."
  *(b) Sam and Sue are husband and wife.
  (c) Both Sam and Sue are married.
  (d) Jones, Smith, and Brown constitute the Board of Directors of the company.
  (e) Mary's mother and father are lawyers.
  (f) (*Colin Powell*) "It [success] is the result of preparation, hard work, learning from failure."
  (g) (*U.S. Const., Art. III, Sec. 1*) "The judicial Power of the United States shall be vested in one Supreme Court, and in such Inferior Courts as the Congress may from time to time ordain and establish."
  *(h) (*U.S. Const., Amendment XIV*) "All persons born or naturalized in the United States and subject to the jurisdiction thereof, are citizens of the United States and of the State wherein they reside."

3. Construct formal proofs of validity for each of the following symbolized arguments:
  (a) A, B ⊢ A & B
  *(b) C & D & E ⊢ D
  (c) F & G ⊢ F & F
  (d) H & I, J & K ⊢ H & K & J & I
  (e) L & M & N ⊢ M & N & L

4. F.R.C.P. Rule 56(c) concerning summary judgment reads in part as follows:

> *The judgment sought shall be rendered forthwith if the pleadings, depositions, answers to interrogatories, and admissions on file, together with the affidavits, if any, show that there is no genuine issue as to any material fact and that the moving party is entitled to a judgment as a matter of law.*

  (a) State what the court is to look at in passing on a motion for summary judgment. Symbolize your statement. (P = The court looks at pleadings, D = The court looks at depositions, I, F, A)
  *(b) State the finding that will result in a summary judgment. Symbolize your statement. (F = There is no genuine issue as to any material fact, E = The moving party is entitled to a judgment as a matter of law)
  (c) A student in a first-year moot-court argument tried to support a summary judgment by showing that there was no genuine issue as to

any material fact. On the basis of your answer to part (b) of this exercise, show the invalid reasoning into which the student had evidently fallen.

§2.2 IF

A statement composed of two constituent statements and the connective 'if . . . then' is called a *conditional*. The component statement that precedes 'then' is called the *antecedent*, and the component following 'then' is called the *consequent*. The antecedent and the consequent may be either simple or compound. A sample conditional:

(S1) If it looks like RAIN I will carry an UMBRELLA.

The statement connective 'if . . . then' is abbreviated by the *arrow* symbol ($\rightarrow$). The arrow is preceded by a letter (or letters) representing the antecedent statement and followed by a letter (or letters) representing the consequent. S1 is symbolized by the formula F1.

(F1) R → U

This formula is read "If R then U" or "R arrow U."
English provides many ways of expressing S2.

(S2) If the PROCESS is properly served, then the court will acquire JURISDICTION.

The sentences in the following list are equivalent to S2:

If the process is properly served the court will acquire jurisdiction.

The court will acquire jurisdiction *if* the process is properly served.

Provided that[2] the process is properly served the court will acquire jurisdiction.

The court will acquire jurisdiction *provided that* the process is properly served.

---

[2]Stylists oppose the use of 'provided' as synonymous with 'if'. See the entry in Fowler, *Modern English Usage*. Nevertheless, we believe the arrow captures fairly well the common use of the term.

Should the process be properly served, the court will acquire jurisdiction.

Proper serving of the process *implies*[3] that the court acquires jurisdiction.

Proper serving of the process will *result in* (*bring about, lead to*, etc.) the court's acquiring jurisdiction.

Each of these statements is correctly symbolized by F2.[4]

(F2) P → J

It is important to realize that none of these statements is correctly rendered by F2X.

(F2X) J → P

The order given to the components of a conditional is critical. We may employ this symbolization guide:

The statement following the word 'if' (or its synonym 'provided that') is the antecedent; accordingly, its abbreviation is placed before the arrow.

If you apply this guide to the statements above that contain 'if' or 'provided that', you will arrive at F2 (not F2X) in each case. (Unfortunately, the guide breaks down when applied to the locution 'only if'. We discuss that expression below.)

Consider this sentence from a newspaper sports story:

A Houston loss to Denver and a Miami victory over New England puts the playoff game in the Orange Bowl.

This amounts to a conditional with a conjunctive antecedent:

---

[3]Logicians often use 'implies' and 'entails' as synonyms, but we attach a weaker meaning to 'implies'. In our usage, P *implies* Q when (and only when) 'If P, then Q' is true, whereas P *entails* Q when (and only when) 'If P, then Q' is necessarily (or logically) true. 'Smith signed the Declaration of Independence' implies (but does not entail) 'Smith is now dead'.

[4]In §5.3 we examine the question whether F2 *fully* captures the meaning of these sentences. At any rate, F2 suffices for most logical purposes.

(S3) If HOUSTON loses to Denver and MIAMI beats New England, then the playoff game will be in the ORANGE Bowl.

This statement is not satisfactorily symbolized as F3X.

(F3X)   H & M → O

The problem with F3X is that it is *amphibolous*. (Amphiboly is ambiguity due to poor grammatical structure.) F3X contains no indication whether it is a conditional or a conjunction; that is, it does not indicate whether the dominant connective symbol is the arrow or the ampersand. Compare F3X with F4 and F5.

(F4)   (H & M) → O
(F5)   H & (M → O)

In F4, we call the arrow the *dominant* symbol, because it applies to the whole formula. Its *scope* is greater than that of the ampersand, which applies only to a part of the formula. In F5, on the other hand, the ampersand is the dominant symbol. It is the placing of the parentheses that makes the difference. Parentheses play the same role in logic that they do in arithmetic. In addition to parentheses we shall use brackets and braces as grouping symbols (or "groupers").

Formula F4 is the correct symbolization of S3. Formula F5 does not have the same content as F4 and is not an acceptable symbolization of S3. Note that 'Houston loses to Denver' follows logically from F5, but does not follow from either F4 or S3. Formula F3X is not even *well formed*; that is, it is not a properly constructed formula. Every formula containing an arrow and an ampersand must also contain a pair of grouping symbols to be well formed. (The rules that determine whether a formula of propositional logic is well formed are stated in Appendix Two.)

Some sentences contain more than one conditional connective; S6 is an example:

(S6) If Ellen is LATE paying this installment, then if the company DEMANDS repayment of the entire loan, she will be required to PAY the balance immediately.

Statement S6 is a *double* conditional. A statement may contain any number of conditional connectives. S6 is symbolized by F6:

(F6)   L → (D → P)

Parentheses are placed around 'D → P' to indicate that the first arrow has greater scope than the second. Shifting the parentheses to the left would change the meaning of the formula to the extent that it would no longer represent S6. Double (or multiple) conditionals always require groupers to identify the dominant arrow. In contrast to multiple conditionals, multiple conjunctions (like 'P & W & I') do not require internal groupers to eliminate amphiboly; nevertheless, groupers may always be added to multiple conjunctions, and adding them can facilitate logical techniques developed in this chapter.

Part of the logical content of the 'if . . . then' connective is captured in this observation:

If a conditional is true and its antecedent is also true, then its consequent will be true.

If you have doubts about this observation, test it with examples. We can base a rule of inference on this observation:

The Arrow Out Rule **(→O): From a conditional and a statement identical to its antecedent derive a statement identical to its consequent.**

This is called the "Arrow Out Rule" because it licenses the move from a formula containing an arrow (and a second formula) to a formula not containing an arrow.

We may use this rule to prove the validity of the following symbolized arguments:

A, A → B, B → C ⊢ C

(1) A   A
(2) A → B   A
(3) B → C   A
(4) B   2,1 →O
(5) C   3,4 →O

D & (D → E) ⊢ D & E

(1) D & (D → E)   A
(2) D   1 &O

(3)   D → E   1 &O
(4)   E 3,2 →O
(5)   D & E   2,4 &I

The Arrow Out Rule corresponds to an argument pattern that has been known since the Middle Ages as "*modus ponens*" (Latin for "the affirming form"):

If *P* then *Q*
P
So, *Q*

This argument pattern is very common in both legal and nonlegal contexts. Because of its simplicity people often use it without recognizing that they are reasoning. Here is a sample *modus ponens* argument from *Ragan* v. *Merchants Transfer & Warehouse Co.*, 337 U.S. 530 (1949):

> Guaranty Trust Co. *v.* York *applied that principle to statutes of limitations on the theory that where one is barred from recovery in the state court, he should likewise be barred in the federal court. It is conceded that if the case were in a Kansas court it would be barred. The theory of* Guaranty Trust Co. *v.* York *would therefore seem to bar it in a federal court.*

We readily extract this *modus ponens* argument:

If the case would be barred in a KANSAS court, then it will be barred in a FEDERAL court.
The case would be barred in a Kansas court.
Therefore, the case will be barred in a federal court.

The argument may be symbolized:

K → F, K ⊢ F

It is important to distinguish *modus ponens* from the following invalid inference pattern:

If *P* then *Q*
Q
So, *P*

The second premise of an argument having this form is identical to the *consequent* of the first premise. This pattern is appropriately labelled *the fallacy of affirming the consequent*. The contrast between the two argument patterns is shown clearly in the following:

$$\text{modus ponens}: P \rightarrow Q, P \vdash Q \text{ (valid)}$$
$$\text{affirming the consequent}: P \rightarrow Q, Q \vdash P \text{ (invalid)}$$

Consider an example of an argument that commits the fallacy of affirming the consequent:

If Sam and Charlie are partners, they share profits and losses.
Sam and Charlie share profits and losses.
Hence, they are partners.

Note that the first premise asserts that partnership implies sharing; it does *not* assert that sharing implies partnership. Thus, when the first premise is coupled with the second, it does not follow that they are partners.

The connective expression 'only if' causes a good deal of confusion. Consider S7 for example:

(S7) The defendant was found GUILTY only if the State's principal witness TESTIFIED.

Does S7 have the content of S8 or S9?

(S8) The defendant was found guilty if the State's principal witness testified.

(F8) $T \rightarrow G$

(S9) If the defendant was found guilty then the State's principal witness testified.

(F9) $G \rightarrow T$

As S8 and S9 do not have the same content, it is important to determine which one is equivalent to S7.

It is a common mistake to believe that S7 and S8 are equivalent in meaning. If they were equivalent in meaning, then the word 'only' would contribute nothing to the meaning of S7, but it obviously does affect S7's meaning.

Suppose that the defendant was found guilty even though the State's principal witness did not testify. Clearly, that outcome would prove S7 to be false; but it would not falsify S8. S8 concerns the consequences of the witness testifying, not the consequences of the witness's failing to testify. So it is not invalidated by what happens when the witness does not testify. If a possible outcome would falsify one sentence and not another, then the two sentences have different content. Note that the possible situation described would falsify S9; this provides some evidence that S7 and S9 are equivalent.

A stronger argument for the equivalence of S7 and S9 centers on their relationship to S10.

(S10) If the State's principal witness failed to testify, then the defendant was not found guilty.

It should be clear that S7 and S10 express the same thought. S9 and S10 also express the same thought (this will be demonstrated in Section 2.3). Therefore, S7 and S9 express the same thought.

Each of the three preceding paragraphs expresses a deductive argument that shows either that S8 is *not*, or that S9 *is*, an adequate reformulation of S7. (Two of these arguments will reappear below as exercises.) So the formula that symbolizes S9 (G $\rightarrow$ T) also represents S7. We can adopt the following guide in symbolizing "only if" sentences:

The expression 'only if' introduces consequents.

A sentence having the form '*P* only if *Q*' is properly symbolized '*P* $\rightarrow$ *Q*'.

Two other problematic conditional connective expressions are 'sufficient condition' and 'necessary condition'. If event (or state of affairs) A is a sufficient condition for event (state) B, then A's occurrence ensures B's occurrence. If event (state) C is a necessary condition for event (state) D, then D cannot occur in C's absence. Since D cannot occur in C's absence, the occurrence of D ensures the occurrence of C; therefore, D is a sufficient condition for C. Similarly, since A's occurrence ensures B's occurrence, A cannot occur in B's absence, and B is a necessary condition for A. The general rule is, therefore, that if E is a sufficient condition for F, then F is a necessary condition for E, and vice versa. So every conditional statement expresses both a sufficient condition (in the antecedent) and a necessary condition (in the consequent).

Sufficient conditions need not be necessary conditions and necessary conditions need not be sufficient conditions. Two examples:

(S11) Fred's being found GUILTY of fraud is a sufficient condition for his being DISBARRED.

(S12) Nancy's SCORING above 175 on the LSAT is a necessary condition for her being ADMITTED to law school.

S11 means approximately the same as S13 and is, therefore, symbolized by F13. S12 means roughly the same as S14 and, so, is symbolized by F14.

(S13) If Fred is found guilty of fraud he will be disbarred.
(S14) If Nancy is admitted to law school, then she scored above 175 on the LSAT.

(F13) $G \rightarrow D$
(F14) $A \rightarrow S$

Note that the phrase 'is a sufficient condition for' precedes the consequent, while the phrase 'is a necessary condition for' precedes the antecedent. Also note that the arrow symbol has only a logical significance and no temporal significance. (The arrow means "then" not "then later.") Achieving a score on the LSAT precedes admission to law school, but that fact provides no reason to place the 'S' before the arrow.

The concept of "necessary condition" can help illuminate the expression 'only if'. S7 and S15 are equivalent and both are correctly symbolized by F15.

(S7) The defendant will be found GUILTY only if the State's principal witness TESTIFIES.
(S15) The State's principal witness's testifying is a necessary condition for the guilty verdict.

(F15) $G \rightarrow T$

The concepts of "sufficient condition" and "necessary condition" also help explain the validity of *modus ponens* and the invalidity of the fallacy of affirming the consequent. The second premise in a *modus ponens* argument affirms the sufficient condition expressed in the first premise. The second premise in the fallacy of affirming the consequent fails to affirm the sufficient condition expressed in the first premise, affirming the necessary condition instead.

A common argument pattern consists entirely of conditionals:

$$A \rightarrow B, B \rightarrow C \vdash A \rightarrow C$$

The consequent of the first premise matches the antecedent of the second; because of this, the premises link together the antecedent of the first premise (which is also the antecedent of the conclusion) and the consequent of the second premise (which is the consequent of the conclusion). The pattern is appropriately called *chain argument*. Chain arguments may have more than two premises. A scene in the movie *Witness* involves a chain argument with three premises. Police detective John Book (Harrison Ford) has been shot trying to protect a young Amish boy from corrupt police who intend to kill him. The boy's mother, Rachel (Kelly McGillis), finds Book injured. This conversation ensues:

> Rachel: *My God, why didn't you get to a hospital?*
> John: *No, no doctor. Gunshot wound—they have to make a report and if they make a report they find me and if they find me they find the boy.*

Part of John's reasoning may be rephrased as follows:

If a doctor is CALLED, the gunshot wound will be REPORTED.
If a report is made, the corrupt police will find ME.
If the police find me, they will also find the BOY.
So, calling a doctor will result in the corrupt police locating the boy.

$C \to R, R \to M, M \to B \vdash C \to B$

It is obvious that this chain argument is valid, but in order to establish this in our proof procedure we will need to adopt a rule of inference for deriving conditionals. The following informal demonstration of the validity of the "Witness" argument suggests the sort of rule we need:

> Look, *suppose* Rachel calls the doctor. It would follow from this supposition and premise one that the doctor will report John's gunshot wound. Then it would follow with help from premise two that the bad cops will find John. And with premise three it will follow that they will find the boy. To summarize: from the assumption of calling the doctor we can conclude the finding of the boy. So, we are warranted in asserting that *if* the doctor is called, *then* the boy will be found.

This is a legitimate way to reason and we will strengthen our proof procedure by adding a rule of inference that sanctions inferences of just this sort.

> **The Arrow In Rule ($\to$I): If from an assumption statement (and perhaps other assumptions) a second statement can be derived, then derive the conditional that has as antecedent the assumption statement and as consequent the second statement.**
>
> From the derivation of *B* from assumption *A* (and perhaps other assumptions) derive $A \to B$.

What is meant by "deriving B from assumption A"? Is it required that the deductive passage from A to B be one step long? No; we may proceed by any number of steps from assumption A to B, and we shall describe that as "a derivation of B from A."

Since the last section, we have been justifying the introduction of premises into proofs by calling them "assumptions." It is necessary now to state formally a rule that covers the introduction of premises, as well as certain other statements.

The Assumption Rule (A, PA): Any statement may be introduced as an assumption at any point in a proof.

This rule may seem inordinately liberal; we shall explain below why it is not.

With the aid of the two rules just introduced—plus the Arrow Out Rule—we can construct a formal proof of the "Witness" argument.

(1)	C → R	A
(2)	R → M	A
(3)	M → B	A
(4)	C	PA
(5)	R	1,4 →O
(6)	M	2,5 →O
(7)	B	3,6 →O
(8)	C → B	4-7 →I

The first four lines are justified by the Assumption Rule. The first three lines are the premises of the argument; let's call them *original* assumptions. The fourth line is a *provisional* assumption made in order that the Arrow In step can be taken at line 8. We identify provisional assumptions with the abbreviation 'PA'. The entry on line 8 of the justification column ('4-7 →I') is short for 'Derived by the Arrow In Rule from the derivation of line 7 from the assumption on line 4'. (The assumption line is always cited first.)

Notice that the Arrow In Rule differs in an important respect from the Arrow Out Rule and the two ampersand rules. Each of those rules sanctions a passage from a *statement* (or two statements) to a statement. The Arrow In Rule sanctions the passage from a *derivation* (or inference) to a statement. That is, it is because we were able to *derive* line 7 from the assumption on line 4 that we are warranted in adding line 8. This difference is reflected in the justification entry for an Arrow In step by the use of a hyphen (rather than a comma) to separate line numbers.

When proofs were first introduced in the last section, a three-column format was provided; now we add a fourth column. This new column, called the "assumption-dependence column," is located to the left of the line-number column. The column indicates for each statement in the proof which assumption(s) it depends upon. The assumption-dependence column enables us to determine that the last line of a proof depends only on original assumptions. If the last line depends on provisional assumptions, then we have not shown that the argument's conclusion follows from the premises alone, but merely that it follows from the premises when they are coupled with the additional assumption.

For each proof rule, we adopt a principle for calculating the assumption dependence of any statement introduced by that rule. The principle governing the Assumption Rule is simplicity itself:

An assumption depends upon itself.

This principle applies to original and provisional assumptions alike. The assumption-dependence principles for the other four rules presented so far:

$\left.\begin{array}{c}\rightarrow O\\ \&I\\ \&O\end{array}\right\}$ The statement derived depends on all of the assumptions on which the premise(s) of the step depend(s).

$\rightarrow I$ The conditional derived depends on all of the assumptions on which the statement corresponding to its consequent depends—less the assumption that corresponds to its antecedent.

(We refer to the first of these two principles as the "standard assumption-dependence principle.")

To illustrate these principles and the modified proof format, we rewrite the proof for the "Witness" argument as follows:

1	(1)	C → R	A
2	(2)	R → M	A
3	(3)	M → B	A
4	(4)	C	PA
1,4	(5)	R	1,4 →O
1,2,4	(6)	M	2,5 →O
1,2,3,4	(7)	B	3,6 →O
1,2,3	(8)	C → B	4-7 →I

Each of the first four lines, being an assumption, depends only on itself. Line 5 depends on the assumptions its premises depend on—namely, 1 and 4. Line 6 is derived from lines 2 (depending on assumption 2) and 5 (depending on assumptions 1 and 4) and, thus, depends on assumptions 1, 2, and 4.

A similar explanation applies to line 7. Line 8 depends on whatever assumptions line 7 depends on (1 through 4) less assumption 4; hence, it depends on 1 through 3. Note that only numbers of assumption lines can appear in the new column. In the above proof, lines 5 through 8 are *derived* lines, not assumption lines. Thus, none of the numbers 5 through 8 could appear in the assumption-dependence column of that proof.

Assumption dependence is reduced by using the Arrow In Rule (and other rules introduced below); for example, the move on line 8 in the proof above eliminates dependence on the provisional assumption on line 4. When you introduce a provisional assumption to a proof you should have some idea how you are going to eliminate it from the set of assumptions on which the final line depends (by using the Arrow In Rule, for example). The requirement that the last line in the proof not depend on any provisional assumption is the check that keeps the Rule of Assumptions from being too liberal. The assumption-dependence column may be included in any proof, but we will use it only in proofs containing provisional assumptions.

Now that the Rule of Assumptions is in our arsenal of rules and we have introduced the notion of "assumption dependence," we can give a better account of what constitutes a formal proof:

A formal proof of symbolized argument S is a list of formulas such that:

(1) each formula is justified by the Rule of Assumptions or is deduced from formulas above it in the list by one of the stated rules of inference,

(2) the last formula in the list is the conclusion of S, and

(3) every assumption on which the last formula depends is a premise of S.

Now consider this sentence:

If the Peruvians SAILED to Polynesia before Columbus and if the only boats available at that time were made of BALSA, then the PERUVIANS sailed to Polynesia in balsa boats.

Should that be symbolized by F16 or F17?

(F16) $S \rightarrow (B \rightarrow P)$
(F17) $(S \mathbin{\&} B) \rightarrow P$

These two formulas are equivalent (they have the same content), so either will do as the symbolization of the sentence displayed above. We can demonstrate the equivalence of F16 and F17 by constructing proofs to show that each entails the other.

A proof that F16 entails F17:

1	(1)	S → (B → P)	A
2	(2)	S & B	PA
2	(3)	S	2 &O
1,2	(4)	B → P	1,3 →O
2	(5)	B	2 &O
1,2	(6)	P	4,5 →O
1	(7)	(S & B) → P	2-6 →I

In constructing this proof, we employed the *Arrow In Strategy*. When we wish to derive a conditional (such as line 7 in this proof), we make a provisional assumption of the antecedent of that conditional (on line 2) and attempt to derive the consequent (line 6).

The Arrow In Strategy is employed twice in this proof that F17 entails F16:

1	(1)	(S & B) → P	A	[1]
2	(2)	S	PA	[3]
3	(3)	B	PA	[5]
2,3	(4)	S & B	2,3 &I	[7]
1,2,3	(5)	P	1,4 →O	[6]
1,2	(6)	B → P	3-5 →I	[4]
1	(7)	S → (B → P)	2-6 →I	[2]

Here are some suggestions designed to simplify the task of constructing proofs; they will be illustrated by reference to this proof. (1) When devising a proof concentrate first on the principal column, which is the list of statements; the other columns can always be added later. (2) Very often, it is helpful to construct this main column by working from both the top and the bottom toward the middle. When you write statements at the bottom while there is a gap in the middle, you are *setting goals* rather than *making deductions*. The order in which we set down the statements when we first worked the proof displayed above is indicated by the bracketed numbers on the extreme right. (This column is not part of the proof format.) At the outset we knew what would be the first and last lines of the proof. Noting that (what we later labelled) line 7 is a conditional, we adopted the Arrow In Strategy of assuming its antecedent on line 2 and aiming for its consequent on line 6. As line 6 is itself a conditional, we employed the Arrow In Strategy again, assuming line 3 and adding line 5 as an additional goal. It didn't take much ingenuity to see that line 4 would plug the gap in the middle of the proof. Having connected the top set of statements with the bottom group, we then added the remaining three columns, working from top to bottom. (3) It may be helpful to view an Arrow In proof as involving a "sub-proof," or proof within a proof. The sub-proof begins with the provisional assumption of the antecedent and concludes with the derivation of the consequent. The proof above contains two sub-proofs. The larger runs from line 2 to 6 and the smaller from line 3 to 5.

Can the Arrow In Rule be applied to *any* pair of lines in a proof? No; these two conditions must be met:

1. **The first line (which is identical to the antecedent of the conditional being derived) must be an** *assumption*.

2. **The other line (which is identical to the consequent of the conditional) must be** *derived from* **the first line (and perhaps other assumptions).**

(A statement **B** is derived from an assumption **A** if and only if **B** is a derived line and **B** depends on **A**.)

EXERCISES

1. Symbolize each statement using the suggested notation.
   (a) (*Dostoevski*) "If God is DEAD, then everything is PERMITTED."
   *(b) (*billboard ad*) "If you have the TIME, we have the BEER."
   (c) (*newspaper*) "Some 7,000 employees of Southern Bell in Dade County would be AFFECTED if a nationwide STRIKE of Bell System workers goes into effect."
   (d) Provided that he does well on the LSAT, John will APPLY to law school.
   (e) The defendant will receive PROBATION provided that she COOPERATES with the district attorney.
   *(f) (*newspaper*) "If PENN State were to win [its final game] and MIAMI were to lose, the Fiesta/Orange OPTION would be Penn State's."
   (g) (*newspaper*) "If she SPEAKS, she BREAKS the spell and has to WAIT for another Midsummer."
   (h) (*tax instructions*) "If MARRIED filing a JOINT return and both spouses worked and had IRAs, figure each spouse's deduction SEPARATELY. . . ." (A = You worked, B = Your spouse worked, C = You had an IRA, D = Your spouse had an IRA, S = Each spouse's deduction should be figured separately)
   (i) (*comic strip*) "If I had some BREAD I could MAKE a ham sandwich if I HAD some ham."

*(j) (*newspaper*) "If PENN State loses [its final game], MIAMI would pick the Orange if NEBRASKA won the Big Eight, and stay with the FIESTA if OKLAHOMA won the Big Eight." (M = Miami chooses to play in the Orange Bowl, F = Miami chooses to play in the Fiesta Bowl)

2. Translate each formula into an English sentence using this "dictionary":
   - A = Miami wins its last regular-season game
   - B = Miami loses its last regular-season game
   - C = New York loses its last regular-season game
   - D = Miami wins the division championship
   - E = Miami is the "wild-card" team

   (a) (A & C) → D
   *(b) A & (C → D)
   (c) A → (C & D)
   (d) (A → D) & (B → E)

3. Symbolize each statement using the suggested notation.
   (a) (*newspaper*) "The bowl in Tempe, Ariz., will be prime time JANUARY 2 only if MIAMI and PENN State win their final games."
   *(b) (*newspaper*) "Should the Dolphins QUALIFY as a runnerup team, they will PLAY at Cleveland Christmas weekend."
   (c) (*bumper sticker*) "In case of RAPTURE, this vehicle will be UNMANNED." (R = The Rapture occurs, U = This vehicle is unmanned)
   (d) Only if YOU come along, will I go to the MOVIES.
   (e) (*billboard ad*) "Like Sports? Love Us. Bill Ray's Sports City." (S = You like sports, B = You will love Bill Ray's Sports City)
   *(f) Smith's CONVICTION is a sufficient condition for a life SENTENCE.
   (g) Having a BATTERY is a necessary condition for my car's STARTING.
   (h) (*Arizona quarterback*) "Our only chance of WINNING [against Miami] is if Gino TORETTA gets hurt and another HURRICANE comes around."

(i) (*TV spot for labor union*) "BUY American and Americans WORK." (B = Americans buy American products, W = There are jobs for Americans)

*(j) (*comic strip*) "Show me a guy who DOESN'T know which side his bread is buttered on and I'll show you a guy with a SLIPPERY sandwich." (D = You show me a guy who doesn't know which side his bread is buttered on, S = I'll show you a guy with a slippery sandwich)

By Permission of Johnny Hart and Creators Syndicate, Inc.

4. Construct formal proofs of validity for each of the following symbolized arguments:
   (a) A & B, A → C ⊢ A & C
   *(b) D → E, F → G, D & F ⊢ E & G
   (c) H → (I → J), H & I ⊢ J
   (d) K → (L & M) ⊢ K → M
   (e) K → L, K → M ⊢ K → (L & M)

Instructions for exercises 5 through 12: symbolize each argument and construct a proof of validity for it.

5. When the American team lost the Davis Cup to the Swedish team in 1984, John McEnroe commented:

   > *I said before the matches if we played badly we were going to lose, and we played bad, and we lost.*[5]

   This is an explanation, cast in the form of an argument:

   If we play BADLY, we LOSE; and we played badly. Hence, we lost.

[5]Richard Finn, "Sweden Walks Off with Davis Cup Title," *USA Today* (December 18, 1984), p. 1C.

*6. Our earlier discussion of 'only if' contained this argument:

> **It should be clear that S7 and S10 express the same thought. S9 and S10 also express the same thought. Therefore, S7 and S9 express the same thought.**

Add this unstated premise:

> If S7 and S10 express the same thought and S9 and S10 express the same thought, then S7 and S9 do also.

(A = S7 and S10 express the same thought, B = S9 and S10 express the same thought, C = S7 and S9 express the same thought)

7. In a 1986 interview, anthropologist Thor Heyerdahl claimed that logic led him in 1947 to sail from Peru to Polynesia in a balsa raft, *Kon-Tiki*. He knew that some Polynesian plants had come from South America before the first Europeans had arrived in South America and he knew that the pre-Columbian Peruvian boats were made of balsa:

> *If experts said this boat or that could not sail in the open sea, I had the evidence of earlier voyages—the plants, say—and if they clearly had made these voyages centuries ago and if they had only one kind of possible boat to make them in, then the logic is clear—they sailed in those boats, no matter what the experts might say about their seaworthiness.*[6]

Heyerdahl's reasoning may be expressed in this argument:

> **Peruvians SAILED to Polynesia before Columbus, and the boats available at that time were all constructed of BALSA. If the Peruvians sailed to Polynesia before Columbus, then if the only boats available at that time were made of balsa the PERUVIANS sailed to Polynesia in balsa boats. Conclusion: the Peruvians sailed to Polynesia in boats made of balsa.**

8. This argument is advanced in panels two through five of the "Doonesbury" strip:

> **Judge Thomas SAID under oath that he had no opinions about *Roe* v. *Wade*, but obviously he did have OPINIONS. If he had opinions about the case but said under oath that he had none, then he committed**

---

[6]Henry Mitchell, "Author's Voyage through History," *The Miami Herald* (May 30, 1986), p. 1B.

PERJURY. If he perjured himself he committed a FELONY. It follows that Judge Thomas committed a felony.

DOONESBURY © 1991 G.B. Trudeau. Reprinted with permission of Universal Press Syndicate. All rights reserved.

9. Sir Arthur Conan Doyle's story, "A Scandal in Bohemia," contains this dialogue:

> Sherlock Holmes: *I am in hopes that she [Irene Adler] does [love her husband].*
> The King of Bohemia: *And why in hopes?*
> Holmes: *Because it would spare your Majesty all fear of future annoyance. If the lady loves her husband, she does not love your Majesty. If she does not love your Majesty, there is no reason why she should interfere with your Majesty's plan.*[7]

Part of Sherlock's reasoning is reflected in this argument:

If Irene Adler loves her HUSBAND, then she does not love the KING. If she does not love the King, then she will not INTERFERE with the King's plans. Therefore, Irene will not interfere with the King's plans, provided that she loves her husband.

(K = Irene does not love the King, I = Irene does not interfere with the King's plans)[8]

---

[7]*The Complete Sherlock Holmes* (New York: Doubleday & Company, Inc., 1956), vol. I, pp.192.
[8]In this section we treat negative sentences as simple statements; beginning with the next section we will treat them as compounds.

*10. Here is an argument critical of Utilitarianism, the moral theory that the right act is the one having the best consequences for all concerned:

> **If UTILITARIANISM is true, then the rightness or wrongness of an act depends on its CONSEQUENCES. We know little about the FUTURE. If the rightness or wrongness of an act depends on its consequences and we know little about the future, then we know little about what is RIGHT and wrong. So, if Utilitarianism is true, then we know little about what is right and wrong.[9]**

(C = The rightness or wrongness of an act depends on its consequences, R = We know little about what is right and wrong)

11. In the *Calvin and Hobbes* strip, Calvin reasons as follows:

> **Mom is taking a SHOWER. That means she is GOING out. She hasn't told me to get CLEANED up. That means I'm staying HOME. Mom's going out while I'm staying home means I'll have a BABY sitter. If I'll have a sitter, it will be ROSALYN. Ergo, Rosalyn will be my baby sitter.**

(C = She hasn't told me to get cleaned up)[10]

Calvin and Hobbes   by Bill Watterson

CALVIN AND HOBBES ©1989 Watterson. Dist. by Universal Press Syndicate. Reprinted with permission. All rights reserved.

12. In the historical musical *1776*, John Adams and Ben Franklin prevail on Virginia's representative to Congress, Richard Lee, to travel home and persuade the House of Burgesses to authorize its congressional delegation to support the cause of independence. As Lee mounts his horse to leave Philadelphia for Williamsburg, he orates grandiloquently:

---

[9]This is a formalized version of an argument advanced by Tom Regan in *The Thee Generation: Reflections on the Coming Revolution* (Philadelphia: Temple University Press, 1991), p. 112.

[10]In this section we treat negative sentences as simple statements; beginning with the next section we will treat them as compounds.

If there's one colony that can get the job done, Virginia—the land that gave us our glorious commander-in-chief, George Washington—will now give the continent its proposal on independence. And when Virginia proposes, the South is bound to follow. And when the South goes, the middle colonies go. Gentlemen, a salute to Virginia and American independence.

Lee's argument:

If VIRGINIA backs independence, then so do the other SOUTHERN colonies. If Virginia and the other southern colonies support independence, then so do the MIDDLE colonies. Conclusion: Virginia's backing independence is a sufficient condition for gaining the support of the rest of the southern colonies as well as that of the middle colonies.

13. It is not obvious how to group the constituent statements in exercise 1(i) above. Here are the alternatives:

    If I had BREAD then (I could MAKE a ham sandwich if I HAD ham).

    $$B \rightarrow (H \rightarrow M)$$

    (If I had bread I could make a ham sandwich) if I had ham.

    $$H \rightarrow (B \rightarrow M)$$

    Prove that the first interpretation entails the second. The same sequence of proof steps would show that the second also entails the first. Hence the two statements are equivalent; so both are acceptable translations of the original.

*14. From the "ABC World News Tonight" telecast of March 31, 1995:

Linda Patillo: . . . *Security is the key to Haiti's entire future, says President Jean-Bertrand Aristide.*
Aristide: *Once we have the breadth of political stability, we can then have the climate which is indispensable to have people feeling safe enough to invest, and by that we will have jobs, and by jobs we'll get money, and by money we'll get food.*

President Aristide offers an argument for the claim that bringing security to Haiti is a sufficient condition for feeding the citizens. (a) Provide a formal restatement of the argument in English. (b) Adopt a symbol dictionary and symbolize the argument. (c) Construct a formal proof for the symbolized argument.

§2.3 NOT

A statement composed of the expression 'It is not the case that' and a constituent statement is called a *negation*. The constituent statement may be either simple or compound. A sample negation:

(S1) It is not the case that Jesse Jackson is a REPUBLICAN.

We introduce the *dash* (−) as an abbreviation for the connective. S1 may be symbolized by the formula F1.

(F1) −R

This formula is read "Not R" or "Dash R."
    The following statements have the same content as S1:

(S2) *It is not true that* Jesse Jackson is a REPUBLICAN.
(S3) *It is false that* Jesse Jackson is a REPUBLICAN.
(S4) Jesse Jackson is *not* a REPUBLICAN.

Accordingly, they may also be represented by F1. From this point forward we will use capital letters to abbreviate affirmative statements, never negative statements.
    To avoid ambiguity when symbolizing negations, we must adopt this principle concerning the use of groupers:

Whenever the constituent of a negation is a compound statement other than another negation, the constituent must be enclosed in parentheses (or brackets, etc.). When the constituent of a negation is either a simple statement or a negation, it is not enclosed in parentheses.

The following formulas illustrate this grouping principle:

(F5) −A
(F6) −−A

(F7) −(B & C)
(F8) −B & −C

Note that F7 symbolizes the negation of a conjunction while F8 represents a conjunction (of the negation of 'B' and the negation of 'C'). Note incidentally that F7 and F8 are not equivalent. F8 is much stronger than F7. F8 claims that both 'B' and 'C' are false, while F7 merely claims that the conjunction formed from those two statements is false. As we saw in Section 2.1, a conjunction can be false when one conjunct is true provided that the other conjunct is false. So the most we can conclude from F7 is that at least one of the statements, 'B' and 'C' is false. F7 is much less informative than is F8.

How should we symbolize this familiar store door sign:

No Shirt. No Shoes. No Service.

The sign is amphibolous. Does it deny service to people who are shirtless (whether or not they are wearing shoes) as well as those who are shoeless (whether or not they are wearing shirts) or does it deny service only to those who are both shirtless and shoeless? No doubt the store owner intends the greater restriction, which is represented by S9 (and F9).

(S9) If it is not the case that you are wearing both a shirt and shoes, then you will not be served.
(F9) −(A & B) → −C

(A = You are wearing a shirt, B = You are wearing shoes, C = You will be served) The less restrictive interpretation of the sign is represented by S10 (and F10).

(S10) If you are not wearing a shirt and not wearing shoes, then you will not be served.
(F10) (−A & −B) → −C

The fact that F9 and F10 are not equivalent makes clear that the dash does not "multiply through" the ampersand. While the dash of logic and the minus of arithmetic are represented by the same symbol, they "behave" differently.

Later in this section we will examine an argument of Plato's containing this sentence:

(S11) Initial learning could not occur before BIRTH unless the soul is IMMORTAL.

S11 has the same content as:

> (S11A) If the soul is not immortal, then initial learning could not occur before birth.

S11A (and so S11 also) are symbolized by this formula:

> (F11) $-I \to -B$

(I = The soul is immortal) 'Unless' means the same as 'if it is not the case that'; sentences containing 'unless' should be symbolized accordingly. (It is sometimes maintained that 'P unless Q' means 'P if *and only if* not Q' rather than the weaker 'P if not Q', but this is incorrect. A friend who tells you, "I'll be there unless I'm ill," doesn't break his or her word by showing up ill.)

Jean Harris, the convicted murderer of the diet doctor, Herman Tarnower, claimed that she killed him accidentally. In an interview with *People* magazine before her sentencing, she said, "If I were going to kill somebody [intentionally], I'd have killed her [Lynne Tryforos, the other woman]."[11] Harris's argument:

If I killed anyone INTENTIONALLY, I would have killed TRYFOROS. I didn't kill her. This shows that I didn't kill anyone intentionally.

> $I \to T, -T \vdash -I$

(I = I killed someone intentionally) This pattern of argument bears the label, "*modus tollens*" (Latin for "the denying form"). It is a pattern that we all employ regularly and it is obviously valid. (Whether Harris's argument has a true first premise and a true conclusion is another question into which we will not enter.) We can give an informal demonstration of the validity of Harris's argument (and by extension any *modus tollens* argument) as follows:

> *Assume* that Harris killed someone intentionally. From this assumption and the first premise it follows that she killed Tryforos. But the second premise contradicts this. So the assumption we made is incompatible with the premises of the argument, and we are entitled to conclude its negation.

[11]"Jean Harris: If I Killed, I Would Have Gotten Rival, Not Beloved Diet Doctor," *Miami News* (March 2, 1981), p. 4A.

We add to our stock of rules one that sanctions reasoning of this sort:

The Dash In Rule (−I): If from an assumption statement (and perhaps other assumptions) a standard contradiction can be derived, then derive the negation of the assumption.

A *standard contradiction* is a conjunction of the form $P \& -P$. We can state the Dash In Rule more perspicaciously with the help of a few symbols:

From the derivation of $B \& -B$ from assumption A (and perhaps other assumptions) derive $-A$.

Note that the Dash In Rule makes no reference to the content of the contradiction involved; any standard contradiction will do as well as any other.

A formal proof for the "Harris" argument:

1	(1)	$I \to T$	A
2	(2)	$-T$	A
3	(3)	I	PA
1,3	(4)	T	1,3 →O
1,2,3	(5)	$T \& -T$	4,2 &I
1,2	(6)	$-I$	3-5 −I

Lines 1 and 2 are original assumptions. Line 3 is a provisional assumption made with the idea of the subsequent application of the Dash In Rule, a move that is made on line 6. '3-5 −I' is short for 'Derived by the Dash In Rule from the derivation of the standard contradiction on line 5 from the assumption on line 3'. (The assumption line is always cited first.) Line 6 of the assumption-dependence column was computed with the aid of this principle:

The statement derived by the Dash In Rule depends on all of the assumptions on which the contradiction depends—less the assumption whose negation is derived.

Thus, line 6 depends on the assumptions of line 5 (1 through 3) less assumption 3; hence, it depends on 1 and 2. The Dash In step "wipes out" the provisional assumption.

Why is it legitimate to deduce '−I' on line 6 of the above proof? The contradiction on line 5 follows from the assumptions on lines 1 through 3. Only a contradictory set of assumptions will yield a contradiction and, thus, lines 1 through 3 constitute a contradictory set. To avoid inconsistency, one of the assumptions must be given up. The assumption-dependence entry on line 6 shows that we are retaining assumptions 1 and 2. We can retain 1 and 2 only by rejecting the third assumption, 'I', which we do on line 6.

The "Sylvia" comic strip revolves around this argument:

If women had played team SPORTS as children, they would FIT smoothly into the corporate world. But since they didn't play team sports, they won't fit into the corporate world.

S → F, −S ⊢ −F

By permission of Nicole Hollander

One who gives this argument a superficial examination may hold that it exhibits the form *modus tollens*. Closer inspection, however, will show that it is the counterfeit of *modus tollens*, the invalid pattern called the *fallacy of denying the antecedent*. Compare the two patterns:

modus tollens: $P \rightarrow Q, -Q \vdash -P$ **(valid)**
denying the antecedent: $P \rightarrow Q, -P \vdash -Q$ **(invalid)**

The second premise of a *modus tollens* argument denies a necessary condition, while the second premise of its counterfeit denies a sufficient condition. In the "Corporate World" argument, playing team sports is identified as a sufficient condition for "fitting in." That is compatible with there being many other sufficient conditions for "fitting in"—having an agreeable personality, for example. A person who does not have one of these sufficient conditions may have another, so you can't validly reason from the absence of one sufficient condition to the absence of that for which it is a condition.

In Section 2.2 we vowed to prove the logical equivalence of S9 and S10.

(S9) If the defendant was found GUILTY then the State's principal witness TESTIFIED.

(S10) If the State's principal witness failed to testify, then the defendant was not found guilty.

(F9) G → T
(F10) −T → −G

Logicians call F10 the "contrapositive" of F9. It may not be important to remember that term, but it is useful to remember that you can switch the constituents in a conditional (and not change its content) if you negate them both.

Here we will present half our proof that this is the case by showing that F9 entails F10; the reverse entailment is the subject of exercise 9 at the end of this section.

1	(1)	G → T	A
2	(2)	−T	PA
3	(3)	G	PA
1,3	(4)	T	1,3 →O
1,2,3	(5)	T & −T	4,2 &I
1,2	(6)	−G	3-5 −I
1	(7)	−T → −G	2-6 →I

This proof contains a Dash In sub-proof (lines 3 through 5) within an Arrow In sub-proof (lines 2 through 6). When the top two lines and the bottom two lines had been determined, we used the *Dash In Strategy* to add line 3:

> **When you aim to derive a negated statement such as '−G' (and a more direct route is not apparent), make a provisional assumption of 'G' and try to derive a standard contradiction.**

The Dash Out Rule is very similar to the Dash In Rule:

> ***The Dash Out Rule (−O):** If from a negative assumption statement (and perhaps other assumptions) a standard contradiction can be derived, then derive the constituent of the negative assumption.*

The rule restated:

> **From the derivation of B & $−B$ from assumption $−A$ (and perhaps other assumptions) derive A.**

The assumption-dependence principle for this rule parallels the one for the Dash In Rule.

We can illustrate the employment of the Dash Out Rule by constructing a proof for this argument from Plato's "Phaedo." The setting for the dialogue is the prison in Athens where Socrates engages in philosophical discussion while awaiting execution.

> 'Besides, Socrates,' rejoined Cebes, 'there is that theory which you have often described to us—that what we call learning is really just recollection. If that is true, then surely what we recollect now we must have learned at some time before; which is impossible unless our souls existed somewhere before they entered this human shape. So in that way too it seems likely that the soul is immortal.'[12]

The argument set out formally and symbolized:

What we call learning is RECOLLECTION. If so, then initial learning occurred before BIRTH. Initial learning could not occur before birth unless the soul is immortal. So, the soul is IMMORTAL.

R, R → B, −I → −B ⊢ I

The Dash Out proof of validity:

1	(1)	R	A
2	(2)	R → B	A
3	(3)	−I → −B	A
4	(4)	−I	PA
1,2	(5)	B	2,1 →O
3,4	(6)	−B	3,4 →O
1,2,3,4	(7)	B & −B	5,6 &I
1,2,3	(8)	I	4-7 −O

The *Dash Out Strategy* is simple:

When you seek to derive some statement such as 'I' (and you do not see a more direct route to 'I'), make a provisional assumption of '−I' and try to derive a standard contradiction.

When is the Dash Out Strategy likely to be useful? When the following three features are all present, the chances are good that the strategy should be employed:

---

[12]Plato, "*Phaedo,*" in *The Last Days of Socrates*, trans. Hugh Tredennick (Baltimore: Penguin Books, 1989), p. 120.

1. The goal (or sub-goal) is affirmative.
2. The premise lines contain one or more dashes.
3. No more direct path from premises to goal is obvious.

The type of reasoning sanctioned by the Dash In and Dash Out rules has traditionally been known as the *Reductio ad Absurdum* method of proof. One reduces some assumption to absurdity (by deriving a contradiction from it) and then concludes its denial. This form of reasoning is common in mathematics, but it works in any discipline, including, of course, the law.

EXERCISES

1. Symbolize each statement using the suggested notation.
    (a) (*bumper sticker*) "SKATEBOARDING is not a crime."
    *(b) (*cracker box*) "This package is sold by WEIGHT, not by VOLUME."
    (c) (*newspaper*) "Tomorrow night's WBC heavyweight title fight between Larry Holmes and Scott Frank will not be unavailable to television viewers in Miami." (W = The Holmes-Frank WBC title fight will be available to Miami television viewers tomorrow night)
    (d) (*lyrics*) "I've waited so long for SCHOOL to be through; PAULA, I can't wait no more for you." (P = I can wait longer for Paula)
    (e) (*Ferrol Sams*) "If MOMMA ain't happy, ain't nobody happy." (M = Momma is happy, S = Some family members are happy)
    *(f) (*comic-strip mother*) "I can't BUY peanut butter and SELL houses, too!"
    (g) (*comic-strip son*) "Hmmpf! She hasn't SOLD any houses, and she hasn't BOUGHT any peanut butter."
    (h) (*comic strip*) "If it wasn't for JUNK mail, I wouldn't get any MAIL at all!!" (J = I receive junk mail, M = I receive mail)
    (i) (*Hans Christian Andersen*) "Now we shall have duck EGGS, unless it's a DRAKE."
    (j) (*television news*) "If Senator Kerrey votes FOR the budget bill it will PASS, and if he votes AGAINST it it will fail."
    *(k) (*Hitler*) "If I do not get the oil of MAIKOP and GROZNY, then I must END this war."
    (l) (*Charles Dickens*) "Unless I SEE it with my own eyes, and HEAR it with my own ears, I never will BELIEVE it."
    (m) (*headline*) "Hamilton can't PRESIDE without MOVING to city." (P = Hamilton presides, M = Hamilton moves to city)
    *(n) (*tax instructions*) "[If you have a] nonresident ALIEN spouse[, then] if you do not file a JOINT return, you can take an EXEMPTION for your . . . spouse only if your spouse had no INCOME from U.S.

sources and is not the DEPENDENT of another person." (E = You are permitted to take an exemption for your spouse)

(o) (*sports page*) "When they [the Notre Dame football team] have a second DOWN and between FOUR and SEVEN yards to go, they never THROW unless it's LATE and they are BEHIND." (F = Notre Dame has four or more yards to go, S = ND has seven or fewer yards to go, T = ND passes)

2. Translate each formula into an English sentence using this dictionary:
   D = The defense witness was granted immunity
   P = The prosecution witness was granted immunity
   (a) −−D
   *(b) −D & −P
   (c) −(D & P)
   (d) −D → −P

3. Sally Swindle filed an income tax return claiming an exemption for a dependent who had been dead for two years. When the IRS held the return to be invalid, she argued with them as follows:

   > *On page 41 of the instructions it says "Form 1040A is not considered a valid return unless you sign it." I signed it. So, it is a valid return.*

   What is wrong with Sally's argument?

4. Construct formal proofs of validity for each of the following symbolized arguments:
   (a) −−A ⊢ A
   *(b) −B → C, −B → −C ⊢ B
   (c) D → −E, E ⊢ −D
   (d) −F ⊢ −(F & G)
   (e) −(H & −I), H ⊢ I

Instructions for exercises 5 through 15: symbolize each argument and construct a proof of validity for it.

5. This argument was employed in Section 2.2 in a discussion of how 'only if' sentences are to be understood.

   > *It is a common mistake to believe that S7 and S8 are equivalent in meaning. If they were equivalent in meaning, then the word 'only' would contribute nothing to the meaning of S7, but it obviously does affect S7's meaning.*

(E = S7 and S8 are equivalent in meaning, A = The word 'only' affects the meaning of S7)

*6. The comic strip character Nancy muses that if she plays on the swings, she'll be happy. The next frame of the strip shows her on the swings obviously unhappy, thus showing that she was mistaken.

> **Nancy is playing on the SWINGS and is not HAPPY. Thus, it was not true that if she plays on the swings she'll be happy.**

7. **The contract was SIGNED and sealed, but it was never delivered. Therefore, it is false that it was signed, sealed, and DELIVERED.**

(A = The contract was sealed)

8. Kingsley Amis presents this argument while discussing rumors that Rudyard Kipling was promiscuous:

> *Nothing contradicts the view that he was an ordinary monogamous man. If there existed real evidence to the contrary, some of it would have been sure to come out. There's no fire without smoke, and in this case there's no smoke.*[13]

Amis's conclusion is that Kipling was monogamous. Paraphrase the argument so that it may be symbolized with this dictionary: M = Kipling was monogamous, E = There is evidence of his promiscuity, K = We know of such evidence. You will need to unpack the metaphor about fire and smoke.

9. We proved on page 52 that S9 entails S10. Now demonstrate that S10 also entails S9 (thereby establishing logical equivalence).

> (S9) If the defendant was found GUILTY then the State's principal witness TESTIFIED.

> (S10) If the State's principal witness failed to testify, then the defendant was not found guilty.

*10. Plato writes in *The Republic*:

> *. . . Pindar and the tragedians . . . allege that Asclepius, although he was Apollo's son, took a bribe to raise to life a rich man already at the point of death,*

---

[13]Kingsley Amis, *Rudyard Kipling* (New York: Thames and Hudson, 1986), p. 37.

and was struck by a thunderbolt for doing so. As we said before, we cannot accept both statements; if he was the son of a god, he was not avaricious . . . .[14]

One analysis of Plato's reasoning:

It is not the case that Asclepius was both the SON of a god and took a BRIBE. Here are the reasons: (1) If Asclepius was the son of a god, he wasn't AVARICIOUS. (2) If he took a bribe, he was avaricious.

11. The last sentence of the quotation from *The Republic* (see previous exercise) could be understood as containing this argument:

    It is not the case that Asclepius was both the SON of a god and AVARICIOUS. So, if he was the son of a god, he wasn't avaricious.

12. In "The Freedom of a Christian," Martin Luther writes:

    The Word of God cannot be received and cherished by any works whatever but only by faith. Therefore it is clear that, as the soul needs only the Word of God for life and righteousness, so it is justified by faith alone and not any works; for if it could be justified by anything else, it would not need the Word, and consequently it would not need faith.[15]

    It is obvious that Luther is reasoning, but not clear (from this translation) exactly how the reasoning goes. Here is one reconstruction:

    If the soul could be justified by something other than faith, it would not need the Word of God. But the soul does need the Word of God. If the soul cannot be justified by anything other than faith, then it cannot be justified by works. Therefore, the soul can be justified by nothing other than faith; it cannot be justified by works.

    Use this dictionary: O = The soul can be justified by something other than faith, N = The soul needs the word of God, W = The soul can be justified by works.

13. In "Million Dollar Bond Robbery," Agatha Christie writes:

    "But if the bonds were thrown overboard, they couldn't have been sold in New York."

---

[14]Plato, *The Republic*, trans. Francis McDonald Cornford (Oxford: Oxford University Press, 1941), p. 98.

[15]John Dillenberger (ed.), *Martin Luther: Selections from His Writings* (Garden City, NY: Doubleday & Company, Inc., 1961), p. 55.

> "I admire your logical mind, Hastings [Poirot replied]. The bonds were sold in New York, therefore they were not thrown overboard. You see where that leads us?"
> "Where we were when we started."
> "Jamais de la vie! If the package was thrown overboard, and the bonds were sold in New York, the package could not have contained the bonds...."[16]

In this passage Hastings and Poirot draw linked inferences. We compress their two arguments into one as follows:

If the bonds were thrown OVERBOARD, they weren't SOLD in New York. The bonds were sold in New York. If the PACKAGE was thrown overboard and the bonds were not thrown overboard, the package did not CONTAIN the bonds. It follows that if the package was thrown overboard, then it did not contain the bonds.

14. (CHALLENGING)[17] The prisoner advances this depressing argument:

If I don't EAT, I'll get SICK. If I do eat, I'll get sick. Hence, I'll get sick.

The proof can be done in ten lines, but it requires *two* provisional assumptions that are eliminated by two applications of the Dash rules.

By permission of Johnny Hart and Creators Syndicate, Inc.

15. (EXTRA CHALLENGING) A newspaper feature recounts the struggles of AIDS patient, Tim Braun.[18] The federal government promised to supply him with marijuana cigarettes to control his nausea and then reneged on the promise. Braun remarked,

---

[16]"Poirot Investigates" in *Triple Threat* (Binghamton, NY: Vail-Ballou Press, Inc., 1923), p. 105.
[17]Exercises marked "challenging" are more difficult than the others. They require ingenuity. If you enjoy a challenge, you will want to tackle them.
[18]Tom Majeski, "AIDS Patient Fights for Marijuana," *Miami Herald* (December 15, 1991), p. 15A.

> *If I buy grass, I can't afford to eat. If I buy food, I can't eat it.*

His dilemma is captured in this argument:

> **If I buy GRASS, then I cannot afford to buy FOOD. If I cannot afford food, then I don't EAT. If I can afford food and don't buy grass, then I don't eat [because of nausea]. So, I don't eat.**

Material surrounded by brackets is meant to clarify or provide background. It should not be symbolized.

§2.4 OR

A statement consisting of two constituent statements joined by the connective 'or' is called a *disjunction*; the component statements are called *disjuncts*. Often, the first disjunct is preceded by the word 'either'. A sample disjunction:

> Either the witness for the DEFENSE is lying or the PROSECUTION witness is lying.

We introduce the *wedge* (v) as an abbreviation for the connective 'or'. The disjunction displayed above is symbolized:

> D v P

This formula is read "D or P" or "D wedge P."

Often when we assert a disjunction we intend to admit the possibility that both disjuncts are true. For example, we would probably not regard the disjunction above as being disproved by the discovery that both witnesses lied. In such a case 'or' is said to be used in the *inclusive* sense. Sometimes, though, when we utter an "or" sentence, we intend to rule out the case where both disjuncts are true. For example, if the boss tells an employee, "You may have Friday off or Monday off," the boss is probably excluding the case where the employee takes both days off. In such a sentence 'or' is said to be used in an *exclusive* sense.[19]

Sometimes when we use 'or' in a sentence, it is not clear which sense we intend.[20] Then, we may have to elaborate in order to avoid the risk of being

---

[19]Latin has distinct words for these two senses of 'or'; 'vel' represents inclusive disjunction and 'aut' exclusive. Logicians chose the first letter of 'vel' to symbolize inclusive disjunction.

[20]It is worth noting that 'or' in the exclusive sense seems to be found more often in regulative or directive language (such as that used by the employer in the above example) than it is elsewhere.

misunderstood. When we wish to make it clear that we are including the possibility that both disjuncts are true, we may use one of these locutions:

A *or* B *or both*.
A *and/or* B.

When we wish to be explicit about excluding that possibility, we may say:

(S1) A *or* B *but not both*.

The wedge symbol abbreviates the inclusive 'or'. We can symbolize exclusive disjunctions with the wedge, ampersand, and dash. F1 will represent S1:

(F1) **(A v B) & −(A & B)**

'Or' is generally used in the inclusive sense. When it is intended in the exclusive sense in this book we will indicate this by appending the phrase 'but not both'.
How shall we symbolize this sentence?

Either the SUSPECT or the WITNESS or the DETECTIVE lied.

Three possibilities:

(F2) **S v W v D**
(F3) **S v (W v D)**
(F4) **(S v W) v D**

Formulas F3 and F4 are logically equivalent (see exercise 15 following this section). This shows that parenthesis placement is not critical in this case; so any of the three formulas is acceptable. F2 is preferable because it is simpler. The situation is different with these sentences and their symbolizations:

(S5) The suspect lied and so did either the witness or the detective.
(S6) Either the suspect and the witness both lied or else the detective lied.

(F5) **S & (W v D)**
(F6) **(S & W) v D**

S5 and S6 are not equivalent statements. One difference between them is that S5—but not S6—flatly asserts that the suspect lied. It is clear that parenthesis placement is critical in the symbolization of sentences like S5 and S6. A formula like F7X is amphibolous.

(F7X) S & W v D

Does it mean the same as F5 or F6? There's no way to tell. We regard F7X as an ill-formed formula and avoid employing it. The sentence S7 is equally amphibolous and should also be avoided.

(S7) The suspect lied and the witness lied or the detective lied.

We can summarize the conventions regarding the use of groupers in wedge formulas as follows:

If a formula contains both a wedge and either an ampersand or an arrow, parentheses (or other grouping symbols) must be included to indicate which connective has greater scope.

Groupers may be omitted in multiple disjunctions.

Part of the logical content of the (inclusive) 'or' connective is captured in this observation:

If either disjunct is true, then the disjunction is true.

We can base a rule of inference on this observation:

The Wedge In Rule **(vI): From a statement derive any disjunction that includes that statement as a disjunct.**

The rule restated:[21]

From A derive either A v B or B v A.

The standard assumption-dependence principle applies to this rule. The Wedge In Rule will permit us to derive from F8 any of the formulas F9 through F11.

---

[21]The first statement of the rule is somewhat broader than the symbolic restatement because it applies to multiple disjunctions that lack internal groupers.

(F8) A
(F9) A v B
(F10) C v A
(F11) D v A v E

Some people's first reaction to the Wedge In Rule is that it is excessively liberal. "You mean you can introduce *any* statement—even one that does not appear elsewhere in the proof—as the other disjunct?" Their concern should be met by this consideration: All we need require of a rule of inference is that it be truth preserving (*i.e.*, that it never permit the derivation of a false statement from a true one). We know that if either disjunct of a disjunction is true, then the disjunction is also true. So, if F8 is true, then each of the other formulas F9 through F11 will also be true regardless of the truth-value of the other disjunct(s). So, in a sense it doesn't matter what the other disjunct is. The disjunction will follow no matter what statement constitutes the other disjunct.

The expression 'neither . . . nor' means "not either . . . or"; so S12 may be reformulated as S13 and symbolized by F13.

(S12) Neither the husband nor the wife requested custody.
(S13) It is not the case that either the husband or the wife requested custody.

(F13) −(H v W)

(H = The husband requested custody, W = The wife requested custody) F13 is logically equivalent to F14, so S12 may also be symbolized with this formula.

(F14) −H & −W

One quarter of the job of establishing the logical equivalence of F13 and F14 is showing that F13 entails the left conjunct of F14. The following proof, which employs the Wedge In Rule on line 3, concerns this entailment.

1	(1)	−(H v W)	A
2	(2)	H	PA
2	(3)	H v W	2 vI
1,2	(4)	(H v W) & −(H v W)	3,1 &I
1	(5)	−H	2-4 −I

A proof with the same structure will establish that F13 also entails the right conjunct of F14.

A sports story about the 1987 Fiesta Bowl (between Penn State and Miami) concerns plans to change the date of the game from the first to the second day of the new year.[22] The story includes this sentence:

A loss by either team and the Fiesta would revert to its original Jan. 1 niche. . . .

This sentence may appear to be a conjunction but is in fact a conditional; it may be symbolized by F15:

(F15) (P v M) → O

(P = Penn State loses a game, M = Miami loses a game, O = The Fiesta Bowl will be played on January 1) How does F15 relate to F16?

(F16) (P → O) & (M → O)

The two formulas are equivalent. With the aid of the Wedge In Rule we can prove that F15 entails F16. The proof begins as follows:

1	(1)	(P v M) → O	A
2	(2)	P	PA
2	(3)	P v M	2 vI
1,2	(4)	O	1,3 →O
1	(5)	P → O	2-4 →I

This part of the proof establishes that F15 entails the left conjunct of F16. The next part shows that F15 also entails the right conjunct of F16.

6	(6)	M	PA
6	(7)	P v M	6 vI
1,6	(8)	O	1,7 →O
1	(9)	M → O	6-8 →I

The proof concludes with this step:

1	(10)	(P → O) & (M → O)	5,9 &I

---

[22]Greg Cote, "Prime Time Hinges on 11–0 Teams," *Miami Herald* (November 18, 1986), p. 1C.

(Exercise 4(b) at the end of the section concerns the entailment of F15 by F16.)

A news story begins:

Democrat Helen Boosalis will fage Republican state Treasurer Kay Orr in the nation's first woman-against-woman campaign for governor, after yesterday's primaries guaranteed that Nebraska will have its first female chief executive.[23]

The quotation involves this simple argument (let's call it "Nebraska Governor"):

**Either BOOSALIS or ORR will win the election for governor of Nebraska.
If Boosalis wins, Nebraska will have its first FEMALE governor.
If Orr wins, Nebraska will have its first female governor.
So, Nebraska will have its first female governor.**

B v O, B → F, O → F ⊢ F

The pattern of inference exhibited by "Nebraska Governor" is one form of the *dilemma*. Logicians recognize several species of dilemma; the two most common forms:

$$\text{simple constructive:} \quad P \vee Q, P \to R, Q \to R \vdash R$$
$$\text{complex constructive:} \quad P \vee Q, P \to R, Q \to S \vdash R \vee S$$

"Nebraska Governor" is a simple constructive dilemma; later in the section we shall consider a complex constructive dilemma. Examples of both types are included in the exercise set at the end of the section.

We can base our Wedge Out Rule on the obviously sound simple constructive dilemma.

The Wedge Out Rule (vO): From $A \vee B$, $A \to C$, and $B \to C$ derive C.

If called upon to justify this rule we would start with the observation that whenever a disjunction is true at least one of its disjuncts is also true. (How would the justification continue?)

We illustrate the employment of the Wedge Out Rule by constructing three proofs, beginning with a proof for "Nebraska Governor":

(1)   B v O   A

---

[23]"Women Vie for Top Nebraska Post," *Miami News* (May 14, 1986), p. 2A.

(2)	B → F		A
(3)	O → F		A
(4)	F		1,2,3 vO

The Wedge Out Rule is applicable because the statement on line 4 matches the consequent of both conditionals and the antecedents of those conditionals match the disjuncts in line 1.

The second proof concerns this simple argument:

Either BOOSALIS or ORR will win the election.
Thus, either ORR or BOOSALIS will win the election.

B v O ⊢ O v B

The proof for this argument will involve both the Wedge Out Rule (because of the disjunctive premise) and the Wedge In Rule (because of the disjunctive conclusion). The dominant pattern in the proof is Wedge Out. In order to apply the Wedge Out Rule at the conclusion of the proof, we will need (in addition to the disjunctive premise) these two conditionals:

B → (O v B)
O → (O v B)

Notice that the antecedent of the first conditional matches the left disjunct of the disjunctive premise and the antecedent of the second matches the right disjunct. The consequent of each conditional matches the conclusion of the argument. The two conditionals will be derived by means of the Arrow In Rule. The proof:

1	(1)	B v O	A
2	(2)	B	PA
2	(3)	O v B	2 vI
	(4)	B → (O v B)	2-3 →I
5	(5)	O	PA
5	(6)	O v B	5 vI
	(7)	O → (O v B)	5-6 →I
1	(8)	O v B	1,4,7 vO

Note that lines 4 and 7 in the proof above depend on *no* assumptions. Line 4, for example, depends upon whatever line 3 depends on (namely, 2) less line 2; that is, nothing. To help explain how a line can depend on no assumptions we introduce the concept of a *logical truth*. A logical truth is a statement whose falsity is logically impossible; it is the opposite of a logical contradiction. An example is 'Either Clinton is a Democrat or he is not a Democrat'. The strength of a logical truth is that it must be true; its weakness is that it conveys no factual information. Logical truths—and only logical truths—can be derived in proofs free of assumptions. The fact that lines 4 and 7 in the above proof are so derived demonstrates that they are logically true.

The standard assumption-dependence principle applies to the Wedge Out Rule. Line 8 depends on all of the assumptions on which lines 1, 4, and 7 depend. A peculiar feature of this proof (and of many Wedge Out proofs) is that the same statement occurs on three lines (3, 6, and 8). However, the lines differ in assumption dependence, and only line 8 is free of dependence on a provisional assumption.

The *Wedge Out Strategy* may be summarized as follows:

If one of the premise lines is a disjunction, search the other premise lines for two conditionals whose antecedents match the disjuncts and whose consequents (both) match some goal line. If you find both conditionals, apply Wedge Out. If you find one of the two conditionals, add the other as a goal line. If you find neither conditional among the premise lines, add both as goal lines.

This strategy is illustrated in a third proof which establishes the validity of a complex constructive dilemma. In a letter to James C. Conkling, Abraham Lincoln writes about the Emancipation Proclamation:

> But the proclamation, as law, either is valid, or not valid. If it is not valid, it needs no retraction. If it is valid, it cannot be retracted, any more than the dead can be brought to life.[24]

By transposing the last two premises and adding the unstated conclusion we reach this formulation of the argument:

The proclamation is either VALID or not valid. If it is valid, it cannot BE retracted. If it is not valid, it does not NEED to be retracted. Therefore, either the proclamation cannot be retracted or it does not need to be retracted.

V v −V, V → −B, −V → −N ⊢ −B v −N

This argument is a *complex*, rather than a *simple*, dilemma. Note that the consequents of the conditionals do not match; instead, each consequent matches

---

[24]Quoted in Irving M. Copi, *Introduction to Logic*, 7th ed. (New York: Macmillan Publishing Company, 1986), p. 259.

one of the disjuncts of the conclusion. In order to complete a Wedge Out proof for this argument, it will be necessary to deduce two conditionals (lines 7 and 11 in the proof below) whose consequents match the conclusion of the argument.

1	(1)	V v −V	A
2	(2)	V → −B	A
3	(3)	−V → −N	A
4	(4)	V	PA
2,4	(5)	−B	2,4 →O
2,4	(6)	−B v −N	5 vI
2	(7)	V → (−B v −N)	4-6 →I
8	(8)	−V	PA
3,8	(9)	−N	3,8 →O
3,8	(10)	−B v −N	9 vI
3	(11)	−V → (−B v −N)	8-10 →I
1,2,3	(12)	−B v −N	1,7,11 vO

EXERCISES

1. Symbolize each statement using the suggested notation.
   (a) (*Plautus dialogue*) "This woman must be either MAD or DRUNK."
   *(b) (*Congressman William Lehman*) "He [Bush] will be an ENVIRONMENTAL president or he won't."
   (c) (*forensic pathologist*) "Either the EYEWITNESSES are mistaken or this is not the body of NATHANIEL Cater." (E = The eyewitnesses are correct, N = This is the body of Nathaniel Cater)
   (d) (*Plato*) "Either [death] . . . is ANNIHILATION, and the dead have no CONSCIOUSNESS of anything, or . . . it is . . . a MIGRATION of the soul. . . ." (C = The dead have consciousness of something)
   (e) (*John Volpe*) "The whole [railroad] system will wind up BROKE and/or NATIONALIZED."
   *(f) (*Bum Phillips*) "If we don't finish at LEAST 8 and 8 I should RESIGN or be FIRED." (L = We finish the season with at least eight wins, R = I should resign, F = I should be fired)
   (g) (*Supreme Court Justice John Harlan*)[25] "Our Constitution is color BLIND and neither KNOWS nor TOLERATES classes among citizens."

---

[25]Dissenting in *Plessy v. Ferguson*, 163 U.S. 537, 552, at 559 (1896).

(h) We will call RACHEL or ALLISON to the stand, but not both.

(i) Zero is either EVEN or ODD or neither.

*(j) (*newspaper*) "If it becomes necessary to PULL equipment off the line for repairs or if an ACCIDENT saps generating capacity, the company could be forced to REDUCE power to all areas or BLACK out some sections of the Gold Coast."

(k) (*tax instructions*) "You will not OWE the [estimated tax] penalty or have to COMPLETE Form 2210 if either 1 or 2 below applies." (C = You are required to complete Form 2210, A = 1 below applies, B = 2 below applies)

2. Translate each formula into an English sentence using this dictionary:
   F = Cruelty to animals is a felony in Florida
   M = Cruelty to animals is a misdemeanor in Florida
   O = Cruelty to animals is a punishable offense in Florida

   (a) $-M \vee -F$
   *(b) $-(M \vee F)$
   (c) $O \rightarrow (M \vee F)$
   (d) $(-M \rightarrow F) \vee -O$

3. The instructions for Form 1040A (for 1992) include:

   If you and your spouse paid JOINT estimated tax but are now FILING separate income tax returns, either of you can claim all of the amount paid.

   Symbolize this sentence once treating the 'either' as inclusive, and again treating it as exclusive. (Y = You are permitted to claim all of the amount paid, S = Your spouse is permitted to claim all of the amount paid) Which of these symbolizations captures the IRS rule? Hint: the next sentence is 'Or you can each claim part of it'.

4. Construct formal proofs of validity for each of the following symbolized arguments:
   (a) $A \& B \vdash A \vee C$
   *(b) $(P \rightarrow O) \& (M \rightarrow O) \vdash (P \vee M) \rightarrow O$
   (c) $-(-D \vee -E) \vdash D$
   (d) $(F \& G) \vee (H \& F) \vdash F$
   (e) $I, (I \vee J) \rightarrow K \vdash J \vee K$

Instructions for exercises 5 through 11: symbolize each argument and construct a proof of validity for it.

5. Billy Hayes, author of *Midnight Express*, told a college audience of his decision to escape from the Turkish prison in which he had been confined for five years:

> *My thoughts were that if I made it, I would be free. If they shot and killed me I would also be free.*[26]

Hayes was advancing a simple constructive dilemma:

> **If I ESCAPE, I will be FREE. If they KILL me, I will also be free. Either I escape or they kill me. Hence, I will be free.**

*6. In Plato's "Apology,"[27] Socrates advances this argument to explain why he does not fear death:

> **Death is either ANNIHILATION or a MIGRATION of the soul. If death is annihilation, it is like a dreamless SLEEP. If it is like a dreamless sleep, then it is a BENEFIT. If, on the other hand, death is a migration of the soul, then it is still a benefit. Consequently, death is a benefit.**

7. In *Woolen* v. *Surtran Taxicabs, Inc.*, 684 F.2d 324 (5th Cir. 1982), a group of class members sought to intervene in an antitrust class action that was being brought on their behalf. They claimed a right to intervene under F.R.C.P. 24(a), because their interests would be affected by the outcome. But the representative class plaintiffs argued that there was no right to intervene, because there is an exception in Rule 24(a) where the existing parties will adequately represent the would-be interveners. But under Rule 23(a)(4), there can be no class action unless the representative parties will adequately represent the absent members. So either the class action should not have been allowed to go forward at all or the additional class members have no right to intervene in it.

> **Either the representative parties will adequately REPRESENT the absent members or they won't. If they will adequately represent the absent members, then the absent members have no right to INTERVENE. On the other hand, if the representative parties will not adequately represent the absent members, then the class action should not have been allowed to go FORWARD at all. So, either the class action should not have been allowed to go forward at all or the additional class members have no right to intervene.**

8. A news story concerns one Willie Dennis who, in exchange for a reduction of charges against him from murder to manslaughter, agreed to

---

[26]Bill Kaczaraba, "No More 'Crazy House' for Hayes," *Miami Hurricane* (October 31, 1978), p. 1.
[27]Plato, "Apology" in *The Last Days of Socrates*, trans. Hugh Tredennick (Baltimore: Penguin Books, 1989), pp. 74–75.

testify in the trial of a second man. However, the testimony Dennis gave in the trial of the other man differed from what he had previously told the state attorney; whereupon the judge ordered Dennis's manslaughter conviction vacated and directed the state to indict him for murder. The newspaper account quotes the judge as giving the following justification of his action:

> "His in-court statement was a confession to first-degree murder and contradicted his story to the state attorney," Sepe said. "He reneged and lied—either to the state attorney or to the jury."
>
> "He did not cooperate with the state, an essential condition to his plea to the lesser charge of manslaughter."[28]

Judge Sepe's reasoning may be paraphrased with this argument:

Dennis's in-court statement CONTRADICTED the story he told the state attorney. If this is so, then he lied either to the JURY or to the state ATTORNEY or perhaps both. Provided that he lied to the jury, he did not cooperate with the STATE. And if he lied to the state attorney, he also did not cooperate with the state. Since Dennis's cooperating with the state was a necessary condition for dropping the charge of MURDER, the state will charge him with murder.

(M = The state charges Dennis with murder)

9. In *Morris* v. *Webber*, Moore K.B. 225, 72 Eng. Rep. 545, 2 Leonard 169, 74 Eng. Rep. 449 (1587), the question was whether one John Bury was illegitimate. John's supposed father, Henry, had been married to another woman before he married John's mother. The first marriage had been annulled on the ground that Henry was incurably impotent. It was argued that if Henry was not in fact impotent, then his first marriage was not legally annulled, and he was not validly married to John's mother. On the other hand, if Henry *was* incurably impotent, he could not have been John's real father. Either way, John could not be legitimate.

Either Henry Bury was incurably IMPOTENT or he wasn't. If he was not incurably IMPOTENT, then his first marriage was not legally ANNULLED. And if Henry's first marriage was not legally annulled, then he was not VALIDLY married to John's mother. If Henry was not validly married to John's mother, then John was not the LEGITIMATE child of Henry. On the other hand, if Henry was incurably impotent, he was not John's real FATHER. If he was not John's real father, then

---

[28]"Murder Suspect-Witness Stripped of Lesser Plea," *Miami News* (October 7, 1971), p. 5A.

John was not legitimate. It follows that John was not the legitimate child of Henry Bury.

*10. The author of *Alice in Wonderland*, Lewis Carroll, was a logician, so it is not surprising to find Alice employing deductive arguments:

> *Soon her eye fell on a little glass box that was lying under the table: she opened it, and found in it a very small cake, on which the words "EAT ME" were beautifully marked in currants. "Well, I'll eat it," said Alice, "and if it makes me larger, I can reach the key; and if it makes me smaller, I can creep under the door; so either way I'll get into the garden, and I don't care which happens!"*[29]

Alice's argument:

The cake will make me either LARGER or SMALLER. If it makes me larger, I can reach the KEY; and if it makes me smaller, I can CREEP under the door. If I reach the key, I'll get into the GARDEN; and if I creep under the door, I'll get into the garden. So [either way], I'll get into the garden.

11. A news story reveals that a number of firms made political contributions to both of the candidates for one seat on the Dade County (Florida) Commission. A political consultant commented:

> *The basic theory is, "If I contribute to both [Phillips and Dusseau], how can I lose? I'll have access to this person or I'll have access to that person."* ...
>
> *It's well known all over the earth. There's a vernacular for it, but since you're a family newspaper, I'll have to call it CYA.*[30]

The contributors must have reasoned:

I will have ACCESS to a commissioner because I am contributing to both campaigns. If I contribute to PHILLIPS'S campaign and she wins, then I have access to a commissioner. And if I contribute to DUSSEAU'S campaign and he wins, then I have access to a commissioner. Obviously, either Phillips or Dusseau will win.

(B = Phillips wins the election, C = Dusseau wins the election)

---

[29] Lewis Carroll, *Alice in Wonderland* (London: Dent, 1961), pp. 8–9.
[30] Luis Feldstein Soto, "Donors Hedged Their Bets," *Miami Herald* (October 10, 1988), p. 1B.

12. (SEMI-CHALLENGING) (a) Symbolize this sentence:

    Fred is forbidden to VISIT or TELEPHONE Wilma.

    (V = Fred is forbidden to visit Wilma, T = Fred is forbidden to telephone Wilma) (b) Prove that 'V v T' is an incorrect symbolization. (c) Explain why this anomaly occurs. (d) Is there any correct symbolization of the sentence employing the dictionary above that utilizes a wedge?

13. (SEMI-CHALLENGING) Symbolize this statement using the suggested notation.

    (*policy proposed by Senator Dole*) ". . . We can't DEPLOY American forces to Haiti without Congressional AUTHORIZATION, unless there's some emergency need to EVACUATE American people or unless there's some national INTEREST and you don't have TIME to go to Congress or unless the President CERTIFIES that the safety [of the military is] involved. . . ."

    (A = Congress authorizes deployment)

14. (CHALLENGING) Congressman Lehman's statement in exercise 1(b) above is a logical truth. Prove this by deducing it free of assumptions. Our proof has two provisional assumptions and is eight lines long.

15. (CHALLENGING) It was claimed above that F3 and F4 are logically equivalent. They are equivalent if and only if each entails the other. Prove that F3 entails F4. (A nearly identical proof would demonstrate the reverse entailment.) The proof involves two applications of the Wedge Out Rule. Our proof has 16 lines.

    (F3) S v (W v D)
    (F4) (S v W) v D

16. (CHALLENGING) In *Harisiades* v. *Shaughnessy*, 342 U.S. 580 (1952), the Supreme Court upheld a 1940 law permitting the deportation of aliens on the ground of having *once* been members of an organization advocating forcible overthrow of the government (read "Communist Party"). Justice Douglas began his dissenting opinion as follows:

    > *There are two possible bases for sustaining this Act:*
    > *(1) A person who was once a Communist is tainted for all time and forever dangerous to our society; or*
    > *(2) Punishment through banishment from the country may be placed upon an alien not for what he did, but for what his political views once were.*

> Each of these is foreign to our philosophy. We repudiate our traditions of tolerance and our articles of faith based upon the Bill of Rights when we bow to them by sustaining an Act of Congress which has them as a foundation. (342 U.S. at 598)

Symbolize Douglas's argument using this dictionary: S = The Act is sustainable, T = A person who was once a Communist is tainted for all time, B = An alien may be banished for what his political views once were. In this context, 'two possible bases' may be interpreted as "only two possible bases," and 'foreign to our philosophy' may be considered tantamount to "false." Construct a proof for the argument.

§2.5 IFF

A statement consisting of two constituent statements joined by the connective 'if and only if' is called a *biconditional*. An example:

(S1) Lynn will GO to law school if and only if she is ACCEPTED.

A biconditional is equivalent to the conjunction of a pair of conditionals. Thus, S1 is logically equivalent to S2:

(S2) Lynn will go to law school if she is accepted, and she will go to law school only if she is accepted.

S2 is, of course, symbolized by F2:

(F2) $(A \rightarrow G) \mathbin{\&} (G \rightarrow A)$

We could employ F2 as a symbolization of S1 but it will prove convenient to have a single symbol to abbreviate the locution 'if and only if'; that symbol is the *double arrow* ($\leftrightarrow$). We symbolize S1 with F1:

(F1) $G \leftrightarrow A$

F1 is read "G if and only if A" or "G double arrow A." The expression 'is a necessary and sufficient condition for' is approximately synonymous with 'if and only if' and is also abbreviated by the double arrow. It is customary among logicians to abbreviate 'if and only if' as 'iff', and we will adopt that custom. The conventions for punctuating formulas containing arrows also apply to formulas containing double arrows.

The two Double Arrow Rules are based on the fact that a biconditional is equivalent to the conjunction of a pair of conditionals.

The Double Arrow In Rule ($\leftrightarrow$I): From $A \to B$ and $B \to A$ derive $A \leftrightarrow B$.

The Double Arrow Out Rule ($\leftrightarrow$O): From $A \leftrightarrow B$ derive either $A \to B$ or $B \to A$.

The standard assumption-dependence principle applies to each rule.

We illustrate the application of these rules by constructing three proofs. The first concerns this simple argument:

Lynn will MOVE to Chapel Hill iff she GOES to law school. So, she goes to law school iff she moves to Chapel Hill.

$M \leftrightarrow G \vdash G \leftrightarrow M$

(1) $M \leftrightarrow G$ A
(2) $M \to G$ 1 $\leftrightarrow$O
(3) $G \to M$ 1 $\leftrightarrow$O
(4) $G \leftrightarrow M$ 3,2 $\leftrightarrow$I

The second proof demonstrates the validity of an argument advanced by the Florida Supreme Court in a case where a woman sought to compel her ex-husband to pay for the college education of their adult daughter:

> While most parents willingly assist their adult children in obtaining a higher education that is increasingly necessary in today's fast-changing world, any duty to do so is a moral rather than a legal one. Parents who remain married while their children attend college may continue supporting their children even beyond age twenty-one, but such support may be conditional or may be withdrawn at any time, and no one may bring an action to enforce continued payments. It would be fundamentally unfair for courts to enforce these moral obligations of support only against divorced parents while other parents may do as they choose.[31]

The court's argument can be formulated as follows:

---

[31]*Grapin v. Grapin*, 450 So.2d 853 (Fla. 1984).

§2.5 Iff

DIVORCED parents are legally obligated to pay for the education of their children who are 21 or older iff MARRIED parents are so obligated. Married parents are under no such legal obligation. Therefore, neither are divorced parents.

D ↔ M, −M ⊢ −D

The proof:

1	(1)	D ↔ M	A
2	(2)	−M	A
3	(3)	D	PA
1	(4)	D → M	1 ↔O
1,3	(5)	M	4,3 →O
1,2,3	(6)	M & −M	5,2 &I
1,2	(7)	−D	3-6 −I

An ad for a TV movie about steroids and athletics included this statement:

(S3) If I TAKE it, I WIN; if I don't take it, I don't win.

S3 is equivalent to this biconditional:

(S4) I win iff I take it.

The proof below shows that S3 entails S4. Exercise 8 treats the reverse entailment.

1	(1)	(T → W) & (−T → −W)	A	[1]
1	(2)	T → W	1 &O	[3]
1	(3)	−T → −W	1 &O	[4]
4	(4)	W	PA	[6]
5	(5)	−T	PA	[8]
1,5	(6)	−W	3,5 →O	[9]
1,4,5	(7)	W & −W	4,6 &I	[10]
1,4	(8)	T	5-7 −O	[7]
1	(9)	W → T	4-8 →I	[5]
1	(10)	W ↔ T	9,2 ↔I	[2]

The column of numbers at the right indicates the order of proof discovery; this column is not part of the proof format.

We have now discussed the five main connectives of propositional logic. We have introduced ten rules of inference—an "In" rule and an "Out" rule for each of these connectives. These rules are displayed together on one page in Appendix Five. These ten rules and the Rule of Assumptions form a set which is both "consistent" and "complete." Calling the set *consistent* means that any argument for which a proof can be completed using only these rules is valid; one cannot complete a proof for an invalid argument. Calling the rule set *complete* means that a proof employing only these rules can be constructed for *any* valid symbolized argument which can be expressed with the symbolic vocabulary we have adopted.

In the next two sections we augment our proof system (§2.6) and present two more logical techniques for evaluating propositional arguments (§2.7).

EXERCISES

1. Symbolize each statement using the suggested notation.
   (a) (*college memo*) "We can emphasize these APPLIED areas most effectively if, and only if, we emphasize also the FUNDAMENTAL areas of research."
   *(b) (*bulletin*) "If and only if it [a motion on racial research] is APPROVED by a majority of the AAA membership will it become an official POSITION of the American Anthropological Association."
   (c) (*Section 2.4*) "They [F3 and F4] are EQUIVALENT if and only if each entails the other." (A = F3 entails F4, B = F4 entails F3)
   (d) (*ABC newscaster Carole Simpson*) "If there *is* a PEACE agreement, and only if, American troops would begin MOVING almost immediately toward Bosnia."
   (e) (*logician W. V. Quine*) "For the TENABILITY of the thesis that mathematics is logic it is not only sufficient but also necessary that all mathematical expressions be capable of DEFINITION on the basis solely of logical ones."
   *(f) (*Winston Churchill, June 1940*) "Provided, but only provided, that the French Fleet is SAILED forthwith for British harbors, His Majesty's Government give their full CONSENT to an armistice for France."
   (g) (*newspaper*) "North Vietnam will MEET with U.S. negotiator Henry Kissinger Monday but only if the U.S. will AGREE to sign the peace agreement Tuesday on schedule."
   (h) (*logician Geoffrey Hunter*) "A set C is a *PROPER* subset of a set D iff there is no member of C that is not a member of D but there is a member of D that is not a member of C." (A = There is a member of

C that is not a member of D, B = There is a member of D that is not a member of C)

2. Translate each formula into an English sentence using this dictionary:
    B = Notre Dame wins its bowl game
    C = Notre Dame wins the national championship
    M = Miami wins the national championship
    F = Notre Dame beats Miami in the season finale
   (a) C ↔ −M
   *(b) −(C ↔ B)
   (c) C ↔ (F & B)
   (d) (C ↔ F) & B

3. From a newspaper sports story appearing on the last day of the major-league season:

    If the BRAVES and GIANTS both win or both lose today, they will play a one-game PLAYOFF Monday night in San Francisco.

   (B = The Braves win) Because tied games are rare in baseball, 'The Braves lose' may be symbolized as '−B'. (a) Provide the literal symbolization of the sentence. (b) Supply an equivalent symbolization that employs neither '&' nor 'v'.

4. Construct formal proofs of validity for each of the following symbolized arguments:
   (a) A ↔ B, B & C ⊢ A & C
   *(b) D → (E & F), E → D ⊢ D ↔ E
   (c) G ↔ H, (H → G) → (I → J), (G → H) → (J → I) ⊢ J ↔ I
   (d) K ↔ L, L ↔ M ⊢ K ↔ M

Instructions for exercises 5 through 10: symbolize each argument and construct a proof of validity for it.

5. A group of black golfers provided a scholarship at the University of Miami for "a promising black golfer who can qualify academically."[32] Some people objected to the terms of the scholarship as embodying a form of "reverse discrimination." They reasoned as follows:

    A scholarship restricted to BLACK golfers is racially discriminatory iff a scholarship restricted to WHITE golfers is discriminatory. Clearly, a scholarship for which only white golfers are eligible is discriminatory. Therefore, a scholarship for which only black golfers are eligible is discriminatory as well.

[32]Charlie Nobles, "Black Golfers on a Treadmill," *Miami News* (February 18, 1972), p. 4B.

*6. Philosopher John Locke writes the following in support of his view that warmth is a "secondary quality" that exists only in our ideas:

> ... *He that will consider that the same fire that, at one distance produces in us the sensation of warmth, does, at a nearer approach, produce in us the far different sensation of pain, ought to bethink himself what reason he has to say—that this idea of warmth, which was produced in him by the fire, is* actually in the fire; *and his idea of pain, which the same fire produced in him the same way, is* not *in the fire.*[33]

Locke is arguing:

The WARMTH is a quality of the fire exactly if the PAIN is a quality of the fire. Since the pain is not a quality of the fire, neither is the warmth.

7. The logicians Hughes and Cresswell write:

> *It should be noted that whenever we have a thesis of the form* Cab *we can always use TR3 to obtain* ⊢ LCab, *and hence, by Def F,* ⊢ Fab. *Moreover, whenever we have* ⊢ Fab *we can, by Def F, substitution of* Cab *for* p *in A5, and Modus Ponens, obtain* ⊢ Cab. *I.e. whenever* Cab *is a thesis, so is* Fab, *and vice versa.*[34]

Their reasoning may be paraphrased:

If 'Cab' is a theorem, then 'LCab' is also, and if 'LCab' is a theorem, then 'Fab' is a theorem. Moreover, if 'Fab' is a theorem, then 'Cab' is too. This proves that 'Cab' is a theorem iff 'Fab' is a theorem.

(C = 'Cab' is a theorem, L = 'LCab' is a theorem, F = 'Fab' is a theorem)

8. We showed above that S3 entails S4.

(S3) If I TAKE it, I WIN; if I don't take it, I don't win.

(S4) I win iff I take it.

Now establish the reverse entailment (thereby proving logical equivalence).

---

[33] John Locke, *An Essay Concerning Human Understanding* (New York: Dover, 1959), vol I: p. 174.
[34] G. E. Hughes and M. J. Cresswell, *An Introduction to Modal Logic* (London: Methuen, 1968), p. 31. We have replaced their logical notation by one that is more readily printed.

9. In a *Star Trek* show, the crew of the starship *Enterprise* is held captive by a powerful computer. The crew escapes after one of them says to the computer, "I am lying to you." This inference blew the computer's circuits:

> **He SAYS that he is lying. If he says that he is lying and he *is* LYING, then he isn't lying. But also, if he says that he is lying and he *isn't* lying, then he is lying. This proves that he is lying iff he isn't lying.**

*10. Some people believe that their bodies will be re-created at some time following their deaths. But would the re-creation of, say, Elvis Presley actually be Elvis or would it merely be an Elvis *replica*? The realization that it is logically possible for there to be two Elvis re-creations (label them "A" and "B") counts against the view that either re-creation would be Elvis. This argument pursues the point.

> **A is Elvis iff B is Elvis. But they can't both be Elvis. So, A is not Elvis and neither is B.**

(A = A is Elvis, B = B is Elvis)

11. Five-year-old Brian Dailey pulled a lawn chair out from under elderly Ruth Garratt as she was sitting down. Ruth fell, broke her hip, and sued Brian for medical expenses (*Garratt* v. *Dailey*, 46 Wash.2d 197, 279 P.2d 1091 (1955)). The appellate court decision includes the following statements:

> *If . . . [Brian] had such knowledge [that plaintiff would attempt to sit down where the chair had been] the necessary intent [to cause the plaintiff's bodily contact with the ground] will be established and the plaintiff will be entitled to recover. . . . If Brian did not have such knowledge, there was no wrongful act by him and the basic premise of liability on the theory of a battery was not established.*

(Each of these two statements may be viewed as the conjunction of two conditionals.) Do these judicial remarks entail:

> Defendant is LIABLE to plaintiff for battery iff defendant KNEW that plaintiff would attempt to sit down where the chair had been.

Use this dictionary: K, I = Defendant intended to cause the plaintiff's bodily contact with the ground, L, W = Defendant committed a wrongful act.

12. (CHALLENGING) In Section 2.4 it was noted that S1 may be symbolized by F1.

>(S1) A or B but not both.

>(F1) (A v B) & −(A & B)

S1 may also be symbolized by the shorter F2.

>(F2) A ↔ −B

Demonstrate the logical equivalence of F1 and F2 by proving that each entails the other.

13. (CHALLENGING) (a) For each of the following formulas produce a logically equivalent formula that uses no connectives other than the dash and the ampersand.

$$P \rightarrow Q$$
$$P \vee Q$$
$$P \leftrightarrow Q$$

(b) For each of the following formulas produce a logically equivalent formula that uses no connectives other than the dash and the wedge.

$$P \rightarrow Q$$
$$P \& Q$$
$$P \leftrightarrow Q$$

The successful completion of this exercise indicates that we could dispense with three of our five connectives (the arrow, the double arrow, and either the wedge or the ampersand).

§2.6 DERIVED RULES

Some proofs that utilize only the eleven rules introduced so far will be complicated—even when the argument involved is simple. Consider, for example, this argument:

> **I either have a TUMOR or I am PREGNANT. But since I can't be pregnant, I must have a tumor.**
>
> T v P, −P ⊢ T

The first premise of this argument was a medical opinion. The second premise was based on the fact that the woman who employed the argument had previously undergone a sterilization procedure. Happily for her the frightening conclusion of this valid argument proved to be false. A valid argument can have a false conclusion only if one or more of its premises are also false. In this instance the second premise was false; the attempted sterilization had failed and she was pregnant.[35]

The proof of validity for this argument:

1	(1)	T v P	A
2	(2)	−P	A
3	(3)	T	PA
2,3	(4)	T & −P	3,2 &I
2,3	(5)	T	4 &O
2	(6)	T → T	3-5 →I
7	(7)	P	PA
8	(8)	−T	PA
2,7,8	(9)	P & −P & −T	7,2,8 &I
2,7,8	(10)	P & −P	9 &O
2,7	(11)	T	8-10 −O
2	(12)	P → T	7-11 →I
1,2	(13)	T	1,6,12 vO

(The fancy footwork on lines 4 and 5 permit the derivation of line 6. With the help of line 9 we make line 10 depend on the assumption on line 8.)

You no doubt agree that this proof represents a lot of effort directed toward demonstrating the validity of a quite simple argument. Constructing proofs should be simpler than this, and we can make proof construction generally simpler if we adopt the seven additional rules of inference displayed on the next page. The standard assumption-dependence principle applies to each of these rules.

Any number of additional rules could be added to our rule set. We chose to keep the set relatively small for the sake of simplicity. Why have we chosen this particular set of rules rather than some other small set? Because the first five rules correspond to natural and common ways to reason and because all seven are particularly useful for abbreviating proofs. We call these rules "derived" rules because they sanction deductions that could be made without their help employing only the original (or "primitive") rules of inference. The proof above, for example, shows how one could dispense with the Disjunctive Argument Rule. Because the set of primitive rules is consistent, we know the expanded set is also.

---

[35]Bill Gjebre, "Unplanned Birth Nearly Cost Marriage," *Miami News* (February 2, 1976), p. 1A.

THE SEVEN DERIVED INFERENCE RULES		
CHAIN ARGUMENT	CH	From $A \to B$ and $B \to C$ derive $A \to C$.
MODUS TOLLENS	MT	From $A \to B$ and $-B$ derive $-A$.
DISJUNCTIVE ARGUMENT	DA	From $A \lor B$ and $-A$ derive B. From $A \lor B$ and $-B$ derive A.
CONJUNCTIVE ARGUMENT	CA	From $-(A \& B)$ and A derive $-B$. From $-(A \& B)$ and B derive $-A$.
DOUBLE NEGATION	DN	From A derive $--A$ and vice versa.
DE MORGAN'S LAW	DM	From $-(A \& B)$ derive $-A \lor -B$ and vice versa. From $-(A \lor B)$ derive $-A \& -B$ and vice versa.
ARROW	AR	From $A \to B$ derive $-A \lor B$ and vice versa. From $A \to B$ derive $-(A \& -B)$ and vice versa. From $-(A \to B)$ derive $A \& -B$ and vice versa.

In Section 2.2 we discussed the argument pattern corresponding to the Chain Argument Rule, and in Section 2.3 we discussed the argument pattern corresponding to the Modus Tollens Rule as well as the pattern that is its invalid counterfeit, "denying the antecedent." We shall now examine the patterns that match the third and fourth derived rules and their associated counterfeit forms.

	VALID PATTERNS	INVALID PATTERNS
DISJUNCTIVE ARGUMENT	$P \lor Q, -P \vdash Q$ $P \lor Q, -Q \vdash P$	$P \lor Q, P \vdash -Q$ $P \lor Q, Q \vdash -P$
CONJUNCTIVE ARGUMENT	$-(P \& Q), P \vdash -Q$ $-(P \& Q), Q \vdash -P$	$-(P \& Q), -P \vdash Q$ $-(P \& Q), -Q \vdash P$

Notice that in a valid disjunctive argument the second premise *denies* a disjunct, while in the invalid pattern the second premise *affirms* a disjunct. Conversely, in a valid conjunctive argument the second premise *affirms* a conjunct (of a negated conjunction) while in the invalid form that premise *denies* a conjunct. Consider simple examples of these four argument patterns:

	VALID ARGUMENT	**INVALID ARGUMENT**
DISJUNCTIVE ARGUMENT	Bill Clinton is an actor or a politician. He is not an actor. So, he is a politician.	Ronald Reagan is an actor or a politician. He is an actor. So, he is not a politician.
CONJUNCTIVE ARGUMENT	Ronald Reagan is not a resident of both California and Ohio. He is a resident of California. So, he is not a resident of Ohio.	Ronald Reagan is not a resident of both Pennsylvania and Ohio. He is not a resident of Pennsylvania. So, he is a resident of Ohio.

Note that each of the invalid arguments has true premises and a false conclusion—a sure indicator of invalidity.

It is fairly obvious why the valid disjunctive argument pattern is valid and also why its counterfeit is invalid. If a disjunction is true, then at least one disjunct must be true; so if it isn't one of the two disjuncts, it must be the other. Both disjuncts of a true disjunction may be true; so you can't reason from the truth of one to the falsity of the other. It is perhaps a little less obvious with the conjunctive argument patterns. A conjunction is true iff both conjuncts are true; it follows that a conjunction is false iff at least one conjunct is false. If it isn't one conjunct that's false, it must be the other. Both conjuncts of a false conjunction may be false; so you can't reason from the falsity of one to the truth of the other.

The validity of the Double Negation Rule rests on the fact that a statement and its negation have opposite truth-values. If 'P' is true, '−P' will be false, and therefore its negation, '−−P', will be true; hence, if 'P' is true, so is '−−P'. The reverse holds as well.

A clarification of the notion of "negation" will prove helpful. The negation of a formula is produced by adding a dash to the formula—not by deleting a dash. Hence, '−P' is the negation of 'P', but 'P' is not the negation of '−P'. There is therefore a difference between the concepts of "negation" and "opposite" (or "contradictory"). The formula 'P' is the opposite (contradictory) of '−P', but it is not its negation. This distinction obviates the absurdity that an affirmative statement may be a negation.

De Morgan's Law (named for the nineteenth-century British logician who propounded it) concerns the equivalence of F1 and F2, and the equivalence of F3 and F4:

(F1) −(A & B)
(F2) −A v −B

(F3) −(A v B)
(F4) −A & −B

Formulas F1 and F2 have this in common: they are true when and only when at least one of the constituent statements ('A' and 'B') is false. Similarly F3 and F4 are true when and only when the constituents are both false. These equivalences are established formally in Section 2.7. Incidentally you should guard against thinking that F1 is equivalent to F4, or that F2 and F3 are equivalent.

The Arrow Rule allows us to convert a conditional into either a disjunction or the negation of a conjunction (and vice versa). We also include a version that treats negated conditionals. The Arrow Rule reflects the fact that there are alternative ways of presenting the information conveyed by a conditional statement—that the antecedent cannot be true if the consequent is false. Each of the following sentences conveys the same information:

> If you eat turnips you'll get sick.
> Either you refrain from eating turnips or you get sick.
> You can't both eat turnips and not get sick.

The validity of the Arrow Rule is the subject of exercise 14 at the end of this section and exercise 3 in the next section.

In order to apply DA, CA, DM, or AR to a disjunction with more than two disjuncts or a conjunction with more than two conjuncts we should reduce the number of disjuncts or conjuncts by introducing internal groupers. For instance, the formula 'A v B v C' will be written as 'A v (B v C)' or else as '(A v B) v C'.

There are several differences between the first four derived rules and the last three. (1) Each of the first four rules involves two premises, while each of the last three involves only one. (2) The last three rules involve logical equivalence and not just entailment; hence they, and only they, are "reversible." For example, the Double Negation Rule sanctions the move from F5 to F6 and also the reverse move.

(F5) P
(F6) −−P

(3) Because the last three rules involve logical equivalence, they may be correctly applied to line *parts* as well as to *whole* lines. (The first four derived rules and all of the primitive rules concern entailment rather than equivalence; so they may *not* be applied to line parts.) Thus, the Double Negation Rule will sanction the move from F7 to F8 (where it is applied to the right conjunct of the antecedent), but the Ampersand Out Rule does not justify the move from F7 to F9.

(F7) (P & Q) → R
(F8) (P & −−Q) → R
(F9) P → R

(It's a good thing that the Ampersand Out Rule does not sanction the move from F7 to F9 because F7 does not entail F9; if it did sanction that move, our rule set would be inconsistent.)

We will illustrate the employment of these derived rules by constructing several proofs. The philosopher Kierkegaard writes:

If you marry, you will regret it; if you do not marry, you will also regret it; if you marry or do not marry, you will regret both; . . .[36]

Kierkegaard appears to be advancing this argument (in symbols):

M → R, −M → R ⊢ R

(M = You marry, R = You will regret your decision about marriage) The proof:

1	(1)	M → R	A
2	(2)	−M → R	A
3	(3)	−R	PA
1,3	(4)	−M	1,3 MT
2,3	(5)	−−M	2,3 MT
1,2,3	(6)	−M & −−M	4,5 &I
1,2	(7)	R	3-6 −O

Consider the "guarantee" offered in the *Far Side* cartoon on page 86. Is the guarantee (S10) equivalent to statement S11?

(S10) Either your IQ is DOUBLED or you will receive no MONEY back.
(S11) Your money will be returned only if your IQ is doubled.

(F10) D v −M
(F11) M → D

---

[36] Soren Kierkegaard, *Either/Or* (Garden City, NY: Doubleday & Company, Inc., 1959), vol I: p. 37.

"Well, I dunno . . . Okay, sounds good to me."

The Far Side cartoon by Gary Larson is reprinted by permission of Chronicle Features, San Francisco, CA. All rights reserved.

Of course, these two statements are equivalent iff each entails the other. With the help of two derived rules we can show that S10 entails S11.

1	(1)	D v −M	A
2	(2)	M	PA
2	(3)	−−M	2 DN
1,2	(4)	D	1,3 DA
1	(5)	M → D	2-4 →I

Our primitive-rules proof for this argument has four provisional assumptions and 14 lines. By applying a couple of derived rules we cut nine lines from the proof. This well illustrates why it was advisable to expand the set of rules.

The step of Double Negation on line 3 was necessary to set up the Disjunctive Argument move on the next line. Note that in a step of DA one line must be *the negation of* one of the disjuncts of another line. Line 3 is the negation of the right disjunct of line 1. Line 2, on the other hand, is not the negation of anything. It would not be correct, then, to apply DA to lines 1 and 2 as they stand.

Now we prove that S11 entails S10. As we have already demonstrated the reverse entailment, this will establish the logical equivalence of the two statements.

(S10) Either your IQ is DOUBLED or you will receive no MONEY back.
(S11) Your money will be returned only if your IQ is doubled.

S10 is a disjunction, and as a general rule disjunctions are difficult to derive. It will help to have strategies for reaching them; here are two:

The *Arrow Strategy:* You can reach a disjunction by setting a conditional sub-goal (which will be reached by Arrow In or Chain Argument). The conditional is then transformed into the disjunction with the help of the Arrow Rule.

The *De Morgan's Law Strategy:* Make a provisional assumption of the negation of the disjunctive goal (with an eye to a subsequent step of Dash Out). Apply De Morgan's Law to the provisional assumption.

Note two things about the Arrow Strategy: (1) When the conditional is transformed into a disjunction, the left disjunct will be the *negation* of the conditional's antecedent. (2) So, if the left disjunct of the goal statement is affirmative, an intermediate step of Double Negation will be required. These two points and the Arrow Strategy are all illustrated in this proof that S11 entails S10:

1	(1)	M → D	A
2	(2)	−D	PA
1,2	(3)	−M	1,2 MT
1	(4)	−D → −M	2-3 →I
1	(5)	−−D v −M	4 AR
1	(6)	D v −M	5 DN

Here are two additional general suggestions for planning strategies for proof construction: (1) Proofs can be constructed from the top down or from the bottom up. Constructing from the bottom up is a matter of listing goals and sub-goals until you reach the premises of the argument. One asks, "What line would I need in order to reach the conclusion of the argument? What line would I need in order to reach that subordinate goal?" etc. Of course, setting sub-goals is not deriving lines; when you reach the top in this chain of goal setting you must turn around and descend again filling in the justification and assumption-dependence columns. Frequently one works a proof partly top-to-bottom and partly bottom-to-top.

(2) Typically a proof line is used exactly once; that is, one and only one deduction is made from that line. (Conjunctions and biconditionals are often exceptions to this generalization; two deductions are frequently made from

these statements.) So, at a given point in devising a proof, it is advisable to concentrate attention on all and only those premise lines that have not yet been used. To facilitate this it may be helpful to place a check mark (or some such sign) by the premise lines as you make deductions from them. Then you can attend to the unchecked lines.

We illustrate these two tips by constructing a proof for an argument advanced by columnist William Safire. During the 1988 presidential campaign, Safire had this to say about the Quayle-National Guard flap:

> *The dilemma is this: If the Bush staff failed to vet the potential running mate aggressively on this issue, it was incompetent, and Bush asks us to let him bring that band of bunglers into the White House. On the other hand, if the Bush staff asked the tough questions and Quayle misled them, then he was duplicitous or naive—and not the best suited to be one heartbeat from the presidency.*[37]

Safire's argument:

> If the Bush staff didn't VET Quayle on the issue, then it was not COMPETENT.
> If they grilled him and Quayle was not FORTHRIGHT with them, then he was DUPLICITOUS or NAIVE.
> If Quayle was duplicitous or naive, then he is not SUITED for the presidency.
> It cannot be both the case that they grilled Quayle and that he was forthright. [suppressed premise]
> It follows that either the Bush staff was incompetent or Quayle is not suited for the presidency.

$-V \to -C$
$(V \& -F) \to (D \lor N)$
$(D \lor N) \to -S$
$-(V \& F)$
$\vdash -C \lor -S$

The second and third premises of the argument are the premises of a step of Chain Argument. Even before developing an overall strategy for constructing the proof we may want to make that derivation.

$$
\begin{array}{lll}
(1) & -V \to -C & A \\
\checkmark\ (2) & (V \& -F) \to (D \lor N) & A \\
\checkmark\ (3) & (D \lor N) \to -S & A \\
\end{array}
$$

[37]"Quayle Issue Tests Bush's Character," *Miami News* (August 23, 1988), p. 11A.

(4)   −(V & F)   A
(5)   (V & −F) → −S   2,3 CH

We "check" lines 2 and 3 to focus attention on the remaining three lines at the top of the proof. No additional familiar patterns emerge so we examine the final goal of the proof (the conclusion of the argument). That goal (let's call it line "z" for now) is a disjunction, so we should adopt either the Arrow Strategy or the De Morgan's Law Strategy. We'll construct proofs illustrating each. In this proof we employ the former strategy and accordingly set down a conditional on line y as a sub-goal. Because the formula on line y is a conditional we use the Arrow In Strategy to set down lines 6 and x. Line x matches the consequent of line 5, so we put down on line w (as a sub-goal) the antecedent of line 5. Lines u and v are added as means of reaching line w.

(6)   C   PA
       .
       .
       .
(u)   V
(v)   −F
(w)   V & −F
(x)   −S
(y)   C → −S
(z)   −C v −S

Of course, lines u through z have not been derived; instead they are goals we need to realize. The gap separating the two halves of the proof is represented by a vertical ellipsis. We have a plan for reaching lines w through z; now we need to make a plan for deriving lines u and v. It's time to look at the so-far unchecked lines at the top of the proof: 1, and 4 through 6. Lines 1 and 6 resemble the premises of a *modus tollens*—except that the consequent of line 1 is the negation of line 6, rather than the other way around. That problem can be solved handily with the help of Double Negation.

✓ (7)   −−C   6 DN
  (8)   −−V   1,7 MT

We see that line u (now 9) follows from 8.

(9)   V   8 DN

After "checking" all the lines from 1 to 9 that have been used to generate further lines we identify these unchecked lines: 4, 5, and 9. Lines 4 and 9 are the premises of a conjunctive argument that will lead to 10 (formerly goal line v).

(10) −F   4,9 CA

Now the two halves of the proof (the assumptions and derivations at the top and the goal lines at the bottom) have been linked together. We complete the proof by making the derivations previously envisioned when we set out goal lines, and adding an assumption-dependence column.

(original goal-line letters)

1	(1)	−V → −C	A	
2	(2)	(V & −F) → (D v N)	A	
3	(3)	(D v N) → −S	A	
4	(4)	−(V & F)	A	
2,3	(5)	(V & −F) → −S	2,3 CH	
6	(6)	C	PA	
6	(7)	−−C	6 DN	
1,6	(8)	−−V	1,7 MT	
1,6	(9)	V	8 DN	(u)
1,4,6	(10)	−F	4,9 CA	(v)
1,4,6	(11)	V & −F	9,10 &I	(w)
1,2,3,4,6	(12)	−S	5,11 →O	(x)
1,2,3,4	(13)	C → −S	6-12 →I	(y)
1,2,3,4	(14)	−C v −S	13 AR	(z)

An alternative proof for this argument employs the De Morgan's Law Strategy for reaching a disjunctive conclusion.

1	(1)	−V → −C	A
2	(2)	(V & −F) → (D v N)	A
3	(3)	(D v N) → −S	A
4	(4)	−(V & F)	A
2,3	(5)	(V & −F) → −S	2,3 CH
6	(6)	−(−C v −S)	PA
6	(7)	−−C & −−S	6 DM
6	(8)	−−C	7 &O
1,6	(9)	−−V	1,8 MT

1,6	(10)	V	9 DN
1,4,6	(11)	−F	4, 10 CA
1,4,6	(12)	V & −F	10,11 &I
1,2,3,4,6	(13)	−S	5,12 →O
6	(14)	− −S	7 &O
1,2,3,4,6	(15)	−S & − −S	13,14 &I
1,2,3,4	(16)	−C v −S	6-15 −O

In this proof we applied the second version of De Morgan's Law (reading from left to right) to line 6 in order to derive line 7. Had we applied the first version (reading from right to left) we would have derived:

− −(C & S)

EXERCISES

1. Construct formal proofs of validity for each of the following symbolized arguments. In each proof employ at least one derived rule.
   (a) −A & (B → A) ⊢ −B
   *(b) C v D, −C ⊢ − −D
   (c) −(E & F), − −E ⊢ −F
   (d) −G v H, −(H & −I) ⊢ G → I
   (e) −(J v K) ⊢ −(K v J)

Instructions for exercises 2 through 13: symbolize each argument and construct a proof of validity for it. In each proof employ at least one derived rule.

2. A TV nature documentary gives this explanation of why each flock of bighorn sheep has a dominant male:

   > If there weren't a DOMINANT ram, all the rams would MATE with the ewes. If that happened the ewes would be SAPPED of energy, and if they were drained of energy their offspring would be WEAK. This shows that if there weren't a dominant ram, the offspring would be weak.

*3. René Descartes argued:[38]

---

[38]*Meditations on First Philosophy* (Indianapolis: The Bobbs-Merrill Co., Inc., 1960), p. 81.

BODY is by nature divisible. If this is so and mind and body are the SAME, then MIND is also divisible. However, the mind is entirely indivisible. Consequently, mind and body are not the same.

4. A Florida Assistant State Attorney gave this explanation (which we have paraphrased as an argument) for dropping rape charges that had been brought against a man who had sex with an elderly victim of Alzheimer's disease:[39]

   The State can OBTAIN a conviction only if it PROVES that the woman did not consent to having sex. The State will prove this only if there is relevant CIRCUMSTANTIAL evidence or the woman TESTIFIES. There is no relevant circumstantial evidence and [because of her illness] the woman will not testify. Therefore, the State cannot obtain a conviction.

   (P = The State proves that the woman did not consent) Material surrounded by brackets is meant to clarify or provide background. It should not be symbolized.

5. From *Alice in Wonderland*:

   Alice noticed, with some surprise, that the pebbles were all turning into little cakes as they lay on the floor, and a bright idea came into her head. "If I eat one of these cakes," she thought, "it's sure to make some change in my size; and, as it can't possibly make me larger, it must make me smaller, I suppose."[40]

   Alice's argument:

   If I EAT the cake, it will make me either LARGER or SMALLER. It can't make me larger. So, if I eat the cake, it will make me smaller.

6. An umpire ruled that Yankees second baseman Bobby Meacham trapped, rather than caught, a line drive in a game in 1988. Manager Billy Martin was ejected from the game after protesting the call. Martin had this to say after the game:

---

[39] Jeff Leen, "Alzheimer's Victim Can't Testify, So Prosecutor Drops Rape Charge," *Miami Herald* (November 6, 1983), p. 28A. The prosecutor's decision seems unfortunate. Under the applicable law, Fla. Stat. § 794.011, the woman's mental condition should make any consent on her part ineffectual.

[40] Lewis Carroll, *Alice's Adventures in Wonderland, Through the Looking-glass,* and *The Hunting of the Snark* (New York: The Modern Library, n.d.), pp. 60–61.

> *He [umpire Scott] first said he didn't see it, and I could have accepted that. But then he said the ball bounced. Either he was a liar the first time or the second time.*[41]

Martin's reasoning:

> **If Scott told the truth the first time, then he did not SEE the play. If he told the truth the second time, he did see the play. Hence, Scott either didn't tell the truth the first time or he didn't tell it the second time.**

(A = Scott told the truth the first time, B = Scott told the truth the second time)

*7. A newspaper story details the plight of Mrs. Bass, an elderly Texas widow caught in the Catch-22 of bureaucratic regulations.[42] Her problems started when the Veteran's Administration notified her that they were awarding her a pension she had not sought; they are rooted in these rules and facts:

> If Mrs. Bass receives the PENSION, then [because of her income level] she will not be eligible for MEDICAID.
>
> If she is ineligible for Medicaid, then she cannot AFFORD to stay in the nursing home.
>
> If she does not receive the pension, then she is also ineligible for Medicaid [because one is required by Medicaid guidelines to take advantage of all available benefits].

Show that these three statements taken together entail that Mrs. Bass cannot afford to remain in the nursing home.

8. Joanie Caucus advances an argument with the sentence beginning in the second panel of the "Doonesbury" strip on page 94. The unstated second premise is:

> Anita Hill is not a delusional SOCIOPATH.

The unstated conclusion of the argument is the left disjunct of the first premise. (H = Hill was telling the truth, L = Thomas is a garden-variety liar covering his guilt with righteous indignation, T = Thomas was telling the truth, S = Hill is a delusional sociopath)

---

[41]"Here's Dirt on Martin: He's in Trouble Again," *Miami Herald* (June 1, 1988), p. 1C.
[42]"Widow, 71, Is Too Rich, Too Poor," *Miami Herald* (January 16, 1986), p. 20A.

Doonesbury BY GARRY TRUDEAU

DOONESBURY ©1991 G. B. Trudeau. Reprinted with permission of Universal Press Syndicate. All rights reserved.

9. A bit of dialogue from *Pride and Prejudice*:

> "An unhappy alternative is before you, Elizabeth. From this day you must be a stranger to one of your parents.—Your mother will never see you again if you do not marry Mr. Collins, and I will never see you again if you do."[43]

(M = Your mother will see you again, F = Your father will see you again, C = You marry Mr. Collins) The argument is presented conclusion first.

10. A newspaper story contains the prediction that the Black Guerrilla Family prison gang would issue a "death sentence" for Tyrone Robinson, the man who gunned down Huey Newton (cofounder of the Black Panthers), on the grounds that he told police that he was a member of the gang.[44] The story contained this reasoning:

> **If Robinson is a MEMBER of the gang, he VIOLATED the rule of invisibility. If he violated this rule, he will be KILLED. If he is not a member, then he was POSING as one. If Robinson was posing as a member, he will be killed. It follows that he will be killed.**

*11. The full-time director of a federal scholarship program was discovered by reporters to be also working at his veterinary clinic where he was the only licensed veterinarian. Florida law requires the full-time presence of a licensed vet at every veterinarian's office. The news article notes:

> *Both Dr. González-Mayo and his wife realized that he was caught in a serious dilemma: If he said he was working at the federal program, then his clinic was being tended by veterinarians who were unlicensed; if he said he*

---

[43] Jane Austen, *Pride and Prejudice* (New York: New American Library, 1961), pp. 97–98.
[44] "Newton Case Suspect May Be Gang Target," *Miami Herald* (August 28, 1989), p. 2A.

was at the clinic supervising them, then he was not fulfilling his role as director of the scholarship program.[45]

Dr. González-Mayo's difficulty may be expressed by this argument:

Dr. González-Mayo cannot both work full-time as the director of the federal scholarship PROGRAM and also work full-time at his veterinary CLINIC. If he is not working full-time at the clinic, the clinic is not operating LEGALLY. Therefore, either his clinic is operating illegally or he is not working full-time as scholarship program director.

12. The following argument was classed as an "extra challenging" exercise in Section 2.3. It is considerably easier to prove when derived rules are available.

    If I buy GRASS, then I cannot afford to buy FOOD. If I cannot afford food, then I don't EAT. If I can afford food and don't buy grass, then I don't eat [because of nausea]. So, I don't eat.

13. (SEMI-CHALLENGING) In the dialogue *Statesman*, Plato writes:

    . . . If there are arts, there is a standard of measure, and if there is a standard of measure, there are arts; but if either is wanting, there is neither.[46]

    Prove that the clause preceding the semicolon entails the clause following it. (A = There are arts, M = There is a standard of measure)

14. (CHALLENGING) The various parts of the Arrow Rule correspond to the following six arguments:

    (a) $P \to Q \vdash -P \vee Q$ (b) $-P \vee Q \vdash P \to Q$
    (c) $P \to Q \vdash -(P \& -Q)$ (d) $-(P \& -Q) \vdash P \to Q$
    (e) $-(P \to Q) \vdash P \& -Q$ (f) $P \& -Q \vdash -(P \to Q)$

    By constructing proofs of validity for these arguments that use only the primitive inference rules we can show that the corresponding inference rules are "derived" (as has been claimed). Skip (c) and (f); they are too easy. Construct proofs for the remaining four arguments employing only primitive rules. Our proofs have these lengths:

---

[45]Louis Salome and Hilda Inclan, "Aid Chief Runs Own Clinic While on U.S. Payroll," *Miami News* (October 4, 1976), p. 4A.
[46]Plato, "Statesman," in *The Dialogues of Plato,* trans. Benjamin Jowett (New York: Random House, 1937), p. 312.

(a) 10 lines   (b) 14 lines
(d) 6 lines   (e) 9 lines

Can you produce more elegant (shorter) proofs?

§2.7 TRUTH TABLES

The logical technique of formal proof is very powerful, but it has one major limitation: it does not establish *in*validity. (Note that failure to complete a proof does not demonstrate invalidity.) In this section we introduce the method of truth tables—a technique that will prove both validity and invalidity.[47]

Every statement—simple or compound—has a *truth-value*; that is, every statement is true or false. The truth-value of a conjunction (billboard ad S1, for example) is determined by the truth-values of its conjuncts (S2 and S3).

(S1) He likes my kid and he drinks Johnnie Walker.
(S2) He likes my kid.
(S3) He drinks Johnnie Walker.

If both S2 and S3 are true, then S1 is true. If either S2 or S3 or both are false, then S1 is false. The same can be said for formulas F1 through F3, the symbolizations of these statements. The truth-value of F1 is entirely determined by the truth-values of F2 and F3.

(F1) L & D
(F2) L
(F3) D

Each of the five connective symbols has this characteristic: the truth-value of a formula in which that symbol is the major connective is determined by the truth-values of the formulas it connects (or, in the case of the dash, the truth-value of the formula to which it is attached).[48] The specific ways in which the truth-values of formulas are determined by the truth-values of their constituents are shown in the following *basic truth table*:[49]

---

[47]Gottlob Frege introduced an early version of the technique of truth tables in *Begriffsschrift* (Halle, 1879).
[48]The ten "In" and "Out" rules adopted in the first five sections of this chapter assign meanings to the connectives that guarantee that they possess this characteristic. One may wonder whether the five connective symbols so interpreted are faithful translations of English connective expressions. This question is addressed in §5.3.
[49]We simplify this section by assuming that multiple conjunctions and disjunctions will be written with internal groupers.

§2.7 Truth Tables

	GUIDE COLUMNS		(1)	(2)	(3)	(4)	(5)
	P	Q	−P	P & Q	P v Q	P → Q	P ↔ Q
(a)	T	T	F	T	T	T	T
(b)	F	T	T	F	T	T	F
(c)	T	F	F	F	T	F	F
(d)	F	F	F	F	F	T	T

Column 3 (to take an example) shows that a disjunctive formula is true when both disjuncts are true (row a), is also true when one disjunct is true (rows b and c), but is false when both disjuncts are false (row d). (We will refer to vertical lines of truth-values as "columns" and horizontal lines as "rows.")

In order to employ the truth-table method one must know the information contained in the basic truth table. Fortunately, that information can be repackaged in the following five principles:

(P1) A statement and its *negation* have opposite truth-values.
(P2) A *conjunction* is true iff both its conjuncts are true.
(P3) A *disjunction* is false iff both its disjuncts are false.
(P4) A *conditional* is false iff its antecedent is true and its consequent is false.
(P5) A *biconditional* is true iff its two components have the same truth-value.

With the aid of these principles we can determine the truth-value of *any* compound statement (or formula) built around the five statement connectives of propositional logic, provided that we know the truth-values of the simple statements it contains. Statement S4 (and formula F4) will serve as an example.

(S4) If Sacramento is the CAPITAL of California, then it is not WEST of the Mississippi.

(F4) C → −W
         T   T
          \\ F /
            F

The simple statements contained in S4 (F4) are true; we record that information by placing 'T's beneath the 'C' and 'W'. By applying principle P1, we determine that the fragment '−W' is false and we record that fact by placing an 'F' under the dash. Then, by applying P4, we see that the entire statement (formula) is false. We record that information by placing an 'F' under the arrow. We can economize on vertical space by writing the 'T's and 'F's on one row:

(F4) $\underline{C \rightarrow -W}$
T F F T

We use this convention: each 'T' ('F') is located under the major symbol in the fragment to which it applies. Accordingly, since the major symbol in '−W' is the dash (the dash governs the 'W', not vice versa), the 'F' located under the dash indicates that '−W' is false. Similarly, the 'F' under the arrow indicates that the whole conditional is false.

We have seen how to determine the truth-value of a statement or formula, given the actual truth-values of its simple constituents. We can also determine the truth-values that a statement or formula *would* have for each of the *possible* assignments of truth-values to its simple constituents. The chart in which these possibilities are worked out is called a *truth table*. A truth table for formula F4 looks like this:

C	W	‖	C	→	−	W
T	T	‖	T	F	F	T
F	T	‖	F	T	F	T
T	F	‖	T	T	T	F
F	F	‖	F	T	T	F
				*		
1	2		3	6	5	4

(The numbers on the bottom row are not part of the truth table; they indicate the order in which the columns were filled in.) There are four possible assignments of truth-values to the simple constituents. A convenient way of listing these combinations is to alternate 'T's and 'F's, singly in the first guide column and in pairs in the second column. The guide columns are separated from the rest of the table by a pair of vertical lines. The 16 entries to the right of the guide columns can be computed on a row-by-row basis or by proceeding from one column to another; the latter procedure is faster. Column 3 is copied from column 1 and column 4 from column 2. As a shortcut, columns 3 and 4 may be omitted from the table; we will adopt this shortcut. Column 5 is computed by applying principle P1 to column 4. Column 6, which applies to the entire formula, is calculated by applying principle P4 to columns 3 and 5. (Column 3 gives values for the antecedent of F4 and column 5 gives values for the consequent.) The column that gives values for the entire formula is marked with an asterisk.

One constructs a truth table for an argument by computing truth-values for each of the premises and the conclusion. For an illustration we devise a truth table for an argument advanced in a study guide for the LSAT:

> *Rule II states that the employer is liable if the employee is negligent. It follows that the employer is not liable if the employee is not negligent.*[50]

The argument may be symbolized:

N → L ⊢ −N → −L

The truth table is easily constructed:

N	L	N → L	⊢	−N	→	−L
T	T	T		F	T	F
✓F	T	T		T	F	F
T	F	F		F	T	T
F	F	T		T	T	T
					*	

The statements composing the argument are separated by a vertical line. Because there is only one column of truth-values for the premise we omit the asterisk for that column. Examine the second row of truth-values; on that row the premise has the value 'T' and the conclusion the value 'F'. We place a check mark on that row because of its critical importance. The table shows that it is logically possible for this argument to lead from truth to falsity; this means that the argument is invalid. Compare this argument with a similar, although valid, argument:

N → L ⊢ −N → −L   (invalid)
N → L ⊢ −L → −N   (valid)

You might describe the valid argument as the equivalent of modus tollens and the invalid one as comparable to the fallacy of denying the antecedent.

---

[50] David Tajgman, *Lovejoy's Shortcuts and Strategies for the LSAT* (New York: Monarch Press, 1985), p. 62.

The error of the invalid pattern is that of supposing that if one state implies a second, then the absence of the first implies the absence of the second. Just because negligence implies liability it doesn't follow that not being negligent implies not being liable.

Let's construct a truth table for the valid argument:

N	L	N → L	⊢	−L → −N
T	T	T		F T F
F	T	T		F T T
T	F	F		T F F
F	F	T		T T T
				*

There is no row in this table where the premise has the value 'T' and the conclusion the value 'F'. Since the truth-values of the premise and conclusion are determined entirely by the truth-values of the simple constituents and since the table includes all the possible combinations of assignments to those constituents, the table shows that it is *impossible* for this argument to lead from truth to falsity. That is, the table shows the argument to be valid.

From our analyses of these two arguments and their truth tables we can formulate the following principle for testing propositional arguments by truth table:

An argument is invalid iff there are any rows on its truth table where all the premises are true and the conclusion is false.

Restated:

An argument is invalid iff there are any "checked" rows on its truth table.

The truth table for an argument with three simple constituents will have eight rows of truth-values; the third guide column will alternate 'T's and 'F's in quartets. In the preceding section we claimed that formula F5 does not entail F6. Now we back that claim up with a truth table.

(F5) (P & Q) → R
(F6) P → R

§2.7 Truth Tables

P	Q	R	(P	&	Q)	→	R		P	→	R
T	T	T	T			T	T		T	T	T
F	T	T	F			T	T		F	T	T
T	F	T	F			T	T		T	T	T
F	F	T	F			T	T		F	T	T
T	T	F	T			F	F		T	F	F
F	T	F	F			T	F		F	T	F
✓T	F	F	F			T	F		T	F	F
F	F	F	F			T	F		F	T	F
						*				*	

(The horizontal line between the fourth and fifth rows of truth-values is an optional device for helping "line up" the entries on a given row.) The "checked" seventh row proves that F5 does not entail F6.

Do the truth-table and formal-proof methods yield consistent results when applied to propositional arguments? Yes, a formal proof of validity can be constructed for an argument iff the argument is assessed as valid by the truth-table technique.

Truth tables can do more than evaluate arguments for validity; they can also be used to determine whether two statements are logically equivalent. Logically equivalent statements must have the same truth-value. So, when we construct a truth table testing for equivalence we "check" any row where there is both a 'T' and an 'F' (in either order) in the asterisked columns; this will be the critical row. We can then use this principle:

Two statements are logically equivalent iff there are no "checked" rows on their truth table.

Consider this truth-table demonstration of the nonequivalence of formulas F7 and F8:

(F7) −(A & B)
(F8) −A & −B

A	B	−(A	&	B)	−A	&	−B
T	T	F		T	F	F	F
✓F	T	T		F	T	F	F
✓T	F	T		F	F	F	T
F	F	T		F	T	T	T
			*			*	

The nonequivalence of F7 and F8 is proof that the dash does not "multiply through an ampersand" the way a minus sign multiplies through a plus operator in an arithmetical equation.

The following table proves the logical equivalence of F7 and F9:

(F7) −(A & B)
(F9) −A v −B

A	B	−(A	&	B)	−A	v	−B
T	T	F		T	F	F	F
F	T	T		F	T	T	F
T	F	T		F	F	T	T
F	F	T		F	T	T	T
			*			*	

The equivalence of F7 and F9 indicates that you can "push" a dash across an ampersand if you replace the ampersand with a wedge. Similarly, a dash may be pushed across a wedge if the wedge is replaced by an ampersand (see exercise 2(a) following this section).

The truth-table method has the virtue of being able to prove both validity and invalidity, but it also has one major practical drawback—the inordinate length of tables for arguments containing four or more simple constituents. Each added constituent doubles the length of the table. The table for an argument with four constituents has 16 rows, for five constituents 32 rows. Safire's argument concerning the Quayle-National Guard flap (considered in the preceding section) has six constituents; its truth table has 64 rows. Computers don't mind doing 64-row truth tables, but humans find them extremely tiresome. It is clear then that the full-truth-table method is impractical for evaluating complex arguments. The method of formal proof provides an alternative way to establish the validity of a complex argument, but it cannot be used to prove invalidity. So, we particularly need another way of demonstrating the invalidity of arguments, especially complex ones.

Happily, there is a shortcut truth-table technique for demonstrating invalidity; we may call it the "brief-truth-table" method.[51] A single "checked" row in a full truth table suffices to establish invalidity; the other rows are inconsequential. The brief-truth-table technique is a way of generating that crucial "checked" row without taking the time to construct the other rows. This is accomplished by computing truth-values in reverse. Instead of beginning with assignments to the smallest fragments and working up to entire statements as is done in the full-truth-table method, we assign the crucial values (that is, truth to each premise and falsity to the conclusion) to the largest units and work backwards to the smallest. By demonstrating that the row of values can be consistently completed one proves invalidity.

---

[51] There is also a brief-truth-table technique for demonstrating validity; it is not introduced in this book.

We demonstrate by applying the brief method to an argument drawn from a "Garfield" comic strip.

If ODIE thinks, he EXISTS. He doesn't think. Therefore, he doesn't exist.

GARFIELD ©1988 PAWS, INC. Reprinted with permission of Universal Press Syndicate. All rights reserved.

The numbers at the bottom of the table below indicate the order in which the 'T's and 'F's were added to the table; they are not part of the table itself.

O	E	O	→	E	−	O	⊢	−	E
✓F	T	F	T	T	T	F		F	T
			*		*			*	
8	5	9	1	6	2	7		3	4

The first three entries are the goal assignments of truth to the premises and falsity to the conclusion. Now we attempt to complete the row of values in a consistent way. In virtue of principle P1, the 'F' in column 3 forces 'T's into columns 4, 5, and 6. By the same principle, the 'T' in column 2 forces 'F's into columns 7, 8, and 9. Principle P4 assures us that the entries forced into columns 9 and 6 are consistent with the goal entry made in column 1. The row of truth-values we have produced is the crucial row of the full truth table for the argument. It is the row that shows that it is logically possible for this argument to have true premises and a false conclusion; that is, it is the row that shows the argument to be invalid. We indicate the successful completion of the row of truth-values by prefixing it with a check mark.

The brief-truth-table procedure for demonstrating invalidity may be summarized as follows:

> **Make a goal assignment of 'T' to each premise of the argument and 'F' to the conclusion. If the remaining assignments are completed consistently, the argument is invalid.**

A useful hint is to concentrate first on those statements in the argument whose truth-value assignments force additional assignments. For example, if the conclusion of the argument is a disjunction, the 'F' assigned to it will force two more assignments (since a disjunction is false only when both disjuncts are false), but if the conclusion is a conjunction, the goal assignment of 'F' forces no additional assignments (since three different assignments to the conjuncts are compatible with the falsity of a conjunction). If you reach a point in devising a brief truth table where the remaining truth-value assignments are not forced, you will need to make an unforced assignment to some guide column letter. If that unforced assignment leads to an inconsistency, you then have to "back up" and make the opposite assignment.

If the brief-truth-table method is employed cleverly it will suffice to establish the invalidity of any invalid propositional argument. How could one use the technique to show the nonequivalence of two statements?[52]

In this chapter we have introduced the five connectives of propositional logic and presented three techniques for evaluating arguments for validity and pairs of statements for logical equivalence. In the next chapter we will develop a more sophisticated system of logic that is rooted in the elements of propositional logic.

EXERCISES

1. Construct full truth tables for each of the following symbolized arguments. Indicate for each argument whether it is valid or invalid.
    (a) M → R, −M → R ⊢ R
   *(b) −N → −L ⊢ N → L
    (c) A ↔ B ⊢ −(−A ↔ −B)

2. Construct full truth tables for each of the following pairs of formulas. Indicate for each pair whether the formulas are logically equivalent.
    (a) −(A v B), −A & −B
   *(b) −(A v B), −A v −B
    (c) S & (W v D), (S & W) v D

3. The Arrow Rule permits one to pass from F1 to either F2 or F3, and vice versa.

   (F1) A → B
   (F2) −A v B
   (F3) −(A & −B)

[52]We assign a 'T' to one statement and an 'F' to the other and show that the row can be completed consistently.

Prove by full truth table (a) that F1 and F2 are equivalent and (b) that F1 and F3 are equivalent.

4. Construct brief truth tables to demonstrate the invalidity of the following symbolized arguments.
   (a) A v P, A ⊢ −P
   *(b) −(U & O), −U ⊢ O
   (c) I → (O → F) ⊢ (I → O) → F

Instructions for exercises 5 through 17: symbolize each argument (or statement pair) and evaluate it by constructing either a full or brief truth table.

5. The transcript of an exchange in a House Judiciary Committee meeting is quoted on pages 9–10. The following statements are involved:

   (S1) There will be a ten-day POSTPONEMENT unless the president fails to give his ASSURANCE.[53]

   (S2) If the president fails to give his assurance, there will be a ten-day postponement.

   (S3) If the president gives his assurance, there will not be a ten-day postponement.

   (S4) There will be a ten-day postponement providing the president gives his assurance.

   (S5) There will be a ten-day postponement unless the president gives his assurance.

   (S6) There will be no ten-day postponement if the president fails to give his assurance.

   (a) Congressman Latta maintains that S2 and S3 are consequences of Congressman McClory's motion (abbreviated by S1). Are they?
   (b) He further suggests that S1 and S4 are not equivalent. Is he right?
   (c) Congressman Mann indicates that S1 and S5 are not equivalent. Is he right?
   (d) Congressman McClory claims that S6 is a consequence of S1. Is it?
   (A = The president gives his assurance)

*6. Extracts from the Supreme Court decision reviewing the murder conviction of Dr. Sam Sheppard:

[53] You may wish to review the discussion of 'unless' on pages 48–49.

> ... We believe that the arrangements made by the judge with the news media caused Sheppard to be deprived of that "judicial serenity and calm to which [he] was entitled."
>
> ...
>
> If publicity during the proceedings threatens the fairness of the trial, a new trial should be ordered.
>
> ...
>
> The case is remanded to the District Court with instructions to issue the writ and order that Sheppard be released from custody unless the State puts him to its charges again within a reasonable time.[54]

(P = Publicity surrounding Sheppard's trial threatened the fairness of the trial, R = Sheppard's conviction should be reversed)

7. Spike (Snoopy's brother) seems to be reasoning as follows in the middle two panels:

I must be DYING because my feet are COLD, and if I'm dying, my feet will be cold.

PEANUTS reprinted by permission of United Feature Syndicate, Inc.

8. From Arthur B. Simon, *Calculus with Analytical Geometry*:

> CONTINUITY THEOREM
> If f is differentiable at x, then f is continuous at x.
>
> It follows from the theorem that if f is not continuous at x, then f is not differentiable at x. ... However, it does not follow that continuity implies differentiability.[55]

Simon is claiming that the first of the following two arguments is valid and the second invalid. Are these two claims correct?

---

[54] *Sheppard v. Maxwell*, 384 U.S. 333, 354, 363 (1966).
[55] (Glenview, IL: Scott, Foresman and Co., 1982), p. 116.

If *f* is DIFFERENTIABLE at *x*, then *f* is CONTINUOUS with *x*. Thus, if *f* is not continuous with *x*, then *f* is not differentiable at *x*.

If *f* is differentiable at *x*, then *f* is continuous with *x*. Thus, if *f* is continuous with *x*, then *f* is differentiable at *x*.

9. Another passage from *Alice*:

> *... This time she found a little bottle ..., and tied around the neck of the bottle was a paper label, with the words "DRINK ME" beautifully printed on it in large letters.*
> *It was all very well to say "Drink me," but the wise little Alice was not going to do that in a hurry. "No, I'll look first," she said, "and see whether it's marked 'poison' or not"; for ... she had never forgotten that, if you drink much from a bottle marked "poison," it is almost certain to disagree with you, sooner or later.*
> *However, this bottle was* not *marked "poison," so Alice ventured to taste it....*[56]

This argument is suggested:

If the bottle is marked "POISON," I shouldn't drink from it. The bottle is not marked "poison." Consequently, it is all RIGHT to drink from it.

*10. An undergraduate argued in a term paper as follows:

> *The only perfect thing on this planet is our ability to destroy ourselves. If something created something else it would protect it. God could not protect us if He was nonexistent in this world. So, God does not exist in this world.*

The argument seems to be:

God does not PROTECT us. If God CREATED us, He would protect us. God cannot protect us if he does not EXIST. So, God does not exist.

11. Philosopher James Rachels writes:[57]

> *Here I will only argue that the two forms of euthanasia [active and passive] are morally equivalent: either both are acceptable or both are not.*

---

[56]Lewis Carroll, *Alice's Adventures in Wonderland, Through the Looking-glass,* and *The Hunting of the Snark* (New York: The Modern Library, n.d.), pp. 30–31.
[57]James Rachels, *The End of Life* (Oxford: Oxford University Press, 1986), p. 108.

Rachels appears to think that these two sentences are logically equivalent. Are they?

> ACTIVE euthanasia is morally acceptable iff PASSIVE euthanasia is morally acceptable.

> Either active and passive euthanasia are both morally acceptable or they are both morally unacceptable.

12. The abortion legislation adopted by Missouri in 1986 includes a clause providing that state laws shall be interpreted to acknowledge on behalf of the unborn child all the rights, privileges and immunities available to other persons, citizens and residents of the state. In August, 1989 a Kansas City lawyer filed a suit against the governor and attorney general for jailing the fetus of a pregnant prisoner.[58] This argument elaborates the point of the suit:

> **If Missouri laws are INTERPRETED to acknowledge on behalf of the unborn child all the rights, privileges and immunities available to other persons, citizens and residents of this state and if OTHER Missourians have the right not to be jailed without due process, then the prisoner's FETUS has that right. Other Missourians do have the right not to be jailed without due process. Therefore, if Missouri laws are interpreted to acknowledge on behalf of the unborn child all the rights, privileges and immunities available to other persons, citizens and residents of this state, then the prisoner's fetus has the right not to be jailed without due process.**

The first premise may be regarded either as a double conditional or as a conditional with a conjunctive antecedent (see next exercise).

13. Determine whether these two statements are logically equivalent:

> If Missouri laws are INTERPRETED to acknowledge on behalf of the unborn child all the rights, privileges and immunities available to other persons, citizens and residents of this state, then if OTHER Missourians have the right not to be jailed without due process then the prisoner's FETUS has that right.

> If Missouri laws are interpreted to acknowledge on behalf of the unborn child all the rights, privileges and immunities available to other persons, citizens and residents of this state and other Missourians have the right not to be jailed without due process, then the prisoner's fetus has that right.

---

[58]James J. Kilpatrick, "If Life Begins at Conception . . . ," *St. Petersburg Times* (September 13, 1989), p. 13A.

*14. In *My Stepmother Is an Alien* (RCA Columbia Pictures), Steve Mills (played by Dan Aykroyd) delivers this speech to Celeste (Kim Basinger):

> *You're everything I ever wanted in a human and in an extraterrestrial and if I send that transmission, which I don't really know how to do, and save your planet, I lose you. If I don't send it, you stay but I kill a whole planet.*

Steve's dilemma formulated as an argument:

If I send the TRANSMISSION, your PLANET survives but you leave. If I don't send it, you STAY but your planet will be destroyed. So, you stay iff your planet is destroyed.

(Treat 'You leave' as the negation of 'You STAY' and 'Your planet is destroyed' as the negation of 'Your PLANET survives'.)

15. Business administration educator Eugene F. Brigham writes:

> *If stocks were negatively correlated, or if there were zero correlation, then a properly constructed portfolio would have very little risk. However, stocks tend to be positively (but less than perfectly) correlated with one another, so all stock portfolios tend to be somewhat risky.*[59]

Add this suppressed premise: 'Stocks are not both POSITIVELY and NEGATIVELY correlated'. Note that 'Stocks have zero correlation' is equivalent to 'Stocks are neither positively nor negatively correlated'. (R = Stock portfolios involve significant risk)

16. This statement is included on an historical marker at the ruins of the nineteenth-century Fort Bowie (Arizona) school:

> (S1) School was not held if: no qualified teacher was available; no school-age children were living at the fort; or no soldiers were interested in attending.

It seems more likely that the statement should read:

> (S2) School was HELD only if there was a qualified TEACHER available, and there were either school-aged CHILDREN living at the fort or soldiers INTERESTED in attending.

---

[59]*Fundamentals of Financial Management* (Hinsdale, IL: Dryden Press, 1978), p. 107.

Symbolize the two statements and test them for logical equivalence; use the truth table to specify the precise difference in content between them.

17. With extraordinary good luck you are dealt this hand in a game of high-low poker: A♥, 2♦, 3♥, 4♥, 6♥, 9♠, J♥ (a heart flush and an excellent low). Naturally, you "declare" both high and low. Two players utter S1 and S2 to remind you of the rules covering this situation.

> (S1) If you tie or lose either way, you lose the entire pot (*i.e.*, both the half reserved for the best high hand and the half reserved for the best low hand).

> (S2) You must win (and not merely tie) both ways in order to win either (the high or the low) half of the pot.

Are your friends expressing the same rule or two rules? Symbolize these statements and test them for logical equivalence. (The statements express the same rule iff they are equivalent.) Use this dictionary: A = Your high hand is better than all the other high hands, B = You take half the pot for high, C = Your low hand is better than all the other low hands, D = You take half the pot for low.

18. (CHALLENGING) The following puzzle appears in *101 Puzzles in Thought and Logic*, by C. R. Wylie, Jr.:

> *The personnel director of a firm in speaking of three men the company was thinking of hiring once said,*
>
>> "We need Brown and if we need Jones then we need Smith, if and only if we need either Brown or Jones[60] and don't need Smith."
>
> *If the company actually needed more than one of the men, which ones were they?*[61]

Assume that the dominant connective in the personnel director's statement is 'if and only if'. (Read no further if you want to solve the puzzle on your own.) One proposed solution to the puzzle: 'The company needs JONES and SMITH, but not BROWN'. Discover whether this solution is correct by evaluating the argument that has it as conclusion and that has as premises the personnel director's statement and the claim that the company needs more than one of the men.

---

[60]Assume that 'we need either Brown or Jones' means the same as 'either we need Brown or we need Jones'. In fact, the two have different meanings, as we will see in Chapter 6.
[61](New York: Dover, 1957), puzzle 42.

19. (EXTRA-CHALLENGING) With one game remaining in the NFL regular season, the *Miami Herald* described the playoff situation for the Dolphins with the help of four statements:[62]

> (S1) If the DOLPHINS win [their last regular-season game], they make the PLAYOFFS.
>
> (S2) If the Dolphins lose, then the Dolphins are out of the playoffs if the JETS win.
>
> (S3) If the Dolphins lose, then if the Jets lose and the Steelers lose, the Dolphins make the playoffs.
>
> (S4) If the Dolphins lose, then if the Jets lose and the STEELERS win, the Dolphins make the playoffs iff the BRONCOS win and Miami ends up ahead of the Raiders in the TIE-BREAKING scheme.

Elsewhere in the same edition, the *Herald* presented a simplified version of the playoff picture with these two claims:[63]

> (S1) If the Dolphins win, they make the playoffs.
>
> (S5) If the Dolphins lose, the only way they make the playoffs is if the Steelers and Jets lose or if the Broncos and Steelers win and the Jets lose.

(a) Do S2 through S4 entail S5? (b) Are S1 through S4 (taken together) equivalent to S6?

> (S6) The Dolphins make the playoffs iff either they win or the Jets lose and either the Steelers lose or the Broncos win and Miami ends up ahead of the Broncos in the tie-breaking scheme.

(Each connective in S6 has greater scope than the next connective.) To simplify symbolizing these statements treat "losing" as equivalent to "not winning." You can solve these problems by constructing huge truth tables or by devising proofs (to show entailment or equivalence) or brief truth tables (to demonstrate nonentailment or nonequivalence).

---

[62]"Playoff Possibilities," *Miami Herald* (December 28, 1993), p. 5D.
[63]"Victories, Praise Suddenly Disappear," *Miami Herald* (December 28, 1993), p. 1D.

3

Predicate Logic

The system of propositional logic set forth in Chapter Two is a powerful instrument, but it is not adequate for the assessment of all deductive arguments. Consider for example the argument suggested by this letter to a newspaper:

Dear Editor,
On November 2nd, I voted in Highlands. I read in your newspaper on November 4th that the Highlands polls recorded no votes for Ed Clark.
Since I voted for Ed Clark for President, either your newspaper is in error or the votes were incorrectly tabulated. . . .[1]

In part the argument of this letter is:

I voted for Ed Clark. So, it is false that no one voted for Ed Clark.

[1]*The Highlander,* Highlands, NC (November 13, 1980), p. 2.

Let's call this argument "The Voter." Propositional logic can provide the following analysis of "The Voter":

$$I \vdash {\sim}{\sim}S$$

(I = I voted for Ed Clark, S = Someone voted for Ed Clark) Now this symbolized argument is invalid, but obviously the argument expressed in English is valid. We can only conclude that propositional logic does not satisfactorily represent the logical form of "The Voter." There is a logical relation between the statements 'I voted for Ed Clark' and 'Someone voted for Ed Clark', but that relation escapes detection in propositional logic because these two statements are *simple* statements and propositional logic can only work by analyzing *compound* statements.[2] Clearly what we need is a logic capable of analyzing simple statements; predicate logic—the subject of this chapter—is such a system. Happily, predicate logic is "grafted" upon propositional logic so your understanding of the symbols and techniques of the latter will give you a head start in grasping this additional branch of logic.

Here is another way to contrast these two areas of logic: Propositional logic is the logic of 'and', 'or', 'if', 'iff', and 'not', while predicate logic is the logic of 'all', 'some', 'none', and related terms. Predicate logic provides the deeper analysis because it can "penetrate" simple propositions. The following section shows how to symbolize statements in the notation of this deeper system of logic.

§3.1 SYMBOLIZATION

The premise of "The Voter" (I voted for Ed Clark) can be analyzed into the two expressions 'I' and 'voted for Ed Clark'.[3] These are ingredients of this simple statement; they cannot be represented in propositional logic, but they can in predicate logic. We will symbolize the premise as

$$Vi$$

where the capital 'V' abbreviates 'voted for Ed Clark' and the lower case 'i' abbreviates the pronoun. By convention we write the capital letter first even though this does not reflect the order of the parts of the English sentence. The conclusion of "The Voter" (It is false that no one voted for Ed Clark) may be symbolized

---

[2] The distinction between simple and compound statements is made in Chapter Two on page 17.
[3] The expression 'voted for Ed Clark' can be further analyzed into 'voted for' and 'Ed Clark', but we do not need to carry the analysis to this level to reveal the structure that validates "The Voter."

$$-(x)-Vx$$

The dash is familiar to us from propositional logic. The complex symbol '(x)' is called the *universal quantifier* and is read "for any *x*." The symbolized argument

$$Vi \vdash -(x)-Vx$$

may be read "I voted for Ed Clark; therefore, it is false that for any *x*, *x* did not vote for Ed Clark" (or, in abbreviated form, "*i V*'s; therefore, it is false that for any *x*, *x* does not *V* "). One more symbol will complete the vocabulary of elementary predicate logic, the *existential quantifier*: (∃x). This symbol is read "there exists an *x* such that." We could use it to symbolize the statement 'Someone voted for Ed Clark':

$$(\exists x)Vx$$

This formula may be read "There exists an *x* such that *x* voted for Ed Clark" (or, in abbreviation, "There exists an *x* such that *x V*'s").

To summarize, these are the five types of symbols employed in elementary predicate logic (in addition to the symbols of propositional logic):

individual constants:	a, b, c, . . . , v
individual variables:	x, y, z
universal quantifiers:	(x), (y), (z)
existential quantifiers:	(∃x), (∃y), (∃z)
predicate letters:	A, B, C, . . . , Z

The meanings assigned to individual constants and predicate letters change from one argument to the next and are assigned by "dictionaries." The meanings given to the other symbols of predicate logic remain fixed. *Individual constants* abbreviate "singular terms," that is, names ('Bill Clinton'), pronouns ('she'), and descriptive phrases that refer to a single individual ('the richest American'). Lower-case letters 'a' through 'v' serve as individual constants. *Individual variables* ('x', 'y', and 'z') are cross-reference devices.[4] They occur in quantifiers and also follow predicates; they are used in "dictionaries" as well. The importance of variables will become more obvious as you proceed through the chapter. Incidentally, by an "individual" we mean in logic any single thing that can be named or referred to; examples of individuals

---

[4]If a symbolization requires more than three variables we can use *w* as a backup variable. If even more are needed, we can create additional variables by affixing prime marks (*x'*, etc.).

are Ralph Nader, Socks (Chelsea Clinton's pet cat), the number 17, Saturn, the Eiffel Tower, the Mississippi River, Boston, the father of Ted Bundy, and the First Amendment to the Constitution.

General terms—expressions that can be applied to several or many individuals—are called "predicates" by logicians.[5] Predicates are often adjectives or plural nouns (with or without copulas). Examples of predicates:

    (is) green
    (is a) conductor of electricity
    (is) Lutheran
    voted for Ed Clark

The capital letters used to abbreviate predicates are called *predicate letters* (or "predicates" for short). Predicates may be divided into *property predicates* and *relational predicates*. For the present we confine our attention to property predicates; we will consider relational predicates in Section 3.5. Note that a *property* predicate letter is always followed by one lower-case letter (either an individual constant or an individual variable), while a *relational* predicate letter is always followed by two or more lower-case letters. A capital letter followed by *no* lower case letter is not a predicate letter at all but the abbreviation of a simple statement. Statement letters may be used in predicate logic just as they are in propositional logic.

Where possible the letter we choose as an individual constant or predicate letter will be the first letter of some prominent word in the singular term or predicate. A word we select to supply a constant will be underlined, while a word indicating a predicate letter will be printed entirely in capital letters. Thus the premise of "The Voter" will be written 'I VOTED for Ed Clark' to specify the symbols 'i' and 'V'. Statement abbreviations will be indicated in dictionaries.

Classical Aristotelian logic (also called "syllogistic logic") deals with arguments composed of statements of the following four kinds:[6]

---

[5]The logician's concept of "predicate" differs from the grammarian's. For instance, the term 'attorneys' in 'All attorneys are lawyers' will be called a "predicate" by a logician.

[6]The scope of predicate logic includes all of Aristotelian logic and extends far beyond. Aristotelian logic deals primarily with "categorical syllogisms." A categorical syllogism is an argument (a) consisting of three statements each of which exhibits one of the four forms displayed on page 117, and (b) including three predicates each of which occurs in two of the three statements.

The four statement forms (in the order of our display) were labeled "A," "E," "I," and "O" by medieval logicians. Each valid syllogism form was assigned a mnemonic whose vowels indicated the forms of the statements composing it and whose consonants identified other formal features of the syllogism. For example, the valid syllogism form composed entirely of *A* statements,

    All M are P; all S are M; so, all S are P,

was called "B*a*r*b*a*r*a*." Medieval students of logic learned to identify the 24 syllogism forms recognized as valid in syllogistic logic by memorizing a verse containing the names of these forms.

Because of their different approach to "existential import" (see §5.5), contemporary logicians recognize as valid only 15 (of the 256) syllogism forms.

TYPE	FORM	EXAMPLE
universal affirmative	All A are B.	All statutes are laws.
universal negative	No A are B.	No statutes are laws.
particular affirmative	Some A are B.	Some statutes are laws.
particular negative	Some A are not B.	Some statutes are not laws.

It is important for two reasons to learn how to symbolize sentences exhibiting these four basic forms: (1) many sentences embody one or another of these forms, and (2) many other sentences resemble the basic forms in some respects. The four basic forms are symbolized as follows:

FORM	EXAMPLE	SYMBOLIZATION
All A are B.	All statutes are laws.	$(x)(Sx \rightarrow Lx)$
No A are B.	No statutes are laws.	$(x)(Sx \rightarrow -Lx)$[7]
Some A are B.	Some statutes are laws.	$(\exists x)(Sx \& Lx)$
Some A are not B.	Some statutes are not laws.	$(\exists x)(Sx \& -Lx)$

These formulas employ the arrow, ampersand, and dash from propositional logic. Note that the universal quantifier ordinarily pairs up with the arrow (as in the first two formulas) and the existential quantifier with the ampersand (as in the other formulas); exceptions to these two rules of thumb are rare. The grouper following the quantifier and the matching one at the end of the formula show the *scope* of the quantifier; the scope consists of the quantifier, the groupers, and all the symbols enclosed within the groupers.[8] Every variable in a properly constructed formula falls within the scope of some quantifier. (The rules that determine whether a formula of predicate logic is well formed are stated in Appendix Two.) The dash in the second and fourth formulas must be placed before the second predicate letter; if the dash is placed elsewhere the formulas will be incorrect symbolizations.[9]

---

[7]This symbolization is also correct: $-(\exists x)(Sx \& Lx)$. As you will discover, the formula displayed in the text is easier to work with in proofs.

[8]When only one predicate follows a quantifier, as in '$(\exists x)Vx$', the quantifier-scope groupers may be omitted. Also, when groupers indicating the scope of a dash (as in '$(x)-(Ax \lor Bx)$') also show the scope of the quantifier, additional groupers may be omitted.

[9]For example, F2 (which is equivalent to F2′) fails to represent S1; instead it symbolizes S2. F4 does not represent S3; it symbolizes S4.

(S1) No statutes are laws.
(S2) Everything is a statute and not a law.

Here are some of the different ways we can express statements exhibiting the basic forms:

ALL A ARE B.	NO A ARE B.	SOME A ARE B.	SOME A ARE NOT B.
Every A is B.	A are not B.	There are A that are B.	There are A that are not B.
Each A is B.	No one is both A and B.	At least one A is B.	At least one A is not B.
Any A is B.	There are no A B's.	A that are B exist.	Not all A are B.
A are B.	A are never B.	A are sometimes B.	A are not always B.

We said above that some sentences resemble statements of the four basic forms studied by Aristotle. Recognizing that similarity can help us determine the symbolization of such sentences. Here is an example from a television nature documentary:

In the Arctic each animal must migrate, adapt or perish.

This statement involves four general terms that can be abbreviated with these predicate letters:

Ax = x is an Arctic animal
Mx = x migrates
Dx = x adapts (to cold)
Px = x perishes

The statement is universal (rather than particular) and affirmative (not negative), so it resembles a statement of the form "All *A* are *B*" and will be symbolized accordingly. The quantifier will be universal and the dominant connective will be the arrow. We reach this symbolization:

(x)[Ax → (Mx v Dx v Px)]

(F2) (x)−(Sx → Lx)
(F2') (x)(Sx & −Lx)
(S3) Some statutes are not laws.
(S4) Something is not both a statute and a law.
(F4) (∃x)−(Sx & Lx)

Here are some additional examples of the symbolization of sentences that resemble one or another of the basic forms:

(*sports story*) "Every HURRICANE who PLAYED got on BASE."	(x)[(Hx & Px) → Bx] or (x)[Hx → (Px → Bx)]
MORMONS are CHRISTIANS but not PROTESTANTS.	(x)[Mx → (Cx & −Px)]
Anyone convicted of premeditated MURDER in a military court is either given the DEATH penalty or sentenced to LIFE imprisonment.	(x)[Mx → (Dx v Lx)]
(*Camus*) "No one can be HAPPY in EXILE or estrangement."	(x)[(Ex v Sx) → −Hx] (Sx = x is estranged)
Some CRIMES are MISDEMEANORS rather than FELONIES.	(∃x)(Cx & Mx & −Fx)

Statements of the above types are called *general* because they make claims about all or some of a group of individuals. General statements are symbolized with the help of a quantifier (either universal or existential). You can often simplify the symbolization of a general statement by adopting a restricted *universe of discourse* (or just "universe"). The universe of discourse is the class of objects over which the quantifiers and variables in the formula are understood to range. Consider how restricting the universe can simplify the symbolization of the first sentence displayed in the table above (Every Hurricane who played got on base.):

UNIVERSE	SYMBOLIZATION
unrestricted	(x)[(Hx & Px) → Bx]
Hurricanes	(x)(Px → Bx)
those who played	(x)(Hx → Bx)
Hurricanes who played	(x)Bx

Restricted universes are often suggested for the exercises in this chapter. Incidentally, you must use the same universe of discourse for each statement in an argument, and a given universe is acceptable for that purpose only if *no* statement in the argument treats individuals falling outside the universe.

Not all statements encountered in predicate logic are general; some are *singular*. A singular statement is about a specific named individual; it is symbolized with an individual constant rather than a quantifier.[10] Examples:

I VOTED for Ed Clark.   Vi
Bill Clinton is not a REPUBLICAN.   −Rb

Some statements are compounded out of general statements, like the following sentence from a newspaper story:[11]

If every American went out today and filled up his gas tank, we would drain every service station dry.

This is a conditional statement whose antecedent and consequent are universal affirmative statements. We can symbolize it with either of these formulas:

(x)(Ax → Fx) → (y)(Sy → Ey)
(x)(Ax → Fx) → (x)(Sx → Ex)

(Ax = x is an American automobile gas tank, Fx = x is full, Sx = x is an American service station tank, Ex = x is empty) Note that the main connective in the formula—the second arrow—does not fall within the scope of a quantifier; this is appropriate because the overall form of the statement is that of a conditional. When a general statement is simple—that is, contains no parts that are statements—any connective in its symbolization will fall within the scope of a quantifier that begins the formula.

In this section we have presented the basic symbolic vocabulary of predicate logic. In the next three sections we will present techniques for evaluating arguments composed of sentences symbolizable in predicate logic.

---

[10]Aristotelian logic, unlike modern symbolic logic, equates singular statements (for example, 'Socrates is mortal') with universal affirmative statements ('All persons identical with Socrates are mortal').

[11]*The Springfield [Ohio] Sun* (May 10, 1979).

EXERCISES

1. Symbolize each statement using the suggested notation.
   - (a) (*newspaper*) "Henry Kissinger is a NATURALIZED citizen." (Nx = x is a naturalized citizen[12])
   - *(b) (*overheard*) "Everybody has WEIRD relatives." (Universe: people)
   - (c) (*prosecutor*) "There is no ski MASK [among the items of evidence]." (Universe: items of evidence)
   - (d) (*Plato*) "All WARS are UNDERTAKEN for the acquisition of wealth."
   - (e) (*newspaper feature*) "Some MARRIAGES are FILLED with anger."
   - *(f) (*Joyce Brothers*) "Any person who makes OBSCENE phone calls needs professional HELP." (Universe: people)
   - (g) (*Jerry Brown on Ross Perot*) "There's no such thing as a BILLIONAIRE POPULIST."
   - (h) (*headline*) "Not all tax LOOPHOLES NEED closing."
   - (i) (*highway sign*) "Sign VANDALISM is a MISDEMEANOR." (Vx = x is an act of sign vandalism, Mx = x is a misdemeanor)
   - *(j) (*proverb*) "All that glitters is not gold." (Lx = x glitters, Ox = x is gold)
   - (k) (*newspaper column*) "Bill Evans is a JEW and a SOUTHERNER."
   - (l) (*FDA official*) "Any liquid which contains no lemon juice is not lemonade." (Universe: liquids; Cx = x contains lemon juice, Lx = x is lemonade)
   - (m) (*feminist Evelyn Young*) "Any WOMAN who ENGAGES in sex [with her husband] for credit cards is a PROSTITUTE." (Ex = x engages in sex with her husband for credit cards)
   - *(n) (*newspaper article*) "[Department of Defense officials said] BLUE and cream-colored blankets do not contain DDT and can continue to be USED." (Universe: blankets supplied by the Department of Defense; Bx = x is blue and cream colored, Dx = x contains DDT, Ux = x is safe to use)
   - (o) (*continuation of above passage*) "Some OLIVE-colored blankets contain DDT, while others do not."

2. Translate each formula into an English sentence using this dictionary:

   Universe: people
   Bx = x accepts bribes
   Cx = x is corrupt
   d = the district attorney
   Gx = x is a government employee
   Jx = x is a judge
   Lx = x is a lawyer
   Px = x is a politician

---

[12]It is customary to use variables in dictionary entries for predicates. This convention does not imply that the symbolization will contain a variable.

(a) Ld
*(b) (∃x)Bx
   (c) −(x)Lx
   (d) (x)(Jx → Lx)
   (e) −(x)(Lx → Jx)
*(f) (∃x)(Jx & Cx)
   (g) (∃x)(Px & −Lx)
   (h) (x)(Jx → −Bx)
   (i) (x)[(Px & Bx) → Cx]
*(j) (x)[Jx → (Lx & Gx)]
   (k) (∃x)(Px & −Bx & −Cx)

3. Symbolize each statement using the suggested notation.
   (a) (*bumper sticker*) "Every MOTHER WORKS."
*(b) (*Harry Reasoner*) "People who are intelligently interested in doing BUSINESS are not WARLIKE." (Universe: people; Bx = x is intelligently interested in doing business)
   (c) (*newspaper column*) "Every MAN is ENTITLED to his own opinion; every WOMAN too."
   (d) (*this text*) "Not all statements encountered in predicate logic are GENERAL; some are SINGULAR." (Universe: statements encountered in predicate logic)
   (e) (*Fichte*) "He who WILLS to do evil to produce a greater good is a godless person." (Universe: people; Wx = x wills to do evil to produce a greater good, Gx = x serves God)
*(f) (*Shakespeare*) "Neither a BORROWER nor a LENDER be." (Universe: people; Bx = x should borrow money, Lx = x should lend money)
   (g) (*newspaper ad*) "To be ELIGIBLE [to participate in the study] you must have DIABETES and a FOOT ulcer."
   (h) (*TV ad*) "Duracell—no regular battery looks like it or lasts like it." (Rx = x is a regular battery, Ax = x looks like a Duracell, Bx = x lasts like a Duracell)
   (i) (*TV ad*) "If they don't PLUMP when you cook 'em, they can't be BALL Park franks." (Universe: frankfurters; Px = x plumps when cooked, Bx = x is a Ball Park frank)
*(j) (*newspaper*) "There was no ASPECT of the game the Americans did not DOMINATE." (Ax = x is an aspect of the game, Dx = x is an area dominated by the Americans)
   (k) (*radio newscast*) "All the AMERICAN POWS held by the IRAQUIS were BEATEN or TORTURED." (Ix = x was held by the Iraquis)
   (l) (*rock lyrics*) "Nobody's RIGHT if everybody's WRONG." (Universe: people)

(m) (*conversation*) "All BEER—and some bread—is KOSHER." (Ax = x is bread)

*(n) (*John 1:3*) ". . . Through him all things came to be; no single thing was created without him." (Universe: created things; Gx = x was created by God)

(o) (*Shakespeare*) "Men have DIED from time to time, and WORMS have eaten them, but not for LOVE." (Universe: men; Wx = x's corpse is eaten by worms, Lx = x died for love)

(p) (*newspaper feature on Mardi Gras*) "Rex . . . has JEWS and ITALIANS, but no BLACKS or WOMEN." (Universe: members of Rex krewe)

4. The humor in this "Wee Pals" comic strip depends on the amphiboly in Wellington's boastful statement. Symbolize the boast's two possible meanings using this dictionary: Qx = x is a question, Wx = Wellington knows the answer to x, Ux = Wellington's uncle knows the answer to x.

By permission of Morrie Turner and Creators Syndicate.

5. (CHALLENGING) Symbolize each statement using the suggested notation.
   (a) (*Camoens*) "Those who SERVE God never lack a perfidious ENEMY." (Universe: people; Ex = x has a perfidious enemy)
   (b) (*newspaper*) "No REPUBLICAN in the HOUSE or SENATE voted FOR the original [budget] bill." (Hx = x is a House member, Sx = x is a Senator)
   (c) (*Kojak*) "You show me a LAWYER who hasn't made some enemies, and I'll show you a NOTARY public." (Ex = x has made some enemies)
   (d) (*Minnesota Fats*) "Shoot HARD and sleep in the STREETS." (Universe: professional pool players)
   (e) (*sign*) "All persons are forbidden to ENTER or TRESPASS upon these grounds." (Universe: people; Ex = x is forbidden to enter these grounds; Tx = x is forbidden to trespass upon these grounds)
   (f) (*Bertrand Russell*) "No man can be a GOOD teacher unless he has feelings of warm AFFECTION towards his pupils and a genuine

DESIRE to impart to them what he himself believes to be of value." (Universe: people; Gx = x is a good teacher)

(g) (*Adam Clayton Powell*) "All of us have a problem until all our problems are solved." (Universe: people, Px = x has a problem)

(h) (*Teddy Roosevelt*) "This country will not be a good place for any of us to live in unless we make it a good place for all of us to live in." (Universe: Americans; Gx = America is a good place for x to live in)

(i) (*newspaper*) "No REPUBLICAN has ever been elected PRESIDENT without carrying OHIO." (Ox = x wins in Ohio)

(j) (*Dickens*) ". . . And it was always said of Scrooge that he KNEW how to keep Christmas well if any man alive possessed the knowledge." (Universe: people)

(k) (*television report*) "If people and animals can't SHARE this place [Martha's Vineyard], they can't share any place." (Sx = people and animals can share place x)

(l) (*T-shirt message*) "There are two kinds of people: SPE's and those who wish they were SPE's." (Universe: people; Sx = x is a member of SPE, Wx = x wishes to become a member of SPE)

(m) (*sign*) "No DOGS ALLOWED (except SEEING-eye dogs)."

(n) (*Hot Rod Hundley's wife*) "If only all of Rod's GIRLFRIENDS BUY the book, we'll get rich." (Universe: people; R = We get rich)

§3.2 PROOFS

The method of formal proof that was developed in the last chapter can be extended to predicate logic simply by adding three rules of inference to the inference rules of propositional logic. Two of the new rules (the "Out" rules) allow us to transform formulas of predicate logic into formulas to which we can apply the propositional inference rules.

Some definitions will simplify the statement of the new rules:

A *quantification* is a formula that begins with a quantifier whose scope is the entire formula. A *universal quantification* is a quantification that begins with a universal quantifier and an *existential quantification* is a quantification that begins with an existential quantifier.

An *instance* of a quantification is a formula that results from deleting the quantifier (and the quantifier-scope groupers) and replacing all of the remaining occurrences of the variable by the same individual constant.

Examples may help clarify the concept of "quantification":

QUANTIFICATIONS	NOT QUANTIFICATIONS
(x)(Ax → Bx)	(F1) Fg
(∃x)(Cx & Dx)	(F2) −(x)(Hx → Ix)
(x)−Ex	(F3) (∃x)(Jx & Kx) & (x)(Lx → Mx)

F1 and F2 are not quantifications because they do not begin with quantifiers (F2 begins with a dash). F3 is not a quantification because the quantifier it begins with does not include the entire formula in its scope; the scope of that quantifier ends before the second ampersand. These examples will elucidate the concept of "instance":

INSTANCES OF '(x)(Ax → Bx)'	NOT INSTANCES OF '(x)(Ax → Bx)'	EXPLANATION
Aa → Ba	Ay → By	Contains no individual constants.
Ab → Bb	Ac → Bx	The last variable was not replaced.
⋮		
Av → Bv	Ad → Be	The variables were replaced with different individual constants.

Now we can state the inference rules of elementary predicate logic. We'll start with the Universal Quantifier Out Rule:

The Universal Quantifier Out Rule (UO): From a universal quantification derive any instance of it.

We can illustrate the use of the rule by constructing a proof for the argument contained in this passage:

> The Constitution of the United States expressly commands the Congress to make no law abridging freedom of speech or press. In violation of this express command, Congress has passed the Sedition Law; and the Sedition Law does exactly what the Constitution forbids. It abridges freedom of speech and of the press. Consequently this law is not constitutional and is void.[13]

[13] Kenneth Roberts, *Lydia Bailey* (Greenwich, CT: Fawcett Publications, Inc., 1947), p. 21.

The argument may be formalized and then symbolized as follows:

All laws that ABRIDGE freedom of speech are unconstitutional. All unconstitutional laws are VOID. The <u>Sedition</u> Law abridges freedom of speech. Consequently this law is not constitutional and is void.

(x)(Ax → −Cx), (x)(−Cx → Vx), As ⊢ −Cs & Vs

(Universe: laws; Cx = x is constitutional) Here is the proof (with interspersed commentary):

(1) (x)(Ax → −Cx)   A
(2) (x)(−Cx → Vx)   A
(3) As   A

We'd like to be able to apply a propositional inference rule (such as Arrow Out) to some of these lines but are prohibited by the fact that lines 1 and 2 are not conditionals (the arrows on those lines fall within the scope of quantifiers). We can solve this problem by applying the Universal Quantifier Out Rule:

(4) As → −Cs   1 UO
(5) −Cs → Vs   2 UO

For obvious reasons we use the constant 's' in the instances on lines 4 and 5. (In future we will describe deriving an instance containing 's' as "instantiating to 's'"). Now we are free to apply Arrow Out:

(6) −Cs   4,3 →O
(7) Vs   5,6 →O
(8) −Cs & Vs   6,7 &I

This completes the proof; the argument has been shown to be valid.

Why is UO a valid rule of inference? Because what is true of *every* individual is true of *each named* individual. If it is true of every current member of the Supreme Court that he or she is a college graduate, then it is true of Justice O'Connor that she is a college graduate. The universal quantification on line 1 of the proof above says of every law that if it abridges freedom of speech it is unconstitutional; the formula on line 4 says of the Sedition Law that if it abridges freedom of speech it is unconstitutional.

It should be clear that we can't phrase the Existential Quantifier Out Rule as simply as the UO Rule because you can't say that what is true of *some* individual is true of *each named* individual. Consider the claim

(S1) Something is green.

Because S1 is true we know that somewhere there is an individual that is green. We may not know the name of that individual, but we can agree to assign it an arbitrary name, say 'a'. Then we can infer 'a is green' from S1. With this idea in mind we state the Existential Quantifier Out Rule as follows:

The Existential Quantifier Out Rule (EO): From an existential quantification derive any instance of it, provided that the individual constant being introduced does not occur in the symbolization of the argument being tested or on any line above the line derived.

The two restrictions in the clause beginning 'provided that' ensure that the individual constant used in the instantiation is truly "arbitrary," that is, not dedicated to any other use in that problem.

A newspaper sports column (written before the cable-TV era) gives this explanation of why an upcoming football game between Texas and Oklahoma won't be televised:

> . . . *The limited-appearance rule strikes again. Because CBS will televise the Oklahoma-Nebraska game Nov. 26, the network has used up its Oklahoma options (each network is permitted to televise a team three times during a two-year period). ABC has used up its Texas telecasts.*[14]

This explanation may be viewed as an argument whose conclusion is the fact to be explained:

No OKLAHOMA game will be broadcast by CBS. No TEXAS game will be broadcast by ABC. Any telecast of a Big Eight Conference game will be on either CBS or ABC. Therefore, there will not be a telecast of a game involving both Oklahoma and Texas.

(x)(Ox → −Cx)
(x)(Tx → −Ax)
(x)(Cx v Ax)
⊢ −(∃x)(Ox & Tx)

(Universe: telecasts of Big Eight Conference games; Ox = x involves the Oklahoma football team; Tx = x involves the Texas football team, Cx = x is a CBS telecast, Ax = x is an ABC telecast) The third premise is an unstated assumption.

---

[14] Richard Rosenblatt, "Doctor Cancels Practice for Flu-bitten Hurricanes," *Miami News* (October 4, 1983), p. 4B.

Because the conclusion of this argument is a negation we employ the Dash In Strategy and make a provisional assumption of the conclusion minus its dash.

1	(1)	$(x)(Ox \to -Cx)$	A
2	(2)	$(x)(Tx \to -Ax)$	A
3	(3)	$(x)(Cx \lor Ax)$	A
4	(4)	$(\exists x)(Ox \mathbin{\&} Tx)$	PA

The next steps are to apply EO to line 4 and UO to lines 1 through 3. Notice the importance of making the EO step first; otherwise we will violate one of the restrictions on the EO Rule. Since no constant appears in the symbolization of the argument or in the first four lines of the proof we can instantiate to any individual constant.

4	(5)	$Ob \mathbin{\&} Tb$	4 EO
1	(6)	$Ob \to -Cb$	1 UO
2	(7)	$Tb \to -Ab$	2 UO
3	(8)	$Cb \lor Ab$	3 UO

Note that the standard assumption-dependence principle applies to EO and UO (as well as the other rules presented in this section). Because the dominant symbols in lines 5 through 8 are connectives from propositional logic we can apply propositional inference rules to reach a standard contradiction. We then complete the proof with the aid of Dash In.

4	(9)	Ob	5 &O
1,4	(10)	$-Cb$	6,9 →O
4	(11)	Tb	5 &O
2,4	(12)	$-Ab$	7,11 →O
1,3,4	(13)	Ab	8,10 DA
1,2,3,4	(14)	$Ab \mathbin{\&} -Ab$	13,12 &I
1,2,3	(15)	$-(\exists x)(Ox \mathbin{\&} Tx)$	4-14 −I

We have stated and illustrated the use of "Out" rules for the two quantifiers, but we have not (yet) introduced an "In" rule for either of these quantifiers. In the absence of such rules can we construct proofs for arguments whose conclusions are either universal or existential quantifications? For an example consider an argument derived from this newspaper item:

Augusta, Maine—A bill in the state legislature reads: "Every person residing in Maine who earns less than $4,000 annually shall be furnished a hearing aid free of charge by the Department of Health and Welfare."

Rep. Robert Soulas of Bangor said, "I guess this bill needs some work" when he was told his measure didn't say you had to be hard of hearing.[15]

The bill's defect is made explicit in this argument (call it "Free Hearing Aids"):

Every person residing in Maine who earns LESS than $4,000 will be FURNISHED a hearing aid free of charge [if the bill is enacted]. Some Maine residents who earn less than $4,000 are not HARD of hearing. Accordingly, [if the bill is enacted] some residents of Maine who are not hard of hearing will be provided free hearing aids.

(x)(Lx → Fx), (∃x)(Lx & −Hx) ⊢ (∃x)(−Hx & Fx)

(Universe: Maine residents) Since (at present) we have no Existential Quantifier In Rule, the proof for this argument must employ the Dash Out Strategy.

1	(1)	(x)(Lx → Fx)	A
2	(2)	(∃x)(Lx & −Hx)	A
3	(3)	−(∃x)(−Hx & Fx)	PA

Now we have a problem. Line 3 is not an existential quantification (because it starts with a dash) and EO cannot be applied to it. At this point we need to introduce a third predicate-logic rule of inference, the Quantifier Exchange Rule:

The Quantifier Exchange Rule **(QE):**

From −(x)Ax **derive** (∃x)−Ax **and vice versa.**
From −(∃x)Ax **derive** (x)−Ax **and vice versa.**

QE embodies an obvious insight. If it is not the case that all trial lawyers are competent, then there must be at least one who is not competent (and vice versa). If it is not the case that there is an excusable homicide, then it must be the case that no homicide is excusable (and vice versa). Because QE involves logical equivalence, it may be applied to line parts as well as to entire lines. The other rules introduced in this section may be applied only to entire lines.

---

[15]"Maine Bill Needs to Be Polished Up" (Associated Press), *Miami News* (March 10, 1969), p. 6A.

QE allows us to derive quantifications from negations of quantifications so that we can then apply UO or EO. With the help of QE we continue the proof:

3	(4)	(x)−(−Hx & Fx)	3 QE

Now we apply EO to line 2 and UO to lines 1 and 4.

1	(5)	Lc & −Hc	2 EO
2	(6)	Lc → Fc	1 UO
3	(7)	−(−Hc & Fc)	4 UO

We may apply the rules of propositional logic to lines 5 through 7:

1	(8)	Lc	5 &O
1,2	(9)	Fc	6,8 →O
1,2,3	(10)	− −Hc	7,9 CA
1	(11)	−Hc	5 &O
1,2,3	(12)	−Hc & − −Hc	11,10 &I
1,2	(13)	(∃x)(−Hx & Fx)	3-12 −O

Many proofs in our system of predicate logic employ the Dash Out Strategy. A typical proof of this sort may be divided into seven stages:

SEVEN STAGES OF A DASH OUT PROOF	
1.	The premises of the argument are assumed.
2.	A provisional assumption is made of the negation of the conclusion.
3.	The QE Rule is applied.
4.	The EO Rule is applied.
5.	The UO Rule is applied.
6.	A standard contradiction is derived by applying propositional inference rules to the lines reached in stages 4 and 5.
7.	The conclusion of the argument is obtained by Dash Out.

Up to this point in the book we have required an assumption-dependence column for every proof that contains a provisional assumption. Now we recommend a relaxation of this requirement: If a proof has only one provisional assumption and that assumption is discharged on the last line of the proof by a step of Dash In or Dash Out, the assumption-dependence column may be omitted from the proof. The primary purpose of that column is to ensure that the last line of a proof (the conclusion of the argument) depends on no provisional assumption. In the case just described that result is assured.

For a more complex argument to validate, consider this passage from a business law text:

> *Contingent claims of the sort questioned here are provable under Sec. 63(a) of the Bankruptcy Act. Sec. 83(d) of the act, however, provides that if a contingent claim is not allowed, it is deemed not provable. Since only provable claims are discharged in bankruptcy, it follows that a contingent claim must be allowable before it can be discharged.*[16]

The argument:

If a CONTINGENT claim is not ALLOWED it is not PROVABLE. Only provable claims are DISCHARGED. It follows that a contingent claim will be discharged only if it is allowable.

(x)[(Cx & −Ax) → −Px][17]
(x)(Dx → Px)
⊢ (x)[Cx → (Dx → Ax)]

(Universe: bankruptcy claims) Note the symbolization of premise two; there is a discussion of "only" statements in Section 5.2. We employ the Dash Out Strategy and then apply QE to the provisional assumption.

(1)   (x)[(Cx & −Ax) → −Px]   A
(2)   (x)(Dx → Px)   A
(3)   −(x)[Cx → (Dx → Ax)]   PA
(4)   (∃x)−[Cx → (Dx → Ax)]   3 QE

Next we instantiate lines 1, 2 and 4 (but 4 first).

---

[16]Rate A. Howell, John R. Allison, and N. T. Henley, *Business Law Alternate Addition* (Hinsdale, IL: The Dryden Press, 1978), p. 766.
[17]Premise one may also be symbolized '(x)[Cx → (−Ax → −Px)]'.

(5) $-[Cd \rightarrow (Dd \rightarrow Ad)]$ 4 EO
(6) $(Cd \ \& -Ad) \rightarrow -Pd$ 1 UO
(7) $Dd \rightarrow Pd$ 2 UO

Now we make propositional logic moves to reach a standard contradiction and apply the $-$O Rule.

(8) $Cd \ \& -(Dd \rightarrow Ad)$ 5 AR
(9) $Cd \ \& \ Dd \ \& -Ad$ 8 AR
(10) Cd 9 &O
(11) $-Ad$ 9 &O
(12) $Cd \ \& -Ad$ 10,11 &I
(13) $-Pd$ 6,12 $\rightarrow$O
(14) $-Dd$ 7,13 MT
(15) Dd 9 &O
(16) $Dd \ \& -Dd$ 15,14 &I
(17) $(x)[Cx \rightarrow (Dx \rightarrow Ax)]$ 3-16 $-$O

Note that the system presented here does not permit applying the rules of propositional logic to formula fragments falling within the scope of a quantifier.[18] For example, we can't apply the Arrow Rule directly to line 4 in the proof above but must first remove the quantifier by EO. The rules of propositional logic can be applied to formulas with quantifiers as long as the relevant propositional connectives are not within the scope of any of the quantifiers. For example, the following derivation is legitimate:

(1) $-[(\exists x)Ax \rightarrow (x)Bx]$ A
(2) $(\exists x)Ax \ \& -(x)Bx$ 1 AR

Let's return again to the two restrictions on the EO Rule:

provided that the individual constant being introduced does not occur (1) in the symbolization of the argument being tested or (2) on any line above the line derived.

Without the first restriction we could construct a "proof" for this invalid argument:

---

[18]The A's and B's used in the statement of the propositional rules mark gaps to be filled by formulas that abbreviate *statements*, not by formula fragments (like 'Ax') that represent *statement parts*.

There are REPUBLICANS. Hence, Bill Clinton is a Republican.

(∃x)Rx ⊢ Rb

(1) (∃x)Rx   A
(2) Rb   1 EO   [violates first restriction]

Without the second restriction we could construct a "proof" for this foolish argument:

There are REPUBLICANS. Hence, everyone is a Republican.

(∃x)Rx ⊢ (x)Rx (Universe: people)

(1) (∃x)Rx   A
(2) −(x)Rx   PA
(3) (∃x)−Rx   2 QE
(4) Re   1 EO
(5) −Re   3 EO   [violates second restriction]
(6) Re & −Re   4,5 &I
(7) (x)Rx   2-6 −I

Obviously a rule set that permitted the construction of "proofs" for invalid arguments would be worthless. The set of rules we have presented here (consisting of UO, EO, QE, the 17 propositional inference rules, and the Rule of Assumptions) is *consistent* and *complete*. Any argument for which a proof is constructed is valid, and a proof can be constructed for any valid argument expressible in our symbolic vocabulary.

We can shorten many proofs for arguments whose conclusions are symbolized by existential quantifications by adding a derived rule that allows the introduction of an existential quantifier:

***The Existential Quantifier In Rule* (EI):** Derive an existential quantification from any instance of it.

With the help of this rule we can shave several lines off the proof for "Free Hearing Aids":

(x)(Lx → Fx), (∃x)(Lx & −Hx) ⊢ (∃x)(−Hx & Fx)

(1) (x)(Lx → Fx)   A
(2) (∃x)(Lx & −Hx)   A

(3)   La & −Ha   2 EO
(4)   La → Fa   1 UO
(5)   La   3 &O
(6)   −Ha   3 &O
(7)   Fa   4,5 →O
(8)   −Ha & Fa   6,7 &I
(9)   (∃x)(−Hx & Fx)   8 EI

The deduction of line 9 from line 8 is correct because the formula on 8 is an instance of the existential quantification on 9.

EI is a "derived rule" because it sanctions deductions that could also be made (in several steps) using only primitive rules. The proof on the right shows how this is accomplished:

(1)   Fa   A   (1)   Fa   A
(2)   (∃x)Fx   1 EI   (2)   −(∃x)Fx   PA
                              (3)   (x)−Fx   2 QE
                              (4)   −Fa   3 UO
                              (5)   Fa & −Fa   1,4 &I
                              (6)   (∃x)Fx   2-5 −O

The use of EI in a proof will typically save four proof lines, as it does in this case and in "Free Hearing Aids."

It would be handy to have a companion UI Rule to abbreviate proofs whose last lines are universal quantifications. Unfortunately, to be valid such a rule would require a number of logical restrictions—so many that the gain in proof brevity would be outweighed by the complexity of the rule and the difficulty of determining in a given instance whether all the restrictions are met. Therefore we will leave our predicate logic rule set asymmetrical, with In and Out rules for the existential quantifier but only the Out rule for the universal quantifier.

EXERCISES

1. Complete the following proofs. Every assumption has been identified.
   (a) (1)   (x)(Ax → Bx)   A
        (2)   (∃x)(Cx & Ax)   A
        (3)   Ca & Aa
        (4)   Aa → Ba

	(5)	Aa	
	(6)	Ba	
	(7)	Ca	
	(8)	Ca & Ba	
	(9)	(∃x)(Cx & Bx)	
*(b)	(1)	(x)(Dx → Ex)	A
	(2)	(x)(Ex → Fx)	A
	(3)	−(x)(Dx → Fx)	PA
	(4)		3 QE
	(5)	−(Da → Fa)	4 EO
	(6)		1 UO
	(7)		2 UO
	(8)		6,7 CH
	(9)		8,5 &I
	(10)		3-9 −O

Instructions for exercises 2 through 14: symbolize each argument and construct a proof of validity for it.

2. In the case of *Yakus* v. *United States*, 321 U.S. 414 (1944), Yakus appealed a conviction for selling beef and veal above the maximum price permitted by the Office of Price Administration (a wartime agency). He appealed the conviction contending that the Emergency Price Control Act "constituted an unconstitutional delegation to the Price Administrator of the legislative power of Congress to control prices." The decision of the Supreme Court is described in one textbook as follows:

> *The Emergency Price Control Act is constitutionally valid. A statute that grants specified authority to an administrative agency constitutes an unlawful delegation of legislative power only if it fails to contain reasonable standards under which the agency is to govern its activities. In this instance, reasonable standards exist.*[19]

(Universe: statutes that grant specified authority to an administrative agency; e = The Emergency Price Control Act, Vx = x is a valid delegation of legislative power, Sx = x contains reasonable standards) This argument is presented conclusion first.

---

[19]Rate A. Howell, John R. Allison, and N. T. Henley, *Business Law* (Hinsdale, IL: The Dryden Press, 1979), pp. 93–94.

*3. A newspaper story on the opening of the Denver Airport includes this paragraph:

> *Portions of the main terminal vibrate noticeably when the automated baggage system runs, which is all the time.*[20]

The reporter encourages the reader to reason:

Whenever the baggage system runs, parts of the terminal vibrate. The baggage system runs all the time. Therefore, parts of the terminal vibrate all the time.

(Universe: moments of time [when the airport exists]; $Bx = x$ is a time when the baggage system runs, $Vx = x$ is a time when parts of the terminal vibrate)

4. The philosopher-novelist Camus writes:

> *How do I know I have no friends? It's very easy; I discovered it the day I thought of killing myself to play a trick on them, to punish them in a way. But punish whom? Some would be surprised, and no one would feel punished. I realized I had no friends.*[21]

The argument:

Any FRIEND of mine would feel PUNISHED by my suicide. But no one would feel punished by my suicide. This proves that I have no friends.

(Universe: people; $Fx = x$ is a friend of mine, $Px = x$ would feel punished by my suicide)

5. Dan Rather, on the "CBS Evening News":

> *If anyone thought AIDS was a disease that struck only drug users and gays, Magic Johnson exploded that myth.*

Rather's reasoning:

---

[20]"Denver Airport Is Ready for Takeoff," *Miami Herald* (February 26, 1995), pp. 1A and 14A.
[21]Albert Camus, *The Fall* (New York: Random House, 1956), p. 74.

It is false that all those infected with the HIV virus are either DRUG users or GAYS, because <u>Magic</u> Johnson has the virus, and he is neither a drug user nor a homosexual.

6. **Every member of the school board voted AGAINST the desegregation proposal. Some board members are BLACK. Thus, some black school board members voted against the proposal.**

   (Universe: school board members)

*7. Historian of philosophy Frederick Copleston summarizes a Platonic argument this way:

   > ... Granted that a man can remember something he has formerly perceived and can know it, even while no longer perceiving it, it follows that knowledge and perception cannot be equated. . . .[22]

   The argument formalized:

   There are cases where we KNOW something we REMEMBER even though we are not (now) PERCEIVING it. It follows that knowing cannot be equated with perceiving.

   (Kx = x is a case of knowing, Rx = x is a case of remembering, Px = x is a case of perceiving)

8. A New York Teamsters Union leader had this to say about unionizing prostitutes:

   > Any person who works for a living deserves to be represented by an organization that can make their livelihood better. If ever there were a procedure where we had legalized prostitution, certainly we would be interested in signing up those people who are earning their living in that manner.[23]

   His argument:

   Anyone who WORKS for a living within the LAW has the right to union REPRESENTATION. PROSTITUTES work for a living. So, any prostitute who works within the law has the right to union representation.

   (Universe: people; Wx = x works for a living, Lx = x works within the law)

---

[22]*A History of Philosophy*, Volume I, Part I (Garden City, NY: Image Books, 1962), p. 169.
[23]"Union Has Proposition for Hookers," *Springfield [Ohio] News and Sun* (March 24, 1979).

9. Dialogue from a children's television cartoon:

> Ultan: *Surrender or the boy [Dorno] will perish.*
> Zandor: *Surrender and we all perish.*

The dialogue suggests this unhappy argument:

Either we surrender or <u>Dorno</u> PERISHES. If we surrender, we all perish. Therefore, Dorno will perish.

(Universe: members of the party under attack; S = We surrender) Note that the symbolized second premise, 'S → (x)Px', is not a quantification and that UO may not be applied to it. Proof hint: use the Dash Out Strategy.[24]

10. The following quotation provides an explanation of why an insane person is not guilty of committing a crime; it does this by providing two premises (expressed in the first two sentences) from which the claim may be deduced.

> *The underlying theory behind the insanity defense has been said to be that a crime requires the joint operation of act and intent. An insane person cannot legally be guilty of any criminal intent. . . . If he is incapable of criminal intent, then he is not in law responsible, since one of the essential ingredients of the crime is necessarily missing.*[25]

(Universe: people; Gx = x is guilty of a crime, Ax = x performs a criminal act, Bx = x has criminal intent, Cx = x is insane)

*11. Newspaper story:

> The Clackamas County chapter of the Oregon Medical Association urged state delegates at a meeting here Sunday to approve a resolution saying all OMA members must be covered by the group insurance carrier, or none will be covered.
> Under the proposal, the group carrier's contract with all physicians would automatically be cancelled if the carrier dropped coverage for any OMA physician.[26]

The reporter's inference:

---

[24]The Wedge Out proof is longer and more cumbersome.
[25]"Criminal Law" § 48, 21 Am. Jur. 2d 167.
[26]"Blazer Team Doctor to Lose Insurance," *Springfield [Ohio] Sun* (April 24, 1979), p. 10.

Either all members are COVERED or none are covered. So, if even one member is not covered, then no members will be covered.

(Universe: members of the OMA) Note that the premise is a disjunction; accordingly UO may not be applied to it. The conclusion is a conditional, so the Arrow In Strategy is indicated for the proof.

12. (SEMI-CHALLENGING) F1 and F2 are logically equivalent formulas; so are F3 and F4. Demonstrate this with four formal proofs.

    (F1) (x)(Ax & Bx)
    (F2) (x)Ax & (x)Bx

    (F3) (∃x)(Ax v Bx)
    (F4) (∃x)Ax v (∃x)Bx

13. (CHALLENGING) The Arab general who conquered Egypt and destroyed the books in the ancient library at Alexandria supposedly "justified" his action with this argument:

    The books consistent with the KORAN are SUPERFLUOUS and the ones inconsistent with the Koran are PERNICIOUS. In either case they should be DESTROYED. Consequently, all the books should be destroyed.

    (Universe: books in the library at Alexandria)

14. (CHALLENGING) This Socratic argument was discussed in Section 1.3:

    If there are supernatural ACTIVITIES, then there are supernatural BEINGS. Any supernatural being is either a GOD or the CHILD of a god. If there are children of gods, then there are gods. So, if there are supernatural activities, then there are gods.

    (Cx = x is a child of a god)

§3.3 PROPOSITIONAL ANALOGUES

The method of formal proof establishes validity, but it does not show invalidity. We need a method for establishing invalidity; two such methods are presented in this section and the next.

We can produce the "propositional analogue" (or just "analogue") of a formula of predicate logic by removing every quantifier, quantifier-scope grouper, variable, and individual constant. The propositional analogue created by this operation will consist of some or all of the following: statement letters, statement connectives, and groupers. Here are some examples:

PREDICATE FORMULA	PROPOSITIONAL ANALOGUE
(x)Ax	A
(∃x)(Bx & −Cx)	B & −C
(x)[Dx → (Ex → Fx)]	D → (E → F)

It is easy to produce the propositional analogue of a symbolized predicate argument, for example:

(x)(Gx → Hx), (∃x)(Ix & −Hx) ⊢ (∃x)(Ix & −Gx)
G → H, I & −H ⊢ I & −G

The formulas in a predicate argument are not ordinarily logically equivalent to their counterparts in the analogous argument. Nevertheless, within a certain restricted range of cases the predicate argument is valid iff the propositional analogue is valid. This provides us with a new approach to testing (some) symbolized predicate arguments—we test them indirectly by testing their propositional analogues. Because the analogues are propositional we may use the truth-table tests.

We need to specify very carefully the range of symbolized predicate arguments to which this technique may be safely applied. We do that in steps one through three of the flow chart on page 141. Step one requires that every formula in the argument be a quantification. Recall that a quantification is a formula that begins with a quantifier whose scope is the entire formula. Note that neither of the following formulas is a quantification:

−(x)Jx
(x)Kx → Lm

Step one also excludes formulas with more than one quantifier and formulas containing individual constants. So neither of the next two formulas satisfies the requirements laid out in step one (even though each *is* a quantification).

(∃x)[Nx v (y)Oy]
(x)(Px v Qr)

> **THE METHOD OF PROPOSITIONAL ANALOGUES
> FOR TESTING ARGUMENTS FOR VALIDITY**
>
> 1. Is each formula in the argument a quantification with only one quantifier and no constant?
>   YES: Go to step 2.
>   NO: Stop the test. Inconclusive result.
>
> 2. Is every formula universal?
>   YES: Go to step 4.
>   NO: Go to step 3.
>
> 3. Are exactly one premise and the conclusion existential?
>   YES: Go to step 4.
>   NO: Stop the test. Inconclusive result.
>
> 4. Create and test the propositional analogue. Is the analogue valid?
>   YES: The original argument is also valid.
>   NO: The original argument is invalid.

Steps two and three further restrict the scope of the method by requiring that the argument being tested either be composed exclusively of universal quantifications or that exactly one premise and the conclusion be existential quantifications. While these conditions exclude some arguments, they include the two typical cases of predicate reasoning. Note that nothing concerning an argument's validity or invalidity follows from its failing to reach stage four in the flow chart. Such arguments simply fall outside the scope of the method of propositional analogues.

Let's test some arguments using this procedure. A newspaper story provides this summary of part of the report of the congressional Joint Committee on Defense Production:

> *The committee said no rational leader would use nuclear force unless he believed his country would emerge from war with its basic social, political and economic institutions intact. And no defense system envisioned could promise that, the report said.*[27]

The committee's argument formalized and symbolized:

No rational leaders will use NUCLEAR force unless they BELIEVE their country will emerge from the war basically intact. No rational leader believes that. Hence, no rational leader will use nuclear force.

[27]"They Say Nobody Wins in a Nuclear War," *Miami News* (May 17, 1977), p. 2A.

(x)(−Bx → −Nx), (x)−Bx ⊢ (x)−Nx

(Universe: rational leaders) This symbolized argument reaches stage four in the flow chart, so we can evaluate it by testing its analogue:

−B → −N, −B ⊢ −N

If we recognize that the analogue is an instance of *modus ponens* (view '−B' and '−N' as units), we know in advance that it is valid, but it may be useful to review the truth-table method of showing validity.

B	N	−B	→	−N	−B	⊢ −N
T	T	F	T	F	F	F
F	T	T	F	F	T	F
T	F	F	T	T	F	T
F	F	T	T	T	T	T

*

The pattern "true premises and false conclusion" appears on no row, so the analogue is valid; and this shows that the predicate argument is also valid.

An LSAT study guide states:

> ... Some overlap exists between the C circle and the F circle (statement 5). Therefore, some portion at least of the C circle must be outside the F circle....[28]

The talk of "circles" refers to a technique for representing the content of general statements; ignoring this we extract the following argument:

Some C are F. Therefore, some C are not F.

(∃x)(Cx & Fx) ⊢ (∃x)(Cx & −Fx)

This argument also reaches stage four in the flow chart, so we produce its analogue:

C & F ⊢ C & −F

---

[28]Karl Weber, *How to Prepare for the New Law School Admission Test* (New York: Harcourt Brace Jovanovich, Publishers, 1983), p. 116.

§3.3 Propositional Analogues 143

We can demonstrate the invalidity of the analogue (and indirectly the invalidity of the predicate argument) by constructing a brief truth table:

C	F	‖	C	&	F	⊢	C	&	−F
✓T	T		T	T	T		T	F	FT
				*				*	

The check-mark signifies that the assignment of truth to the premise and falsity to the conclusion was consistently carried through, thus proving the invalidity of the argument.

While some predicate arguments fail to reach stage four in the flow chart, quite a large number do. We can extend the class of treatable arguments even further if we rewrite formulas that are negations of quantifications. The QE Rule tells us that F1A and F1B are logically equivalent, as are F2A and F2B.

(F1A) −(x)Ax
(F1B) (∃x)−Ax

(F2A) −(∃x)Bx
(F2B) (x)−Bx

Therefore, if a predicate argument fails to reach stage four only because one or more of the formulas in it are negations of quantifications, we can bring the argument into the fold simply by rewriting the recalcitrant formulas in accordance with these equivalences.

For an example consider an argument advanced in a book review, which we paraphrase as follows:[29]

> **It is false that any MACHINES are FREE. It is false that any machines can DETECT Gödel statements. Some PEOPLE can detect Gödel statements. So, at least some people are free.**
>
> −(∃x)(Mx & Fx), −(∃x)(Mx & Dx), (∃x)(Px & Dx) ⊢ (∃x)(Px & Fx)

If we rewrite the first two premises as quantifications (using the equivalence of F2A and F2B noted above), the resulting argument can be evaluated with this technique:

[29]*Choice*, June 1971, p. 560.

$(x)-(Mx \& Fx), (x)-(Mx \& Dx), (\exists x)(Px \& Dx) \vdash (\exists x)(Px \& Fx)$

We test the propositional analogue by brief truth table:

M F D P	$-(M \& F)$	$-(M \& D)$	P & D	$\vdash$ P & F
✓F F T T	T F F F *	T F F T *	T T T *	T F F *

The invalidity of the analogue establishes the invalidity of the original argument.

The scope of the technique can be extended further if we adopt a method for avoiding employing individual constants in symbolizations. A sentence whose symbolization would normally include a constant may be recast so that a predicate letter does the work of the constant. This is accomplished with the help of the locution 'is identical to'. S3, for example, may be rewritten as S4, and S5 may be recast as S6:[30]

(S3) The <u>Sedition</u> Law ABRIDGES freedom of speech.   (F3) As

(S4) Any law identical to the SEDITION law ABRIDGES freedom of speech.   (F4) $(x)(Sx \rightarrow Ax)$

(S5) <u>Bill</u> Clinton is not a REPUBLICAN.   (F5) $-Rb$

(S6) No one identical to BILL Clinton is a REPUBLICAN.   (F6) $(x)(Bx \rightarrow -Rx)$

(Sx = x is a law identical to the Sedition Law, Bx = x is a person identical to Bill Clinton) In many cases an argument blocked by step one of the flow chart because of an individual constant may be transformed with the help of this device into one that will reach stage four.

The method of propositional analogues may be used to determine whether two predicate formulas are logically equivalent provided that they reach stage three in the following flow chart:

---

[30] Note this difference between F3 and F4: F3 entails, while F4 does not entail, '$(\exists x)Ax$'. (See §5.5 for elaboration.) This difference does not constitute a practical problem; there is little chance that it would bear on the validity of an argument advanced in actual legal discourse.

§3.3 Propositional Analogues

**THE METHOD OF PROPOSITIONAL ANALOGUES
FOR TESTING STATEMENTS FOR EQUIVALENCE**

1. Are both formulas quantifications with only one quantifier and no constant?
   YES: Go to step 2.
   NO: Stop the test. Inconclusive result.

2. Are both formulas universal or both existential?
   YES: Go to step 3.
   NO: Stop the test. Inconclusive result.

3. Create the propositional analogues and test them for equivalence. Are the propositional formulas equivalent?
   YES: The predicate formulas are also equivalent.
   NO: The predicate formulas are not equivalent.

In the first section of this chapter we suggested two acceptable symbolizations for one statement:

(x)[(Hx & Px) → Bx]
(x)[Hx → (Px → Bx)]

As these formulas reach stage three of the equivalence flow chart, we can prove their equivalence by constructing a truth table for their analogues:

H	P	B	(H	&	P)	→	B	H	→	(P	→	B)
T	T	T			T		T				T	T
F	T	T			F		T				T	T
T	F	T			F		T				T	T
F	F	T			F		T				T	T
T	T	F			T		F				F	F
F	T	F			F		T				T	F
T	F	F			F		T				T	T
F	F	F			F		T				T	T
							*				*	

The two analogous statements have the same truth-value on every row; this proves that they are logically equivalent and (indirectly) that the two quantifications are also equivalent.

Failure of a pair of formulas to reach stage three in the flow chart is not proof of nonequivalence. Regarding failure to satisfy step two, however, it is worth knowing that only in the rarest of cases will a universal and an existential quantification be equivalent.

Aristotelian logicians called a statement "validly convertible" if the statement that results from switching the grammatical subject and predicate (or as we would say, switching the two predicates) is logically equivalent to the original. Which of the following pairs are equivalent?

ORIGINAL	CONVERSE
All A are B.	All B are A.
No A are B.	No B are A.
Some A are B.	Some B are A.
Some A are not B.	Some B are not A.

We prove here that the statements in the first pair are not equivalent by symbolizing them and testing their analogues. One can establish the nonequivalence of two propositional formulas with the brief truth-table method by showing that the two formulas can have opposite truth-values.

predicate symbolization: (x)(Ax → Bx), (x)(Bx → Ax)
propositional analogue: A → B, B → A

A B	A → B	B → A
✓T F	T F F	F T T
	*	*

The equivalence or nonequivalence of the remaining pairs of statements in the table above is the subject of exercise 12 in the exercise set following this section.

Another important relation between statements is that of being *contradictory* (or exactly opposite in content). Two statements are contradictory iff on logical grounds they must have opposite truth-values. (Contradictoriness and equivalence are connected: one statement is the contradictory of a second iff the first and the negation of the second are logically equivalent.) It is worth knowing that the contradictory of 'All A are B' is 'Some A are not B' (rather than 'No A are B') and that the contradictory of 'No A are B' is 'Some A are B'. These relations are the subject of exercise 11.

Two further points regarding the method of propositional analogues: Recall that the brief truth-table test presented in this book does not demonstrate either validity or logical equivalence, but only invalidity and nonequivalence. The full truth-table test may be used to prove all of these results. Secondly, because of their size, full truth tables are not practical for arguments containing more than three (or perhaps four) statement letters. If a propositional analogue contains four or more statement letters you can prove it invalid by the brief truth-table method or valid by constructing a formal proof.

How can we be sure that the method for assessing arguments explained in this section delivers the correct answer in every case? And why do we impose the restrictions in the first three steps of the flow chart? A formal justification of the method is possible, but it is too long to include here.[31] The general idea is that the Dash Out proof for an argument that reaches stage four on the flow chart will have the same form as the one for its propositional analogue, because all the variables can be instantiated to the same constant. That is not always the case with arguments rejected before the fourth stage of the chart.

EXERCISES

1. Each of the following symbolized arguments is rejected at one of the first three stages of the flow chart for testing arguments. Identify the feature(s) of the argument that results in rejection. Can any of these arguments be rewritten in an equivalent form that reaches stage four? If so, rewrite it and test its propositional analogue by formal proof or full or brief truth table.
    (a) −(x)Ax ⊢ (∃x)−(Ax & Bx)
    *(b) (x)Cx & (x)Dx ⊢ (x)(Cx & Dx)
    (c) (x)Ex v (x)Fx ⊢ (x)(Ex v Fx)
    (d) −(x)(Gx → Hx), −Gi ⊢ −Hi
    (e) (x)Jx ⊢ (∃x)Jx
    *(f) (x)Kx, (∃x)Lx ⊢ (x)(Kx v Lx)

2. Demonstrate the validity of the arguments in these exercises following Section 3.2 by constructing full truth tables (for (a) and (b)) or formal proofs (for (c) and (d)) for their propositional analogues.
    (a) Exercise 4
    *(b) Exercise 6
    (c) Exercise 8
    (d) Exercise 10

Instructions for exercises 3 through 10: symbolize each argument and test by constructing a full or brief truth table for its propositional analogue.

---

[31]See Howard Pospesel, "The Method of Propositional Analogues," *Teaching Philosophy*, XVI (June 1993), 157-163.

3. The following arguments have been culled from philosophy examinations—they are not "manufactured."
   (a) No CAUSED acts are FREE.
       Some HUMAN acts are not caused.
       So, some human acts are free.
       (Universe: acts)
   *(b) We HUMANS are all ACTIVE. But since ROBOTS are not human, they are not active.
   (c) Some EVENTS are not CAUSED.
       All ROBOT acts are events.
       So, some robot acts are not caused.
       (Universe: events)
   (d) All FREE acts are uncaused.
       All HUMAN acts are not CAUSED.
       So, human acts are free.
       (Universe: acts)
   (e) All HUMAN acts are CAUSED.
       Some caused acts are FREE.
       Thus some human acts are free.
       (Universe: acts)
4. In *United States* v. *Lee*, 106 U.S. 196, 220 (1882) the Supreme Court argued:

> *No man in this country is so high that he is above the law. No officer of the law may set that law at defiance, with impunity. All the officers of government, from the highest to the lowest, are the creatures of the law and are bound to obey it.*

Their argument, reduced to the core:

No American is ABOVE the law. Therefore, no American government OFFICIAL is above the law.

(Universe: Americans)

*5. Psychiatrist Carl Jung writes:[32]

> *If, however, we restrict the psyche to acts of the will, we arrive at the conclusion that psyche is more or less identical with consciousness, for we can hardly conceive of will and freedom of choice without consciousness. This apparently*

[32]"On the Nature of the Psyche," in *The Basic Writings of C. G. Jung*, ed. Violet Staub de Laszlo (New York: Random House, 1959), p. 54.

brings us back to where we always stood, to the axiom psyche = consciousness.

Perhaps Jung is advancing this argument:

Anything is PSYCHE iff it is WILL. Whatever is not CONSCIOUSNESS is not will. Conclusion: anything is psyche iff it is consciousness.

6. A government tax publication states:

Depreciation can be claimed only on assets that the taxpayer owns. Property that is leased is never subject to an allowance for depreciation by the lessee. . . .[33]

This may be viewed as an argument with the unstated premise 'No LEASED properties are OWNED assets'. (Dx = depreciation may be claimed on x) The first premise is equivalent to 'Any asset on which depreciation may be claimed is owned.'

7. A passage from *Brave New World*:

. . . "*You can only be independent of God while you've got youth and prosperity; independence won't take you safely to the end! Well, we've now got youth and prosperity right up to the end. What follows? Evidently, that we can be independent of God.* . . ."[34]

The argument formalized:

All people who can be INDEPENDENT of God are YOUNG and PROSPEROUS. All people are young and prosperous. It follows that all of us can be independent of God.

(Universe: people)

8. The Irish philosopher Berkeley writes:

. . . *Because intense heat is nothing else but a particular kind of painful sensation; and pain cannot exist but in a perceiving being; it follows that no intense heat can really exist in an unperceiving corporeal substance.*[35]

---

[33]*Fundamentals of Tax Preparation* (Department of the Treasury publication 796, revised July, 1977), p. 11-1.
[34]Aldous Huxley (New York: Bantam Books, 1953), p.159.
[35]George Berkeley, *Three Dialogues Between Hylas and Philonous* (Indianapolis: The Bobbs-Merrill Company, Inc., 1954), p. 16.

Berkeley is advancing this argument (to bolster his thesis that there is no material world):

All intense HEATS are PAINS. No pains can exist in unperceiving CORPOREAL substance. It follows that no intense heats can exist in unperceiving corporeal substance.

(Cx = x can exist in unperceiving corporeal substance)

*9. In *The Praise of Folly*, Erasmus writes:

> *Christ himself in the Gospel denies, that anyone is to be called good but one, and that is God. And then if he is a fool that is not wise, and every good man according to the Stoics is a wise man, it is no wonder if all mankind be included under folly.*[36]

Erasmus appears to be reasoning as follows:

No men are GOOD. He that is not wise is a FOOL. Every good man is a WISE man. Hence, all men are fools.

(Universe: men)

10. Bertrand Russell states:

> *Matter is only a certain way of grouping events, and therefore where there are events there is matter.*[37]

He seems to be arguing:

A thing is MATTER iff it is a complex of events GROUPED in a matter-constituting way. Therefore, any complex of events is matter.

(Universe: complexes of events; Mx, Gx = the events in x are grouped in a certain matter-constituting way) We give Russell the benefit of the doubt by treating the premise of his argument as a quantified *bi*conditional.

Instructions for exercises 11 through 14: symbolize these statements and test for logical equivalence by constructing a full or brief truth table for their propositional analogues.

---

[36]Trans. John Wilson (Ann Arbor, MI: The University of Michigan Press, 1958), p. 129.
[37]*Religion and Science* (New York: Henry Holt and Company, 1935), p. 147.

11. (a) Prove that S1 and S2 are contradictories by demonstrating that S1 and S3 are logically equivalent.

   (S1) All A are B.
   (S2) Some A are not B.
   (S3) It is false that some A are not B.

   Rewrite the symbolization of S3 (and S6) as a quantification.

   *(b) Prove that S4 and S5 are contradictories by demonstrating that S4 and S6 are logically equivalent.

   (S4) No A are B.
   (S5) Some A are B.
   (S6) It is false that some A are B.

12.

ORIGINAL	CONVERSE
All A are B.	All B are A.
No A are B.	No B are A.
* Some A are B.	Some B are A.
Some A are not B.	Some B are not A.

We have already established that the statements in the first pair are not logically equivalent. Test the three other pairs for equivalence.

13. The *contrapositive* of a general statement is formed by converting it and then affixing the prefix 'non-' to each of its predicates.

ORIGINAL	CONTRAPOSITIVE
All A are B.	All non-B are non-A.
* No A are B.	No non-B are non-A.
Some A are B.	Some non-B are non-A.
Some A are not B.	Some non-B are not non-A.

Test each of these pairs for equivalence.

14. Sign posted behind the counter of a rock and gem shop:

> **There Are Three Kinds of Work: Cheap–Good–Fast**
>
> **Cheap Good Work Is Not Fast.**
> **Fast Cheap Work Is Not Good.**
> **Fast Good Work Is Not Cheap.**

Do the last three sentences on the sign make one, two, or three different claims? Test them for logical equivalence by the method of propositional analogues. (Universe: instances of work; Cx, Gx, Fx)

§3.4 INTERPRETATIONS

The method of propositional analogues is fine as far as it goes, but there are many predicate arguments to which it cannot legitimately be applied. Consider, as an example, this argument:

If every jury member has been BRIBED, then all of them will vote ACQUITTAL. Thus, every jury member who has been bribed will vote acquittal.

(x)Bx → (x)Ax ⊢ (x)(Bx → Ax)

(Universe: members of the jury) This argument advanced in an LSAT study guide provides another example:[38]

All of those on the CRICKET team are also on either the slow-pitch SOFTBALL team or the LACROSSE team, but not both. So some cricket players are slow-pitch softball players.

(x){Cx → [(Sx v Lx) & −(Sx & Lx)]} ⊢ (∃x)(Cx & Sx)

The first of these arguments (call it "Bribed Jurors") escapes the method of propositional analogues because the symbolized premise is not a quantification; the first quantifier has only part of the formula within its scope. The argument also escapes the method because the symbolized premise contains two quantifiers. The statements making up the second argument ("Cricket")

---

[38]Karl Weber, *How to Prepare for the New Law School Admission Test* (New York: Harcourt Brace Jovanovich, Publishers, 1983), pp. 69 and 116.

are all quantifications containing one quantifier apiece, but the argument fails to reach stage four of the flow chart because it has an existential quantification for a conclusion but no existentially quantified premise. We need a method of demonstrating invalidity that applies to arguments such as these. In this section we will develop a method that can demonstrate the invalidity of *any* invalid symbolized predicate argument.

A valid argument cannot have true premises and a false conclusion. So any argument that has this combination of truth-values is bound to be invalid. Suppose that the predicates in "Bribed Jurors" had these meanings:

Bx = x is male
Ax = x is wifeless

(Universe: people) In that case the argument would run as follows:

If all people are male, then all people are wifeless. (T)
Thus, every person who is male is wifeless.   (F)

This argument has a true premise and a false conclusion, so it is invalid. But it has the same logical form as "Bribed Jurors"; and since validity is a matter of logical form, "Bribed Jurors" must be invalid as well.

We call this technique for establishing invalidity the method of "interpretation" because logicians call assigning meanings to symbols "interpreting" them. We can summarize the method as follows:

To establish the invalidity of a symbolized predicate argument adopt a universe of discourse and interpret the predicate letters, statement letters, and individual constants so that the reinterpreted argument has true premises and a false conclusion.

Of course, predicate letters must be interpreted with predicates, statement letters with statements, and individual constants with names or expressions that denote an individual. This interpretation establishes the invalidity of "Cricket":

Universe: animals
Cx = x is a bear
Sx = x is a reptile
Lx = x is a mammal

Any bear is either a reptile or a mammal, but not both. (T)
So, some bears are reptiles.   (F)

We indicate on the right the crucial pattern of truth-values.

There are a few logical and epistemic requirements that must be met when this technique is used. The *logical* requirements are (1) that the universe of discourse selected must have at least one member, and (2) that the name used to interpret an individual constant must refer to an individual who belongs to that universe of discourse. (On the other hand, it is not required that the predicates be interpreted so that they are true of some member of the universe.) The *epistemic* requirement is that the truth-values of the reinterpreted statements be known to the people for whose benefit the interpretation is being constructed.

Here are two hints that may make the construction of interpretations easier: (1) Choose a universe that is partitioned into subclasses in a way known to most people. Examples are *people* (a class that subdivides nicely by religion, politics, gender, etc.), the *animal kingdom*, and the set of *integers* or counting numbers (which divides into even, odd, greater than 7, etc.). (2) First attempt to make the conclusion false by appropriately interpreting the symbols occurring there; then make the premises true by suitable interpretation of the remaining symbols.

As a final example of the method, we refute this argument advanced by A. J. Ayer in *Language, Truth and Logic*:[39]

No proposition that has FACTUAL content can be NECESSARY. Accordingly either all MATHEMATICAL propositions lack necessity or they all lack factual content.

$(x)(Fx \to -Nx) \vdash (x)(Mx \to -Nx) \lor (x)(Mx \to -Fx)$

(Universe: propositions) This interpretation does the trick:

Universe: integers
Fx = x is even
Nx = x is odd
Mx = x is prime

No even integer is odd. (T)
Thus, either no primes are even or no primes are odd. (F)

Why is the conclusion false? Well, 2 is an even prime integer, so the left disjunct of the conclusion is false, and 3 is an odd prime, so the right disjunct is false. And a disjunction is false if each of its disjuncts is false.

The method of interpretation is the application to symbolic logic of an ancient technique called by logicians "refutation by logical analogy." Beth uses that method in this conversation:

---

[39] (Harmondsworth, Middlesex: Penguin Books, Ltd., 1971), p. 97. The argument can be transformed into a valid one by adding this premise: If even one mathematical proposition is necessary, then all are necessary. Ayer may have taken this premise for granted.

Arnold: *Anyone admissible to N.Y.U. Law School has an LSAT score above 175. My score is above 175. So, I'm admissible.*
Beth: *Whoa! That's like reasoning, 'Any cat is a mammal; Lassie is a mammal; so, Lassie is a cat.'*

The refutation works because Beth's argument has the same logical structure as Arnold's and is obviously invalid (because it has true premises and a false conclusion). Which version of this logical device is better—the informal refutation by logical analogy or its more rigorous development in symbolic logic, the method of interpretation? Refutation by logical analogy has the advantage of being persuasive for those lacking formal logical training (such as a typical jury member). The method of interpretation has the advantage of ensuring that the two arguments have the same logical form; furthermore it can assist you in devising the refuting argument. Of course, the two techniques are not incompatible; you can employ the method of interpretation to invent an argument and then use that argument in an informal refutation by logical analogy.

The method of interpretation will also establish that two statements are not logically equivalent. All that is required is an interpretation under which either statement is true and the other false. For instance, let's establish the nonequivalence of these statements:

All creatures with HEARTS have KIDNEYS.   (x)(Hx → Kx)
All creatures without hearts lack kidneys.   (x)(−Hx → −Kx)

(Universe: creatures) This interpretation suffices to prove nonequivalence:

Universe: animals
Hx = x is a snake
Kx = x is a reptile

All snakes are reptiles.   (T)
All nonsnakes are nonreptiles. (F)

The second statement is false because alligators (for example) are reptilian nonsnakes.

Two concluding observations about the method of interpretation: First, its scope is even greater than predicate logic. It may be used in propositional logic or indeed *any* branch of symbolic logic. Second, failure to come up with a refuting interpretation is <u>not</u> proof of validity (or logical equivalence) any more than failure to complete a formal proof is a demonstration of invalidity (or nonequivalence).

In this section and the two preceding sections we have presented three methods for evaluating symbolized predicate arguments. Some of the differences among these techniques are noted in this chart:

METHOD	APPLIES TO	DEMONSTRATES	REQUIRES
PROOF	all valid arguments	validity only	ingenuity
PROPOSITIONAL ANALOGUES	only some arguments	both validity and invalidity	no ingenuity
INTERPRETATION	all invalid arguments	invalidity only	ingenuity

In the next two sections we extend the scope of predicate logic to include relational arguments (§3.5) and identity arguments (§3.6). The methods of proof and interpretation can be applied to arguments embodying either or both of these extensions; the method of propositional analogues cannot.

EXERCISES

1. Demonstrate by the method of interpretation the invalidity of each of the arguments in exercise 3 following Section 3.3.

Instructions for exercises 2 through 7: symbolize each argument. If it is valid, show this by formal proof or propositional analogue. If it is invalid, demonstrate this by the method of interpretation.

2. The comic strip on page 157 suggests this argument:

> **Some FOOL will be KISSED by F. M is no fool. So, M will not be kissed by F.**

(Kx = x will be kissed by F, m = the male)

*3. Dialogue in Plato's *Phaedo*:

> Socrates: *... Tell me, what must be present in a body to make it alive?*
> Cebes: *Soul.*
> Socrates: *Is this always so?*
> Cebes: *Of course.*

By permission of Johnny Hart and Creators Syndicate, Inc.

Socrates: *So whenever soul takes possession of a body, it always brings life with it?*
Cebes: *Yes, it does.*[40]

Socrates advances, and Cebes accepts, this argument:

Every live BODY has a SOUL. This proves that every body that has a soul is ALIVE.

(Ax = x is alive, Bx = x is a body, Sx = x has a soul)

4. The meditative bureaucrat depicted in the Oliphant cartoon on page 158 may be reasoning:

   Only ESSENTIAL federal employees REPORT for work. No federal employees report for work. Hence, no federal employees are essential.

   (Universe: federal employees) The first premise amounts to 'All federal employees who report for work are essential'.

5. Huck Finn reasons:

   Jim said bees wouldn't sting idiots; but I didn't believe that, because I had tried them lots of times myself, and they wouldn't sting me.[41]

   His argument appears to be:

---

[40]Plato, "Phaedo," in *The Last Days of Socrates*, trans. Hugh Tredennick (Baltimore: Penguin Books, 1989), p. 167.
[41]Mark Twain, *Adventures of Huckleberry Finn* (Berkeley and Los Angeles: University of California Press, 1985), p. 55.

Bees won't STING me. I am not an IDIOT. So, it is false that bees won't sting idiots.

(Universe: people; Sx = bees will sting x, h = Huck Finn, Ix = x is an idiot) The second premise is unstated.

OLIPHANT ©1979 UNIVERSAL PRESS SYNDICATE. Reprinted with permission. All rights reserved.

6. From an I.R.S. publication:

> *Every taxpayer . . . is entitled to at least one personal exemption of $750. . . .*
> *In addition to the regular exemption, a taxpayer 65 years old or older on the last day of the tax year may claim a second exemption. . . . A taxpayer who is blind on the last day of the taxable year, also is entitled to an additional exemption.*
> *Therefore, a taxpayer 65 or over and blind is entitled to two additional exemptions along with the regular exemption. . . .*[42]

(Universe: taxpayers; Rx = x is entitled to a regular personal exemption; Sx = x is 65 years old or older, Ax = x is entitled to an additional exemption for age, Bx = x is blind, Cx = x is entitled to an additional exemption for blindness)

---

[42]*Fundamentals of Tax Preparation* (Department of the Treasury publication 796, revised July, 1977), p. 2–1.

*7. Charles Ives:

> We like the beautiful and don't like the ugly; therefore, what we like is beautiful, and what we don't like is ugly.[43]

Adopt "objects in our experience" as the universe and use these symbols: Lx = x is liked by us, Bx, Dx = x is disliked by us, Ux. In this passage 'don't like' means "dislike." In other cases what is 'not liked' includes what is neutral as well as what is disliked.

8. Formula F1 does not entail F2, and F3 does not entail F4. Show these results by the method of interpretation.

   (F1) $(x)(Ax \lor Bx)$
   (F2) $(x)Ax \lor (x)Bx$

   (F3) $(\exists x)Ax \,\&\, (\exists x)Bx$
   (F4) $(\exists x)(Ax \,\&\, Bx)$

9. Demonstrate by the method of interpretation the nonequivalence of the statements in each of the following lettered pairs.

   CONVERSE
   (a) All A are B.   All B are A.
   *(b) Some A are not B.   Some B are not A.

   CONTRAPOSITIVE
   (c) Some A are B.   Some non-B are non-A.
   (d) No A are B.   No non-B are non-A.

10. (SEMI-CHALLENGING) Demonstrate by the method of interpretation the invalidity of the arguments in these exercises following Section 3.3.
    (a) Exercise 5
    (b) Exercise 7
    (c) Exercise 9
    (d) Exercise 10

11. (CHALLENGING) The author of an LSAT study guide believes that statements 1 through 7 together entail 8.[44] If he's correct, show this with a formal proof. If he's wrong, refute him with an interpretation.

---

[43]*Essays before a Sonata, the Majority, and Other Writings*, ed. Howard Boatwright (New York: W. W. Norton & Co., Inc., 1962), p. 77.
[44]Karl Weber, *How to Prepare for the New Law School Admission Test* (New York: Harcourt Brace Jovanovich, Publishers, 1983), pp. 69, 70 and 116.

(1) None of those on the LACROSSE team are also on the field HOCKEY team.
(2) All of those on the POLO team are also on the field hockey team.
(3) Some of those on the field hockey team are also on the BADMINTON team.
(4) None of those on the lacrosse team are also on the badminton team.
(5) All of those on the CRICKET team are also on either the slow-pitch SOFTBALL team or the lacrosse team, but not both.
(6) Some of those on the cricket team are also on the field hockey team.
(7) All of those on the badminton team are also on the polo team.
(8) Some of those on the cricket team are not on the polo team.

(Universe: members of a college's athletic teams)

12. (EXTRA CHALLENGING) If any of these three statements entails another, demonstrate this by constructing a formal proof. If any fails to entail another, demonstrate this by the method of interpretation.

(S1) Anyone who AGREES with Lucy is RIGHT.
(S2) If everybody agrees with Lucy, everybody is right.
(S3) If somebody agrees with Lucy, somebody is right.

(Universe: people; Ax = x agrees with Lucy) This exercise was inspired by a *Peanuts* calendar cartoon in which Lucy asserts smugly, "If everybody agreed with me—they'd all be right."

§3.5 RELATIONS

St. Thomas Aquinas quotes St. Augustine on suicide and the fifth[45] commandment:

> ... *Augustine states, "It remains that the precept, Thou shalt not kill, refers to man. And this means both other men and oneself. For nobody but a man is killed when a person commits suicide."*[46]

Augustine's argument:

---

[45]This numeration is employed by Catholics and Lutherans. Jews and most Protestants number this commandment "six."
[46]*Summa Theologica* (London: Blackfriars, 1975), 2a2ae, Question 64, article 5, p. 33. See Augustine, *The City of God*, bk. I, chapter 20.

Whoever kills a person breaks the fifth commandment. So, whoever commits suicide breaks this commandment.

We might symbolize this argument (which we will call "Suicide"):

(x)(Kx → Bx) ⊢ (x)(Sx → Bx)

(Universe: people; Kx = x kills a person, Bx = x breaks the fifth commandment, Sx = x commits suicide) The English argument is valid, but this symbolization is not. However, we can transform the symbolization into a valid argument by adding this supplemental premise:

Whoever commits suicide kills a person.

(x)(Sx → Kx)

A deeper analysis of "Suicide"—one that will reveal the link between killing oneself and killing someone—would certify the validity of the argument without appealing to any additional premise. Relational logic, the subject of this section, provides that deeper analysis.[47]

In the following pages we discuss symbolizing statements in relational logic and proving the validity and invalidity of arguments containing such statements. We will reexamine Augustine's argument when we have equipped ourselves for the job.

Symbolization

All of the predicates treated so far in this chapter have been *property* predicates. A property predicate denotes a characteristic or quality that an individual may possess. Examples of properties are "greenness," "being a felon," and "having acne." A property predicate is symbolized by a predicate letter followed by a single individual variable or constant. A relational predicate denotes a relation that can hold between two or more individuals. Examples of relations are "killing," "suing," and "being taller than." A relation is symbolized by a predicate letter followed by two or more individual symbols (variables or constants). A two-place relation (like "killing") is symbolized by a predicate letter followed by two individual symbols (Kxy), a three-place relation (like "being between") is symbolized by a predicate letter followed by

---

[47]Traditional logic could not handle relational statements. Charles Peirce and Gottlob Frege developed the first successful ways of doing so.

three individual symbols (Bxyz = x is between y and z), and so on. We indicate the letter to be used to represent a relational predicate by capitalizing a prominent word in the predicate and appending a subscripted 'R'.

Here are some sample symbolizations:

Axel LOVES$_R$ Brenda.	Lab
Someone loves Brenda.	(∃x)Lxb
Brenda loves everyone.	(x)Lbx
Brenda and Axel have a CHILD$_R$.	(∃x)Cxba

(Universe: people; Cxyz = x is a child of y and z) Note that by convention 'Lxy' symbolizes "x loves y" and not "y loves x" or "x is loved by y." Obviously the order of individual symbols following a relational predicate letter is critical, as this example shows:

Someone KILLED$_R$ Edith.	(∃x)Kxe
Edith killed someone.	(∃x)Kex

Here the order of letters makes all the difference between Edith's being victim and perpetrator.

How shall we symbolize S1?

(S1) Someone killed someone.

F2 won't do the job because it symbolizes the entirely different statement S2.

(F2) (∃x)Kxx
(S2) Someone killed himself or herself.

F3X is unacceptable because it is not properly formed.

(F3X) (∃x)(∃x)Kxx

In Section 3.1 we stated the requirement that every variable fall within the scope of some quantifier; now we add the requirement that no variable fall within the scope of two quantifiers employing that variable. F3X violates this requirement. The correct symbolization of S1 is F1:

(F1) (∃x)(∃y)Kxy

Not only is the order of individual symbols following a relational predicate letter crucial, in many cases the order of quantifiers is also critical, as these symbolizations make clear.

(S4) Each person LOVES$_R$ someone or other.
(F4) (x)(∃y)Lxy

(S5) There is a person who is loved by everyone.
(F5) (∃y)(x)Lxy

(Universe: people) It is customary to use *x* in the first quantifier in a formula, *y* in the second, etc.; we breach that custom in symbolizing S5 in order to establish the point about the importance of quantifier order. The only difference between F4 and F5 is quantifier order, yet they represent very different sentences. (While S4 and S5 are not equivalent, S5 does entail S4.) The order of contiguous quantifiers is critical when and only when one is universal and one existential. The topic of quantifier order is covered more fully in Section 5.4.

Consider the amphibolous S6; it can mean S7 or S8.

(S6) Somebody loves everybody.

(S7) There is a person who loves everyone.
(F7) (∃x)(y)Lxy

(S8) Each person is loved by someone or other.
(F8) (x)(∃y)Lyx

It is an interesting and significant fact that there is no formula in predicate logic that preserves the amphiboly found in S6.

Statements often contain both property and relational predicates. Some common patterns are displayed in the box on page 164. In relational logic, as in property logic,[48] when a formula has more than one predicate falling within the scope of a quantifier, groupers are required to show its scope. However, when two quantifiers are side by side (as in the first symbolization in the box) only one pair of quantifier-scope groupers is needed. As these symbolizations illustrate, the customary pairing of universal quantifier with arrow and of existential quantifier with ampersand continues to hold true in relational logic. When a formula begins with several quantifiers followed by a

---

[48]By "relational logic" we mean predicate logic including relational predicates; by "property logic" we mean predicate logic involving only property (that is, one-placed) predicates.

quantifier-scope grouper, expect this pairing to obtain between the connective and the quantifier nearest it.

Some POLITICIAN KNOWS$_R$ some LOBBYIST.	$(\exists x)(\exists y)(Px \ \& \ Ly \ \& \ Kxy)$
There is a politician who knows every lobbyist.	$(\exists x)[Px \ \& \ (y)(Ly \rightarrow Kxy)]$
Every politician knows some lobbyist or other.	$(x)[Px \rightarrow (\exists y)(Ly \ \& \ Kxy)]$
Every politician knows every lobbyist.	$(x)(y)[(Px \ \& \ Ly) \rightarrow Kxy]$ *or* $(x)[Px \rightarrow (y)(Ly \rightarrow Kxy)]$
There is a politician who does not know every lobbyist.	$(\exists x)[Px \ \& -(y)(Ly \rightarrow Kxy)]$ *or* $(\exists x)(\exists y)(Px \ \& \ Ly \ \& -Kxy)$
There is a politician who does not know any lobbyist.	$(\exists x)[Px \ \& \ (y)(Ly \rightarrow -Kxy)]$ *or* $(\exists x)[Px \ \& -(\exists y)(Ly \ \& \ Kxy)]$
No politician knows every lobbyist.	$(x)[Px \rightarrow -(y)(Ly \rightarrow Kxy)]$ *or* $(x)[Px \rightarrow (\exists y)(Ly \ \& -Kxy)]$ *or* $-(\exists x)[Px \ \& \ (y)(Ly \rightarrow Kxy)]$
No politician knows any lobbyist.	$(x)(y)[(Px \ \& \ Ly) \rightarrow -Kxy]$ *or* $(x)[Px \rightarrow (y)(Ly \rightarrow -Kxy)]$ *or* $-(\exists x)(\exists y)(Px \ \& \ Ly \ \& \ Kxy)$

If the symbolization of relational sentences seems difficult at first, try symbolizing in stages, like this:

1. There is a POLITICIAN who does not KNOW$_R$ any LOBBYIST.
2. There is an x such that x is a politician and x does not know any lobbyist.
3. $(\exists x)[Px \ \& $ for any y, if y is a lobbyist, x does not know y$]$
4. $(\exists x)[Px \ \& \ (y)(Ly \rightarrow -Kxy)]$

Some symbolizations of sentences exhibiting less common forms are displayed in the box on page 165.

We are now able to symbolize "Suicide" in relational logic:

Whoever KILLS$_R$ a person BREAKS the fifth commandment. So, whoever commits suicide breaks this commandment.

(*Columbo*) "Nobody's ACCUSING$_R$ anybody of anything."	(x)(y)(z)−Axyz Axyz = x accuses y of doing z
(*Milton*) "<u>Belial</u> CAME last than whom a spirit more LEWD$_R$ fell not from heaven."	Cb & (x)−Lxb Universe: fallen spirits Cx = x came last Lxy = x is more lewd than y
(*Michigan Supreme Court ruling*)[49] Assisting a suicide is murder.	(x)(y)(Hxy → Mxy) Hxy = x helps y commit suicide Mxy = x murders y
A statement and its NEGATION$_R$ have opposite truth-values.	(x)(y)[Nxy → (Tx ↔ −Ty)] Universe: statements Nxy = x is the negation of y Tx = x is true
(*editorial*) "<u>CBS</u>[50] spends part of every corporate DAY KNUCKLING$_R$ under to somebody about something."	(x)[Dx → (∃y)(∃z)Kcyzx] Dx = x is a business day Kxyzw = x knuckles under to y about z on day w[51]
(*Kipling*) "Those who KILL$_R$ SNAKES get killed by snakes."	(x)[(∃y)(Sy&Kxy) → (∃y)(Sy&Kyx)]

(x)[(∃y)Kxy → Bx] ⊢ (x)(Kxx → Bx)

(Universe: people) Could the left grouper in the symbolization of the premise be placed after the existential quantifier?

(x)(∃y)(Kxy → Bx)

This is an incorrect symbolization—one that represents a distinct sentence. Two features of this incorrect symbolization should lead one to suspect its accuracy: (1) it pairs an existential quantifier with an arrow, and (2) it unneces-

---

[49] *People* v. *Roberts*, 211 Mich. 187, 178 N.W. 690 (1920). In *People* v. *Kevorkian*, 447 Mich. 436, 527 N.W.2d 714 (1994), *Roberts* was overruled, and assisting a suicide was declared to be a separate offense.

[50] *Newsweek*, April 23, 1973, p. 53.

[51] Variables are written in the customary order (*xyzw*) in dictionary entries. There is no implication that they will follow the predicate letter in the same order in the symbolization. The order in which variables follow a predicate letter in the symbolization is dictated by the content of the sentence being symbolized and by the convention that the first quantifier in the formula employs 'x', etc.

sarily extends the scope of the existential quantifier across a part of the formula (→ Bx) that lacks the variable of quantification (y). While there are statements that can be correctly symbolized in this way, they are uncommon. This matter is discussed further in Section 5.4.

Formal Proofs

With minor changes in the definition of 'instance', the formal proof mechanism set out in Section 3.2 may be extended to include proofs for relational arguments. The changes are required because relational formulas often contain more than one quantifier. The additions to the definition are italicized:

> **An instance of a quantification is a formula that results from deleting the *initial* quantifier (and *its* quantifier-scope groupers) and replacing all of the remaining occurrences of the variable *in that quantifier* by the same individual constant.**

Remember that the quantifier being eliminated by EO or UO or introduced by EI must be the first symbol in the formula and must have the rest of the formula within its scope. By contrast, QE may be applied to a quantifier in the interior of a formula.

We will provide several relational proofs as examples. *Beachcomber Coins, Inc.* v. *Boskett*, 166 N.J. Super. 442, 400 A.2d 78 (App. Div. 1979),[52] concerns the sale for $500 of a 1916 dime believed by both parties to be genuine. When the buyer, Beachcomber Coins, discovered that the dime was counterfeit, the seller, Boskett, refused to take back the coin and refund the purchase price. The buyer sued, lost, and appealed. The appellate judge ruled for the plaintiff saying that "the case falls under the general rule that a contract entered into under a mutual mistake of fact may be set aside by either party." The judge's argument symbolized:

(x)(y)(z)[(Cxyz & Mxyz) → (Ryx & Rzx)], Cabc & Mabc ⊢ Rca

(Cxyz = x is a contract between y and z, Mxyz = x involves the same mistake of fact on the part of both y and z, Rxy = x may rescind y, a = the contract to purchase the dime, b = Boskett, c = Beachcomber Coins) The proof:

---

[52]Discussed in Rate A. Howell, John R. Allison, and N. T. Henley, *Business Law* (Chicago: The Dryden Press, 1982), p. 210.

(1)	(x)(y)(z)[(Cxyz & Mxyz) → (Ryx & Rzx)]	A
(2)	Cabc & Mabc	A
(3)	(y)(z)[(Cayz & Mayz) → (Rya & Rza)]	1 UO
(4)	(z)[(Cabz & Mabz) → (Rba & Rza)]	3 UO
(5)	(Cabc & Mabc) → (Rba & Rca)	4 UO
(6)	Rba & Rca	5,2 →O
(7)	Rca	6 &O

The instantiating constants used on lines 3 through 5 were chosen carefully with an eye on the second premise and the conclusion. Any other choice of constants would prevent the completion of the proof.

Let's adopt this shortcut: when a formula begins with two or more contiguous universal quantifiers, all may be removed in one step of UO. Had we employed this shortcut in the proof above we would have cut two lines from the proof (lines 3 and 4). Because of the restrictions on the EO Rule, this shortcut does not apply when any of the quantifiers to be instantiated is existential.

A newspaper story some years ago hailed the end of smallpox on the occasion of the recovery of the last smallpox victim on earth. The story runs in part:

> *Though the search for other cases of smallpox will continue, there is scant chance any will be found. The reason lies in the nature of the virus that causes smallpox.*
>
> *Unlike the viruses of polio, flu, and other diseases, the variola (smallpox) virus must be transmitted from human to human only. When that chain is broken, the virus cannot survive, and Ali's recovery without a further spread of infection broke the chain.*[53]

The argument:

Anyone who CONTRACTS smallpox gets it from someone who HAS smallpox. No one has smallpox. So, no one will contract smallpox.

(x)[Cx → (∃y)(Hy & Iyx)], (x)−Hx ⊢ (x)−Cx

(Universe: humans; Cx = x contracts smallpox, Hx = x has smallpox, Ixy = x infects y) The second premise and the conclusion may also be symbolized as negated existential quantifications.

---

[53]"Science Ends Years-long Search, Finds the Last Victim of Smallpox," *Miami News* (April 20, 1978), p. 2A.

The proof (with interpolated comments):

(1) $(x)[Cx \rightarrow (\exists y)(Hy \,\&\, Iyx)]$ A
(2) $(x)-Hx$ A
(3) $-(x)-Cx$ PA
(4) $(\exists x)--Cx$ 3 QE
(5) $--Ca$ 4 EO

The next step will be by UO, but should the rule be applied to line 1 or 2? It makes a difference. We know we want to instantiate line 1 to *a* (in order to forge a link with line 5) but we don't know that about line 2. Also, we want to make EO moves as early as possible to avoid violating the restrictions on that rule, and there is an existential quantification on line 1 that will have to be dealt with. Each of these points tells us to apply UO to line 1 now (rather than to 2).

(6) $Ca \rightarrow (\exists y)(Hy \,\&\, Iya)$ 1 UO
(7) Ca 5 DN
(8) $(\exists y)(Hy \,\&\, Iya)$ 6,7 $\rightarrow$O

We could not apply EO to line 6 because 6 is not an existential quantification; we can apply EO to line 8 since it is an existential quantification. Notice that we are prohibited from instantiating to *a*.

(9) $Hb \,\&\, Iba$ 8 EO

Now we know to instantiate line 2 to *b*. (Had we instantiated line 2 to *b* earlier, we would have been prevented from instantiating line 8 to *b*.) The rest of the proof is cut and dried:

(10) $-Hb$ 2 UO
(11) Hb 9 &O
(12) $Hb \,\&\, -Hb$ 11,10 &I
(13) $(x)-Cx$ 3-12 $-$O

The most challenging aspect of relational proofs lies in the selection of instantiating constants. Incidentally, nothing in the statement of the UO and EO rules prohibits instantiating the same quantifier twice (on different lines, of course), and occasionally this tactic is required to complete a proof.

The proof for "Suicide" is unusual in that it requires instantiating all quantifiers to the same constant. This is possible because EO is employed only once in the proof.

(1)	(x)[(∃y)Kxy → Bx]	A	
(2)	−(x)(Kxx → Bx)	PA	
(3)	(∃x)−(Kxx → Bx)	2 QE	
(4)	−(Kaa → Ba)	3 EO	
(5)	(∃y)Kay → Ba	1 UO	
(6)	Kaa & −Ba	4 AR	
(7)	Kaa	6 &O	
(8)	(∃y)Kay	7 EI	
(9)	Ba	5,8 →O	
(10)	−Ba	6 &O	
(11)	Ba & −Ba	9,10 &I	
(12)	(x)(Kxx → Bx)	2-11 −O	

You may want to check the definition of 'instance' on page 166 to satisfy yourself that the formula on line 7 is an instance of the existential quantification on line 8.

Two-place or *dyadic* relations themselves have properties. Four properties of logical import are *symmetry, asymmetry, transitivity*, and *intransitivity*; they may be defined as follows:

Relation R is *symmetrical* iff whenever x bears R to y, y bears R to x.

> *Examples of symmetrical relations: being sibling of, being married to, logical equivalence.*

Relation R is *asymmetrical* iff whenever x bears R to y, y does *not* bear R to x.

> *Examples of asymmetrical relations: being older than, being mother of, being north of.*

Relation R is *transitive* iff whenever x bears R to y and y bears R to z, x bears R to z.

> *Examples of transitive relations: being ancestor of, being taller than, entailment.*

Relation R is *intransitive* iff whenever x bears R to y and y bears R to z, x does *not* bear R to z.

Examples of intransitive relations: being father of,[54] being two inches taller than, being the immediate successor of.

These four properties of dyadic relations are readily symbolized:

symmetry	(x)(y)(Rxy → Ryx)
asymmetry	(x)(y)(Rxy → −Ryx)
transitivity	(x)(y)(z)[(Rxy & Ryz) → Rxz]
intransitivity	(x)(y)(z)[(Rxy & Ryz) → −Rxz]

Occasionally it is necessary to add one of these formulas as a supplementary premise in order to capture the validity of a relational argument.[55] An example:

Anne is **OLDER**$_R$ than **Ben**, and he is older than **Carla**. So, Anne is older than Carla.

Oab & Obc ⊢ Oac

To transform this symbolized argument into a valid one we must add the supplemental premise that "being older than" is a transitive relation:

(x)(y)(z)[(Oxy & Oyz) → Oxz]

With the aid of this premise the proof is easily constructed.

Interpretations

The method of propositional analogues is not available in relational logic, but the method of interpretation is. The one new "wrinkle" is that you must be sure that one-place (*i.e.*, property) predicate letters are interpreted with property predicates, two-place predicate letters with two-place relations, etc. We will use the method of interpretation to prove the invalidity of several arguments.

A simplistic version of the cosmological argument for the existence of God runs as follows:

---

[54] While "being father of" is intransitive, "being parent of" is not—as the plot of *Oedipus Rex* shows.
[55] Exercise 22 following §3.6 is such an argument.

Every event in the SENSIBLE world has a CAUSE$_R$. Therefore, there is one cause for all the events in the sensible world.

(x)[Sx → (∃y)Cyx] ⊢ (∃x)(y)(Sy → Cxy)

(Sx = x is an event in the sensible world) This is a common fallacious pattern. In its simplest form ((x)(∃y)Cyx ⊢ (∃y)(x)Cyx) it may be described as an "illicit quantifier shift"; it involves moving an existential quantifier across a universal quantifier from right to left. A refuting interpretation of the cosmological argument displayed above:

Universe: integers
  Sx = x is positive
Cxy = x is greater than y

Every positive integer is exceeded by some integer.   (T)
So, there is an integer greater than all positive integers. (F)

The conclusion is false because the set of integers is infinite, and also because the "greatest number" postulated by the conclusion (if positive—and it could hardly be negative) would have to be greater than itself. We've just advanced a relational argument:

No integer is GREATER$_R$ than itself. Hence, it is false that there is an integer that is greater than all integers.

(x)−Gxx ⊢ −(∃x)(y)Gxy

(Gxy = x is greater than y) This argument is evaluated in exercise 9 at the end of this section.
Biologist E. E. Stanford writes:

> *One of the most comprehensible and one of the most firmly held of biological laws was long ago expressed in the Latin, "Omne vivum ex vivo"—"All life comes from life," or "Life springs only from life."*[56]

Stanford's two English statements of the law (we'll call them 'S9' and 'S10') can be symbolized as F9 and F10.

(F9)   (x)[Lx → (∃y)(Cxy & Ly)]

---

[56]*Man and the Living World*, 2nd ed. (New York: The Macmillan Company, 1951), p. 18.

(F10) $(x)[Lx \rightarrow (y)(Cxy \rightarrow Ly)]$

(Lx = x is alive, Cxy = x comes from y) He must think his English statements equivalent (since he speaks of one law—not two), but they are so far from being equivalent that neither entails the other. We show here that S9 does not entail S10 and leave the other demonstration to you as an exercise (number 20).

Universe: people
Lx = x lives in Cleveland
Cxy = x lives within 50 miles of y

Every Clevelander lives within 50 miles of some Clevelander. (T)
So, it is true of each Clevelander that everyone he or she lives within 50 miles of is a Clevelander. (F)

The philosopher Descartes[57] advanced this argument[58] to support a metaphysical principle (the argument's conclusion) which he employed in further reasoning:

There is no SUBSTANCE that POSSESSES$_R$ no ATTRIBUTES. Therefore, every attribute belongs to some substance.

$-(\exists x)[Sx \& (y)(Ay \rightarrow -Pxy)] \vdash (x)[Ax \rightarrow (\exists y)(Sy \& Pyx)]$

The argument is invalidated by this interpretation:

Universe: people
Sx = x is male
Pxy = x is child of y
Ax = x is female

There is no male who is child of no female (*i.e.*, every male has a mother). (T)
Thus, every female has a male child. (F)

The method of interpretation requires ingenuity—especially in relational logic—but finding an invalidating interpretation for a complex relational argument can be as satisfying as solving a crossword puzzle.

---

[57]René Descartes (1596–1650) was one of the founders of modern philosophy, as well as an important mathematician (Cartesian coordinates) and scientist (Cartesian diver). The fact that he could make this mistake should encourage you not to be too concerned if you make such mistakes from time to time.

[58]*The Philosophical Works of Descartes*, 2 vols., trans. Elizabeth S. Haldane and G. R. T. Ross (New York: Dover Publications, Inc., 1955), I, 240.

EXERCISES

1. Symbolize each statement using the suggested notation.
   (a) (*Chief Seattle*) "All things are CONNECTED$_R$."
   *(b) (*newspaper*) "Nobody doesn't LIKE$_R$ Cary Grant." (Universe: people; Lxy = x likes y)
   (c) (*newspaper*) "Not everyone LOVES$_R$ John Travolta." (Universe: people; Lxy = x loves y)
   (d) (*court decision*) "There was no contractual OBLIGATION$_R$ between Jimmy Don Bowman with [sic] anyone." (Universe: people)
   (e) (*song title*) "Everybody's somebody's FOOL$_R$." (Universe: people; Fxy = x is a fool for y)
   *(f) (*Garrett Hardin*) "We are all the DESCENDANTS$_R$ of THIEVES." (Universe: people; Dxy = x is descended from y)
   (g) (*overheard*) "Everybody has WEIRD RELATIVES$_R$." (Universe: people; Rxy = x is a relative of y)[59]
   (h) Every VARIABLE falls within the SCOPE$_R$ of some QUANTIFIER. (Sxy = x includes y in its scope)
   (i) (*adage*) "There is a FLY IN$_R$ every OINTMENT." (Ox = x is a container of ointment)
   *(j) Some JAPANESE person is RICHER$_R$ than any AMERICAN. (Universe: people; Rxy = x is richer than y)
   (k) (*naturalist*) "The elephant and the manatee have a common ANCESTOR$_R$." (Universe: animal species; Axy = x is an ancestor of y)
   (l) (*James Herriot*) "No MAN is as strong as a COW." (Lxy = x is at least as strong as y)
   (m) No MAN is as strong as a COW. (Sxy = x is stronger than y)
   *(n) (*pirate adage*) "DEAD MEN TELL$_R$ no tales." (Sx = x is a tale, Txy = x tells y)

2. Translate each formula into an English sentence using this dictionary:

   Universe: people
   Lxy = x loves y
   Hxy = x hates y
   t = Mother Teresa

   (a) (x)Lxx
   *(b) (x)(y)Lxy
   (c) (x)Ltx
   (d) (x)(∃y)Lxy
   (e) (x)(∃y)Lyx

---

[59]This sentence served also as an exercise in property symbolization (see exercise 1(b) on page 121). The two symbolizations illustrate how analysis can be carried to different levels.

*(f) (∃x)(∃y)(Lxy & Lyx)
(g) (x)(y)(Lxy → −Hxy)
(h) (∃x)(∃y)(Hxy & Lxy)
(i) (∃x)(y)(Lxy → Lyx)
*(j) (x)(∃y)(Lyx & Lxy)
(k) (x)[(∃y)Lyx & (∃y)Lxy]
(l) (x)[Lxx → (∃y)Lxy]

3. Symbolize each statement using the suggested notation.
   (a) Every philosophical PAPER BEGINS$_R$ with an unproved ASSUMPTION. (Px = x is a philosophical paper, Ax = x is an unproved assumption)
   *(b) Someone has DESTROYED$_R$ all of the EVIDENCE. (Px = x is a person, Ex = x is a piece of evidence)
   (c) There is a GULL ON$_R$ every PILING. (Oxy = x is on y)
   (d) (*short story*) "The Lord ACCEPTS$_R$ all who accept Him." (Universe: persons)
   (e) (*lyrics*) "To KNOW$_R$, know, know you is to LOVE$_R$, love, love you." (Universe: people; s = song subject)
   *(f) (*slogan*) "FRIENDS$_R$ don't LET$_R$ friends drive DRUNK." (Fxy = x and y are friends, Lxy = x allows y to drive, Dx = x is drunk)
   (g) (*Lincoln*) "No man is GOOD$_R$ enough to govern another man without that other man's CONSENT$_R$." (Universe: people; Gxy = x is good enough to govern y, Cxy = x consents to being governed by y)
   (h) (*detergent ad*) "No DETERGENT GETS$_R$ out everything." (Gxy = x gets out y, Sx = x is soil)
   (i) (*lyrics*) "Everybody loves somebody sometime." (Px = x is a person, Lxyz = x loves y at time z)
   (j) (*Irving Robbins*) "Not everyone LIKES$_R$ all our FLAVORS." (Px = x is a person, Fx = x is a Baskin-Robbins flavor)
   *(k) (*newspaper*) "[MALE] JOCKEYS always marry TALL WOMEN." (Axy = x marries y)
   (l) (*sign on rear of 18-wheeler*) "If you can't SEE$_R$ [any of] my MIRRORS, I can't see you." (a = you, Mx = x is one of my mirrors, Sxy = x can see y)
   (m) (*logic text*) "Any CLOSURE$_R$ of a closure of a formula is a closure of that formula." (Universe: formulas; Cxy = x is a closure of y)
   *(n) (*statute*) "Any person who pays MONEY upon the event of any WAGER$_R$ prohibited is ENTITLED$_R$ to sue and recover the money from the winner." (Wxyz = x loses y to z in a prohibited wager, Exyz = x is entitled to sue and recover y from z)

4. Humorist Sam Levinson:

   Somewhere on this globe every ten seconds, there is a woman giving birth to a child. She must be found and stopped.[60]

   The humor depends upon the amphibolous first statement. Symbolize both possible meanings using this dictionary: Tx = x is a ten-second period, Wx = x is a woman, Bxy = x gives birth during period y.

5. Complete the following proofs. Every assumption has been identified.

   (a) (1) (∃x)(∃y)Lxy   A
       (2)   1 EO
       (3)   2 EO
       (4)   3 EI
       (5) (∃y)(∃x)Lxy   4 EI

   *(b) (1) (x)(y)Lxy   A
        (2) −(y)(x)Lxy   PA
        (3) (∃y)−(x)Lxy
        (4) −(x)Lxa
        (5) (∃x)−Lxa
        (6) −Lba
        (7) Lba
        (8) Lba & −Lba
        (9) (y)(x)Lxy

   These proofs suggest (what is true) that quantifier order is inessential when both quantifiers are existential (proof a) and when both are universal (proof b).

Instructions for exercises 6 through 15: symbolize each argument and construct a proof of validity for it.

6. The reasoning in Pleading Form 9 (see page 2) may be analyzed as a relational argument:

   Whoever NEGLIGENTLY$_R$ drives a motor vehicle against another person and causes damages is LIABLE$_R$ to that other person. The <u>defendant</u> negligently drove a motor vehicle against the <u>plaintiff</u> and caused damages. Therefore, the defendant is liable to the plaintiff.

   (Universe: people; Nxy = x damages y by negligently driving a motor vehicle against y, Lxy = x is liable to y)

[60]*Orlando Sentinel Tribune* (June 5, 1991), p. A2.

*7. In Section 1.3 we formalized a Socratic argument as two deductive arguments. This was the second:

> **Socrates BELIEVES$_R$ that there are supernatural <u>activities</u>. Socrates has PROVED$_R$ that the existence of supernatural activities implies that there are <u>gods</u>. One who believes a claim and proves that it implies another claim also believes the second claim. Therefore, Socrates believes that there are gods.**

(Bxy = x believes y, a = the proposition that there are supernatural activities, Pxyz = x proves that y implies z, g = the proposition that there are gods)

8. Child-welfare workers in Connecticut removed a baby from her mother on grounds of child abuse because she injected herself with cocaine several hours before the baby was born. The mother sued, contending that her rights were wrongly terminated (In re *Valerie D.*, 223 Conn. 492, 613 A.2d 748 (1992)). In its ruling on the case, the State Supreme Court interpreted Conn. Gen. Stat. § 45a-717 (f)(2), which allows termination of the parent's rights when:

> *The child has been denied, by reason of an act of parental commission or omission, the care . . . necessary for his . . . well-being. . . .*

The Court ruled for the mother, noting that the fetus was not a "child" within the meaning of the above language, and that the action of the mother was not "parental" because she was not (at the time of the injection) a parent. The Supreme Court's reasoning may be formulated as follows:

> **To be guilty of an act of parental COMMISSION$_R$ (or omission) toward a person, one must be a parent of that person. One can be a PARENT$_R$ only of a CHILD. One can be a child only if one has been BORN. [At the time of the injection] <u>Valerie</u> had not been born. Consequently, the <u>mother</u> was not guilty of an act of parental commission (or omission) toward Valerie.**

(Cxy = x is guilty of an act of parental commission (or omission) toward y, Bx = x has been born)

9. The following argument was employed in the text of this section:

> **No integer is GREATER$_R$ than itself. Hence, it is false that there is an integer that is greater than all integers.**

(Universe: integers)

10. Second Timothy, chapter two, verse 13:

   If we are faithless, he keeps faith, for he cannot deny himself.

   The verse suggests (to one of us) this relational argument:

   People are FAITHFUL iff they BELIEVE$_R$ in <u>Jesus</u>. All of us believe in ourselves. So, Jesus is faithful.

   (Universe: persons; Bxy = x believes in y)

*11. Sonya Holloway and Michael Cresse filed for divorce, but Michael died in a construction accident on the day the decree was signed. Sonya contended that he died before the decree was signed, and asked to have the divorce nullified so that she could collect his workers compensation and life insurance benefits.[61] Her reasoning:

   Two people can obtain a DIVORCE$_R$ only if both are alive. <u>Michael</u> was not alive. So, <u>Sonya</u> and Michael did not obtain a divorce.

   (Universe: people; Dxy = x and y obtain a divorce, Ax = x is alive)

12. Prove that S5 entails S4.

   (S4) Each person LOVES$_R$ someone or other.
   (S5) There is a person who is loved by everyone.

   (Universe: people; Lxy = x loves y)

13. One of the arguments Larry Adams used in his successful appeal from his conviction for rape was based on expert testimony that "it would be unlikely for a woman not to have contracted gonorrhea after having sex with a diseased man."[62] Adams had an advanced case of gonorrhea at the time the rape occurred, and the victim did not develop the disease. This is actually an inductive argument (concerning what is probable), but we can transform it into a deductive argument without much distortion.

   Anyone RAPED$_R$ by a person who has GONORRHEA will also have the disease. <u>Adams</u> has gonorrhea, but the <u>victim</u> does not. This proves that Adams did not rape the victim.

---

[61]"Widow or Divorcee? Court to Rule," *Miami News* (April 6, 1981), p. 4A.
[62]*Adams v. State*, 417 So.2d 826 (Fla. App. 1982).

(Universe: people; Rxy = x rapes y)

14. When Johnny Carson asked Judith Martin (Miss Manners) if etiquette isn't just common sense, she replied that it isn't because although common sense is the same everywhere, etiquette is different in Japan from what it is in America. Her argument formalized:

> A requirement of COMMON sense will be in FORCE$_R$ everywhere. Some requirements of ETIQUETTE are in force in <u>Japan</u> but not <u>America</u>. So, it is false that every requirement of etiquette is a requirement of common sense.

(Cx = x is a requirement of common sense, Fxy = requirement x is in force in place y, Ex = x is a requirement of etiquette)

*15. From a health newsletter:

> "Thank God my kid isn't hooked on drugs!"
> If he's hooked on drinking, he's hooked on drugs. . . . It's time we stopped pretending it [alcohol] isn't a drug.[63]

The argument:

> <u>Alcohol</u> is a DRUG. Thus, any PERSON who is HOOKED$_R$ on alcohol is hooked on a drug.

Instructions for exercises 16 through 21: symbolize each argument and demonstrate its invalidity by the method of interpretation.

16. One version of the cosmological argument for God's existence appears to include this inference:

> Each CONTINGENT thing does not EXIST$_R$ at some time. Therefore, there is a time at which no contingent things exist.

(Exy = x exists at time y)

*17. Whoever commits suicide BREAKS the fifth commandment. Hence, whoever KILLS$_R$ a person breaks that commandment.

(Universe: people; Kxy = x kills y)

---

[63]*Life Letter* (University of Miami Lifelines Program), November, 1981.

18. **Every THOUGHT is a BRAIN process. Nothing CAUSES$_R$ itself. It follows that no thought is caused by a brain process.**

   (Cxy = x causes y)

19. **For any consistent formal SYSTEM of arithmetic there is some truth of ARITHMETIC that is not a THEOREM$_R$ in that system.[64] Therefore, there is some truth of arithmetic that is not a theorem of any consistent formal system of arithmetic.**

   (Sx = x is a consistent formal system of arithmetic, Ax = x is a truth of arithmetic, Txy = x is a theorem of y)

20. Prove that S10 does not entail S9.

   (S9) All LIFE COMES$_R$ from life.
   (F9) (x)[Lx → (∃y)(Cxy & Ly)]

   (S10) Life springs only from life.
   (F10) (x)[Lx → (y)(Cxy → Ly)]

*21. Berkeley advances an argument for the existence of God that is based on his philosophical position (that matter does not exist):

   Every sense idea has some CAUSE$_R$. Everything that exists is either SPIRIT or IDEA. Ideas do not cause anything. This shows that there is a spirit [God] who causes all sense ideas.

   (Ax = x is a sense idea)

22. (CHALLENGING) Symbolize each statement using the suggested notation.
   (a) (*W. H. Auden*) "GOOD PEOPLE, if they KEEP$_R$ one, have good DOGS." (Px = x is a person, Gx = x is good, Kxy = x keeps y, Dx = x is a dog)
   (b) (*sheriff, quoted in newspaper*) "Unless they're in your HOUSE$_R$, you're not allowed to SHOOT$_R$ anyone unless you're in imminent DANGER$_R$." (Hxy = x is in y's house, Sxy = x is allowed to shoot y, Dxy = x poses an imminent danger to y)
   (c) (*philosopher*) "Whatever is not CAUSED$_R$ by anything is not caused by me." (Cxy = x causes y)

---

[64]Established by Kurt Gödel.

(d) (*La Rochefoucauld*) "We always LOVE$_R$ our ADMIRERS$_R$, but not always those whom we admire." (Universe: people; Lxy = x loves y, Axy = x admires y)

(e) (*comic-strip dialogue*) "Mailmen have mailmen." (Mxy = x is y's letter carrier)

(f) (*newspaper*) "Marriages are PERMITTED$_R$ only between a MAN and a WOMAN." (Pxy = x is permitted to marry y)

(g) (*adage*) "Any LAWYER who represents himself has a FOOL for a CLIENT$_R$." (Cxy = x is a client of y)

(h) (*lyrics*) "Everybody LOVES$_R$ a lover." (Universe: people; Lxy = x loves y) (Use no additional predicate letter.)

(i) (*adage*) "Nothing VENTURED$_R$, nothing GAINED$_R$." (Vxy = x ventures y, Gxy = x gains y)

(j) (*TV newscast*) "Everyone who OWNS$_R$ at least one SHARE HAS$_R$ a PORTFOLIO." (Ax = x is a person, Ox = x owns y, Sx = x is a share, Hxy = x has y, Px = x is a portfolio)

(k) (*Katherine Hepburn*) "If you OBEY$_R$ all the RULES . . . you MISS$_R$ all the FUN."

(l) (*federal law*) "If two or more[65] persons CONSPIRE$_R$ . . . to commit any offense against the United States . . . and one or more of such persons shall do any ACT$_R$ to effect the object of the conspiracy, each shall be FINED not more than $10,000 or IMPRISONED not more than five years, or both." (Cxyz = person x conspires with person y to commit offense z against the United States, Axy = x acts to effect y)

(m) (*Mark Twain*) "Everyone . . . has a DARK$_R$ side which he never SHOWS$_R$ to anybody." (Px = x is a person, Dxy = x is the dark side of y, Sxyz = x shows y to z)

(n) (*Ann Landers*) "Some ALCOHOLICS DRINK$_R$ only BEER." (Ax = x is an alcoholic, Dxy = x drinks y, Bx = x is beer, Cx = x contains alcohol)

(o) (*Elliott Sober*) ". . . You can rationally justify a proposition to someone only if you agree with that person about something." (Cx = x is a person, Dx = x is a proposition, Jxyz = x rationally justifies y to z, Bxy = x believes y)

(p) (*legal decision*[66]) "Conflict of INTEREST$_R$ exists if client and attorney are being PROSECUTED$_R$ by the same OFFICE." (Ixy = x's representation of y constitutes conflict of interest, Cxy = x is legal client of y)

(q) (*Bill Clinton*) "The U.S. will not be the first nation to RESUME$_R$ nuclear testing." (Rxy = nation x resumes nuclear testing at time y, Bxy = time x is before time y)

---

[65] For the sake of simplicity, disregard both occurrences of 'or more'.
[66] *United States* v. *McLain*, 823 F.2d 1457, 1463-64 (11th Cir. 1987).

(r) (*Thomas Jefferson*) "The man who READS$_R$ nothing at all is BETTER$_R$ educated than the man who reads nothing but NEWSPAPERS." (Px = x is a person, Rxy = x reads y, Bxy = x is better educated than y)

(s) (*Justice Byron R. White*[67]) "If the STATE PERPETUATES$_R$ policies . . . TRACEABLE to its prior system [of segregated higher education] that continue to have segregative EFFECTS . . . and such policies are without sound educational JUSTIFICATION and can be practicably eliminated, the state has not satisfied its BURDEN of proving that it has dismantled its prior system." (Ax = x is a policy, Tx = x is traceable to a prior system of segregated higher education, Jx = x has sound educational justification, Cx = x can be practically eliminated, Bx = x has satisfied the burden of proving that it has dismantled its prior system of segregated higher education)

(t) No SYSTEM which has either an all-black or an all-white school is UNITARY. (Axy = x is part of y, Bx = x is black, Wx = x is white, Pxy = x is a pupil at y, Cx = x is a school)

Instructions for exercises 23 through 26: symbolize each argument. If it is valid, show this by formal proof. If it is invalid, demonstrate this by the method of interpretation.

23. (SEMI-CHALLENGING) In *Yniguez v. Arizonans for Official English*, 63 F.3d 920, 935 (9th Cir. 1995), Judge Stephen Reinhardt argues that one's choice of language is as much protected by the First Amendment as the content of one's speech. He writes, "language is . . . speech, and the regulation of any language is the regulation of speech." Determine whether the second conjunct of Judge Reinhardt's statement follows from the first. Use this dictionary: Lx = x is language, Sx = x is speech, Rxy = x is a regulation of y.

24. (CHALLENGING)

> **Each COCONSPIRATOR has had BUSINESS$_R$ dealings with some JUDGE in the county. County judges who have had business dealings with any coconspirator will WITHDRAW$_R$ from that person's trial. "Having business dealings with" is a symmetrical relation.[68] It follows that for at least one of the coconspirators, all the judges in the county will withdraw from the trial.**

(Cx = x is a coconspirator, Bxy = x has had business dealings with y, Jx = x is a judge in the county, Wxy = x will withdraw from the trial of y)

25. (CHALLENGING) A patient (call him "A") in a Florida institution for the retarded killed another patient by shoving a broomstick into his

---

[67]*United States v. Fordice*, 505 U.S. 717, 731 (1992).
[68]That is, if x has business dealings with y, then y has business dealings with x.

body. Authorities suspected that he was also responsible for a similar death (of "B") four years earlier, but although they charged A with the current death, they did not charge him with the earlier death, for this reason:[69]

> [Under the Florida statute of limitations,] no CHARGE$_R$ except first degree MURDER will be brought more than TWO years after the event. The killing of B occurred more than two years ago. No RETARDED person will be charged with first-degree murder [because a retarded person is incapable of forming the requisite intent]. Individual A is retarded. Thus, no charge will be brought against A for the killing of B.

(Cxyz = charge x is brought against individual y for event z, Tx = x happened more than two years ago, Mx = x is a charge of first-degree murder, k = the killing of B, Rx = x is retarded, a = individual A)

26. (CHALLENGING) The "Momma" comic strip suggests at least one of the following arguments:

(a) All people's CHILDREN$_R$ AGGRAVATE$_R$ them. So, any person who has no children who aggravate that person doesn't have children.

(b) All people's CHILDREN$_R$ AGGRAVATE$_R$ them. So, any person who has a child who doesn't aggravate that person doesn't have children.

(Universe: people; Cxy = x is child of y, Axy = x aggravates y) Evaluate both arguments.

MOMMA

By permission of Mel Lazarus and Creators Syndicate.

[69] "Sunrise Youth Is Indicted in Fellow Patient's Death," *Miami News* (March 16, 1972), p. 8A.

27. (CHALLENGING) Prove that S1 is a logical contradiction by deriving a standard contradiction from it.

(S1) There is a BARBER who SHAVES$_R$ all and only those people who do not shave themselves.

(Universe: people)

§3.6 IDENTITY

Muhammed Ali proclaimed S1:

(S1) I am the greatest [boxer].

We can symbolize S1 in property logic with F1A:

(F1A) Ga

(Gx = x is the greatest boxer, a = Muhammed Ali) Can we symbolize it in relational logic with F1X?

(F1X) (x)Bax

(Universe: people; Bxy = x is a better boxer than y) F1X is unsatisfactory because it entails the absurd statement S2 (which is not entailed by S1).

(S2) Muhammed Ali is a better boxer than Muhammed Ali.
(F2) Baa

We might try to skirt this problem by using F1Y:

(F1Y) (x)−Bxa

F1Y avoids the difficulty attaching to F1X, but it is open to another objection. F1Y is compatible with there being other boxers whose abilities are on a par with Ali's; S1 is not compatible with that state of affairs.

In order to provide a correct relational symbolization of S1 we must employ the relation of identity:

(F1B) (x)(−Ixa → Bax)

(Ixy = x is identical to y) F1B makes the claim that Ali is a better boxer than everyone *else*. It does not entail S2, and it also rules out the case of the equally good boxers.

Symbolization

The relation of identity is so important in logic that it is given its own symbol, a symbol that is familiar to you from mathematics: '='. Unlike other relational symbols, the identity symbol is written *between* (instead of before) individual constants or variables. The order in which letters flank the identity symbol is not crucial. Here are some sample symbolizations of statements involving identity:

(S3) Lew <u>Alcindor</u> is identical with Kareem Abdul <u>J</u>abbar.
(F3) a = j

(S4) Anyone who DIED$_R$ before <u>Shakespeare</u> died was not Shakespeare.
(F4) (x)[Dxs → −(x = s)]

(Universe: people; Dxy = x died before y) We may abbreviate '−(a = b)' as 'a ≠ b', so F4 may also be expressed as F4A.

(F4A) (x)(Dxs → x ≠ s)

Because '=' ('≠') is a relational symbol and not a statement connective, we are not required to place parentheses around expressions constructed with it. However, we may use parentheses on occasion to make a formula easier to read. (F4 is an example.)

There are at least two relations of identity, which may be labelled *qualitative* and *numerical* identity. Qualitative identity holds between things that have the same properties (or qualities). Two 1997 red Geo convertibles powered by the same size engine and outfitted with the same optional equipment are qualitatively identical.[70] Numerical identity holds between individual A and individual B when A and B are *one* individual. Sentence S5 is ambiguous; it can be synonymous with S6 or S7.

---

[70] Strictly speaking, no two individuals share all their properties (for example, one Geo has a paint blemish in a spot where the other does not), but we call them qualitatively identical when they share the relevant properties.

(S5) Tom's <u>gun</u> and the weapon used in the <u>crime</u> are identical.
(S6) Tom's gun is just like the weapon used in the crime.
(S7) Tom's gun is the weapon used in the crime.

S6 is a statement of qualitative identity while S7 asserts numerical identity. The identity symbol represents numerical identity. S6 and S7 may be symbolized as follows:

(F6) Sgc
(F7) g = c

(Sxy = x and y share all their properties, g = Tom's gun, c = the weapon used in the crime) A statement of numerical identity (S3 or S7 for example) is true iff the two singular terms occurring in it refer to one individual.

Not all statements containing 'is' are identity statements. S8 is an identity statement; S9 is not.

(S8) <u>Alcindor</u> is <u>Jabbar</u>.
(F8) a = j

(S9) Alcindor is TALL.
(F9) Ta

(Tx = x is tall) We call the copula in S8 the "*is* of identity" and the one in S9 the "*is* of predication." This criterion may be used to identify the *is* of identity:

The 'is' (occurring in a given sentence) is an *is* of identity iff the sense of the sentence is not altered by inserting the words 'the same individual as' after the 'is'.

Many sentences can be adequately symbolized only with the help of the identity sign. The chart on page 186 provides several examples. Notice that the sentence in the chart about Oliver Parker expresses two claims: (a) that Oliver took the exam, and (b) that no one else did. That is why its symbolization is a conjunction (or a quantified *bi*conditional).

Formal Proofs

We may extend our proof procedure to encompass "identity logic" by adding "In" and "Out" rules for the identity symbol. Identity Out is the more frequently used rule:

Only <u>Oliver</u> Parker took the final EXAMINATION.	Eo & (x)(Ex → x = o) *or* (x)(Ex ↔ x = o)
<u>Nixon</u> and <u>Reagan</u> were the only natural-born citizens over 35 who were not ELIGIBLE to be candidates for president in 1992.	−En & −Er & (x)[−Ex → (x=n v x=r)] *or* (x)[−Ex ↔ (x = n v x = r)] Universe: natural-born citizens over 35
Everyone LOVES$_R$ someone else.	(x)(∃y)(Lxy & x ≠ y) Universe: people
Some person LOVES$_R$ all other people.	(∃x)(y)(y ≠ x → Lxy) Universe: people
(*Walt Whitman*) "Whoever DEGRADES$_R$ another degrades me."[71]	(x)[(∃y)(Dxy & x ≠ y) → Dxa] Universe: people; a = Whitman
There is exactly one GOD.	(∃x)[Gx & (y)(Gy → y = x)]
There is at most one DEVIL.	(x)(y)[(Dx & Dy) → x = y]
At least two JUDGES have been IMPEACHED.	(∃x)(∃y)(Ix & Iy & x ≠ y) Universe: judges

The Identity Out Rule (=O): From $a = b$ (or $b = a$) and Fa derive Fb.

(Let 'Fa' represent any formula containing one or more occurrences of some constant and let 'Fb' represent the result of replacing some or all of those occurrences with occurrences of another constant.) The standard assumption-dependence principle applies to this rule.

We illustrate the Identity Out Rule with the story of the Abbé and the nobleman recounted by the philosopher Bosanquet:

> *The Abbé, talking among friends, has just said "Do you know, ladies, my first penitent was a murderer"; and a nobleman of the neighborhood, entering the room at the moment, exclaims, "You there, Abbé? Why, ladies, I was the Abbé's first penitent, and I promise you my confession astonished him!"*[72]

The ladies gasped, for each drew this inference:

---

[71]Whitman's aphorism has two possible meanings: (1) Whoever degrades someone *other than the degrader* degrades Whitman, and (2) Whoever degrades someone *other than Whitman* degrades Whitman. The formula in the chart expresses the first meaning; the second meaning is symbolized: (x)[(∃y)(Dxy & y ≠ a) → Dxa].

[72]Bernard Bosanquet, *Implication and Linear Inference* (London: Macmillan and Co., Ltd., 1920), p. 26, n. 1. Bosanquet attributes the story to Thackeray.

The Abbé's first <u>penitent</u> was a MURDERER. The <u>nobleman</u> was the Abbé's first penitent. Hence, the nobleman is a murderer.

Mp, n = p ⊢ Mn

(This is the first predicate argument we have encountered whose symbolization involves no quantifier.) The validity of the argument is established by this brief proof:

    (1)   Mp   A
    (2)   n = p   A
    (3)   Mn   2,1 =O

Line 3 is generated from line 1 by substituting 'n' for 'p'. It is obvious why the argument is valid (and why the inference rule is sound). If (as the second premise alleges) the nobleman and the Abbé's first penitent are one and the same individual, then any characteristic of the former must be a characteristic of the latter, and vice versa. If the Abbé's first penitent has the property of being a murderer, then so too must the nobleman possess that property if they are the same individual.

A common pattern of legal argument shows that individuals A and B must be distinct (that is, nonidentical) because there is some property that one has and the other lacks. This argument (quoted, reformulated, and symbolized) is typical:

> *Daniels . . . served 14 years in a mental hospital after being arrested for a rape that Thomas Woods, a legislative investigator, said he did not commit. The rape victim said her attacker was black. Daniels is white.*[73]

The <u>rapist</u> was BLACK. <u>Daniels</u> is not black. Thus, Daniels is not the rapist.

Br, −Bd ⊢ d ≠ r

The proof employs the Identity Out Rule:

    (1)   Br   A
    (2)   −Bd   A
    (3)   d = r   PA
    (4)   Bd   3,1 =O

---

[73] "8,000 Voted Kin in False Rape Case," *Miami News* (April 17, 1974), p. 10A.

(5)   Bd & −Bd   4,2 &I
(6)   d ≠ r   3-5 −I[74]

We adopt also a rule for introducing identity formulas:

The Identity In Rule (=I): $a = a$ may be introduced at any point in a proof.

(Let '$a = a$' represent any formula consisting of the identity sign flanked by two occurrences of the same individual constant.) This rule is sound because every individual is identical to itself. In fact, it is a logical truth that each individual is identical to itself; therefore a formula introduced into a proof by this rule is free of assumptions.

We use the Identity In Rule in establishing the validity of the following argument:

Anyone identical to the <u>nobleman</u> is a MURDERER. Consequently, the nobleman is a murderer.

(x)(x = n → Mx) ⊢ Mn

1	(1)	(x)(x = n → Mx)	A
1	(2)	n = n → Mn	1 UO
	(3)	n = n	=I
1	(4)	Mn	2,3 →O

Although no assumption-dependence column is required for this proof, we include one in order to show that no entry is made in that column for a formula justified by the Identity In Rule. The reason the justification entry for a step of Identity In indicates no line number is that the formula is neither an assumption nor an inference from a previous line.

Interpretations

We can extend the method of interpretation to identity arguments very easily. We need only note that the identity sign is a *logical* symbol (not a descriptive symbol) and that its meaning may not be altered during reinterpretation. We illustrate the technique on two arguments.

---

[74]Note that 'd ≠ r' is merely an abbreviation for '−(d = r)'. That is why we can derive it by Dash In.

§3.6 Identity   189

Al is not identical with either Bob or the rapist. Therefore, Bob is not the rapist.

$-(a = b \lor a = r) \vdash b \neq r$

This interpretation demonstrates invalidity:

Universe: positive integers
a = 1
b = 2
r = the even prime

One is not identical with either two or the even prime.   (T)
So, two is not the even prime.   (F)

We show by interpretation that S10 does not entail S11:

(S10) If anyone can TURN the Georgia Tech football program around, Pepper Rogers can do it.   (F10) $(\exists x)Tx \to Tr$

(S11) Only Pepper Rogers can turn around the Georgia Tech football program—if indeed anyone can do it.   (F11) $(x)(Tx \to x = r)$

Universe: people
Tx = x is male
  r = Kevin Costner

If anyone is a male, Kevin Costner is a male.   (T)
So, only Kevin Costner is a male—if indeed anyone is a male. (F)

With the completion of this section on identity logic we have laid out the elements of standard symbolic logic in four stages:[75]

propositional logic   (Chapter Two)
property logic   (Sections 3.1 through 3.4)
relational logic   (Section 3.5)
identity logic   (Section 3.6)

---

[75]There are further developments of standard logic not covered in this volume, such as the introduction of property variables. However, we have presented enough of standard logic to treat most deductive legal inferences one is likely to encounter.

The analysis at each succeeding stage is carried out at a deeper level. Just how deep should one take the symbolic analysis in a given case? Our recommendation is to use the simplest treatment that reveals the essential structure of the inference under scrutiny. Consider, for example, this argument, and its symbolic analysis on four levels:

Fran is the sole owner of the Galley Diner.
So, Fran is the sole owner of something.

(1) A ⊢ B
(2) Cg ⊢ (∃x)Cx
(3) Dfg ⊢ (∃x)Dfx
(4) (x)(Exg ↔ x = f) ⊢ (∃x)(y)(Eyx ↔ y = f)

(Cx = Fran has sole ownership of x, Dxy = x has sole ownership of y, Exy = x owns y)

Obviously the first treatment is inadequate, but each of the remaining analyses suffices to reveal the structure on which the argument's validity depends. We prefer treatment (2) because of its simplicity. Notice that in another argumentative context, the statements above may require a deeper analysis.

As we developed our system of logic we showed how it may be applied to legal argumentation. In the balance of the book we increase this emphasis on legal applications. In the next section, for example, we present a technique for simplifying the treatment of inferences in which a legal principle is applied to a case.

EXERCISES

1. Symbolize each statement using the suggested notation.
   (a) (*Inspector Clousseau*) "Sir <u>Charles</u> and the <u>Phantom</u> are one and the same."
   *(b) Al <u>Gore</u> is identical to no REPUBLICAN.
   (c) <u>Fred</u> LIVES in the Burns mansion by himself. (Universe: people; Lx = x lives in the Burns mansion)
   (d) (*Joyce Kilmer*) "Only <u>God</u> can MAKE a tree." (Mx = x can make a tree)
   (e) (*Guinness Book of Records*) "The oldest museum in the world is the <u>Ashmolean</u> Museum. . . ." (Universe: museums; Oxy = x is older than y)
   *(f) (*football aphorism*) "[On any given Sunday] any NFL team can BEAT$_R$ any other NFL team." (Universe: NFL teams; Bxy = x can beat y)

(g) (*sign on store door*) "In God we trust; all others, strictly cash!" (Cx = we will extend credit to x)

(h) (*resident of Union, South Carolina*) "With a town like this, everybody is some kind of relation of everybody else or knows who's who." (Universe: residents of Union; Rxy = x is related to y, Kxy = x knows who y is)

(i) (*Samuel Johnson*) "No GOOD and worthy man will INSIST$_R$ on another man's drinking wine." (Universe: men; Gx = x is good and worthy; Ixy = x insists on y's drinking wine)

*(j) (*John Collins Bossidy*) "The CABOTS TALK$_R$ only to God." (Cx = x is a Cabot)

(k) One president was IMPEACHED. (Universe: presidents)

(l) (*Johnny Carson*) "No man is an ISLAND—except Orson Wells." (Universe: men)

(m) (*Karl Britton*) ". . . BETWEEN$_R$ any two real numbers there is another real number. . . ." (Universe: real numbers; Bxyz = x is greater than y and less than z)

2. Translate each formula into an English sentence using this dictionary:
   Universe: people
   Lxy = x loves y
     h = Leona Helmsley
     t = Mother Teresa

(a) h ≠ t
*(b) (∃x)(∃y)(Lxy & x ≠ y)
(c) (∃x)(y)(x ≠ y → Lxy)
(d) (x)(y)(x ≠ y → Lxy)
(e) (∃x)(Ltx & x ≠ t)
*(f) (x)(x ≠ t → Ltx)
(g) (x)(Ltx ↔ x ≠ t)
(h) (x)(y)[(x ≠ y & Lxy) → Lxx]
(i) (x)(y)[(Lxx & Lyy) → x = y]
*(j) (∃x)[Lxx & (y)(Lyy → y = x)]

3. Complete the following proofs. Every assumption has been identified.

(a) (1)   Rab   A
    (2)   −Rcb   A
    (3)   PA
    (4)   3,1 =O
    (5)   4,2 &I
    (6)   a ≠ c   3-5 −I

*(b) 1 (1)   −(x)(x = x)   PA
     1 (2)   (∃x)(x ≠ x)
     1 (3)   a ≠ a

(4) a = a
1 (5) a = a & a ≠ a
(6) (x)(x = x)

Proof (b) establishes that '(x)(x = x)' is a logical truth by deriving it free of assumptions.

Instructions for exercises 4 through 12: symbolize each argument and construct a proof of validity for it.

4. Bertrand Russell tells about speaking with a man "who did not agree that Julius Caesar is dead, and when . . . asked why, replied: 'Because I am Julius Caesar.' "[76] The man may have been insane, but his inference was valid:

   I am not DEAD. Therefore, Julius Caesar is not dead—because I am Julius Caesar.

*5. In *The Pink Panther Strikes Again*, Inspector Dreyfus goes to a "dentist" (actually Inspector Clousseau posing as a dentist) to have a painful tooth extracted. After Clousseau removes a tooth, Dreyfus shouts to his assistants:

   He has pulled the wrong tooth! There is only one man who would pull the wrong tooth. It's Clousseau. Kill him!

   Dreyfus reasons:

   The "dentist" has pulled the WRONG tooth. Only Clousseau would do that. So, the "dentist" is Clousseau.

6. The philosopher G.E. Moore reasons:

   . . . If my sense-datum[77] is identical with the surface we are both of us seeing, his [sense-datum] must be identical with it also. My sense-datum can, therefore, be identical with this surface only on condition that it is identical with his sense-datum. . . .[78]

7. Moore continues building his case against the view that a sense-datum is

---

[76] *The Autobiography of Bertrand Russell* (Boston: Little, Brown and Company, 1968), II, 202.
[77] "Sense-datum" is a philosopher's concept. A sense-datum is a mental entity, the direct object of sensation.
[78] "A Defense of Common Sense," in J. H. Muirhead (ed.), *Contemporary British Philosophy*, 2nd series (London: George Allen and Unwin, 1925), p. 220.

identical with the surface being seen with this argument about the phenomenon of double images:

Double images <u>a</u> and <u>b</u> are distinct. Thus, they cannot both be identical with the <u>surface</u> being seen.

8. In "The Veiled Lady," Agatha Christie writes:

"The shoes were wrong," said Poirot dreamily, while I was still too stupefied to speak. "I have made my little observations of your English nation, and a lady, a born lady, is always particular about her shoes. She may have shabby clothes, but she will be well shod. Now, this *Lady Millicent [i.e., the veiled lady] had smart, expensive clothes, and cheap shoes. . . ."*[79]

Poirot's unstated conclusion is that the <u>veiled</u> lady is not the real Lady <u>Millicent</u>. Add the suppressed premise that Lady Millicent is a LADY. Adopt "Englishwomen" as the universe and use these symbols: Lx, Sx = x wears good shoes, v, m.

*9. In their children's book of Greek myths, Ingri and Edgar Parin d'Aulaire tell how King Minos discovered that Daedalus was in the palace of the King of Sicily:

. . . The King of Sicily hid Daedalus and denied that he had him in his service. Slyly King Minos sent a conch shell up to the palace, with a message that, if anyone could pull a thread through the windings of the conch, he would give him a sack of gold as a reward. The King of Sicily asked Daedalus to solve the problem. Daedalus thought for a while, then he tied a silken thread to an ant, put the ant at one end of the conch shell and a bit of honey at the other end. The ant smelled the honey and found its way through the conch, pulling the thread along with it. When King Minos saw this, he demanded the immediate surrender of Daedalus, for now he had proof that the King of Sicily was hiding him. Nobody but Daedalus could have threaded the conch![80]

King Minos' inference:

Nobody but <u>Daedalus</u> could THREAD the conch. Someone in the King of Sicily's PALACE threaded it. This proves that Daedalus is in the palace.

(Universe: people; Tx = x can thread the conch, Px = x is in the palace)

---

[79]"Poirot Investigates" in *Triple Threat* (Binghamton, NY: Vail-Ballou Press, Inc., 1923), pp. 247–48.
[80]*Book of Greek Myths* (Garden City, NY: Doubleday & Company, Inc., 1962), p. 154.

10. News story:

> *Buckingham Palace officials came to the defense yesterday of the Duke of Clarence, great uncle of Queen Elizabeth II, disputing beliefs that he may have been Jack the Ripper, ghoulish murderer of the last century.*
>
> *The duke could not have killed at least two of the Ripper victims because he was away from London at the time, officials said after delving through the palace archives.*[81]

A formalization of this argument:

Jack the Ripper must have been in LONDON$_R$ whenever a "Ripper" MURDER was committed. The duke was out of London when some of those murders occurred. Consequently, the duke was not Jack the Ripper.

(Lxy = x is in London at time y, Mx = a "Ripper" murder is committed at time x)

11. A newspaper story advances this argument to predict the promotion of Lt. Leroy Smith of the Miami Police Department:[82]

[By order of the City Commission] some BLACK will be PROMOTED to major. [By departmental regulations] only LIEUTENANTS and up can be promoted to major. Smith is the only black who is at least a lieutenant. It follows that Smith will be promoted to major.

(Universe: City of Miami police personnel; Lx = x is at least a lieutenant)

12. When a teenager applying for a marriage license tested positive for syphilis, a health officer persuaded her to reveal her sexual contacts by advancing this argument:[83]

Annie has SYPHILIS. Anyone who has syphilis has had INTERCOURSE$_R$ with someone who has syphilis. Therefore, if Tony [Annie's fiancé] does not have syphilis, Annie has had intercourse with someone besides Tony.

---

[81]"Ripper Royalty? Palace Says No," *Miami News* (November 5, 1970), p. 3B.
[82]Bill Gjebre, "Rules Put Black Officer in Line for Major," *Miami News* (December 6, 1973), p. 6A.
[83]Al Volker, "Annie's Story: Typical Teenager Tries to Hide Shame," *Miami News* (September 1, 1973), p. 13A.

Instructions for exercises 13 through 16: symbolize each argument and demonstrate its invalidity by the method of interpretation.

13. A logic manuscript presents this as a complete valid argument:

    The <u>inventor</u> of the gnomon held that reality was UNLIMITED. <u>Anaxagoras</u> held that reality was MIND. So, Anaxagoras was not the inventor of the gnomon.

    (Ux = x holds that reality is unlimited, Mx = x holds that reality is mind)

*14. **Every prosecutor DESPISES$_R$ every other prosecutor. Hence, every prosecutor despises every prosecutor.**

    (Universe: prosecutors)

15. Some Aristotelians reasoned:

    All HEAVY bodies STRIVE$_R$ toward the <u>center</u> of the universe. All heavy bodies strive toward the <u>earth</u>. This shows that the earth is the center of the universe.

16. When Gerald Ford was searching for a vice-presidential nominee in 1974, Michigan Senator Robert Griffin, appearing on ABC's "Issues and Answers," used this argument to deny that he was a candidate for the position:

    No Republican senator from MICHIGAN is a potential CANDIDATE [because the Constitution doesn't allow the vice president and the president to be from the same state]. <u>I</u> am the only Republican senator from Michigan. So, I am the only Republican senator who is not a potential candidate.

    (Universe: Republican senators)

17. (CHALLENGING) Symbolize each statement using the suggested notation.
    (a) There are two FEMALE Supreme Court Justices. (Universe: Supreme Court Justices)
    (b) (*newspaper*) "Only <u>California</u>, <u>Arizona</u> and <u>Maryland</u> still have GAS chambers." (Universe: states)
    (c) (*newspaper ad*) "<u>Pan</u> American and <u>Aviateca</u> are not the only airlines that FLY from Miami to San Pedro Sula." (Universe: airlines; Fx = x flies from Miami to San Pedro Sula)
    (d) Among SOUTHEASTERN universities only <u>Duke</u> does more federally funded RESEARCH$_R$ than <u>Miami</u>. (Sx = x is a southeastern university, Rxy = x does more federally funded research than y)

(e) (*newspaper*) "All but one of the Rosenkowitz sextuplets have GAINED weight." (Universe: the Rosenkowitz sextuplets)

(f) (*Catch-22*) "The only GOOD colonel, he [Colonel Cathcart] decided, was a DEAD colonel, except for himself." (Universe: colonels)

(g) (*newspaper*) "There are at least two sky MARSHALS aboard if any at all." (Universe: people on board)

(h) (*Geoffrey Hunter*) "Each thing that has a FATHER$_R$ has one and only one father." (Fxy = x is father of y)

(i) (*Newsweek*) "[Hunter] Thompson was one of the two reporters who SMOKED marijuana on the GOP candidate's press bus." (Universe: reporters on the Republican candidate's press bus)

(j) (*comic strip*) "Show me a man who only CARES about himself, and I'll show you a man that nobody else cares about." (Universe: people; Cxy = x cares about y)

By permission of Johnny Hart and Creators Syndicate, Inc.

(k) (*adage*) "Lightning never STRIKES$_R$ twice in the same place." (Lx = x is a bolt of lightning, Sxy = x hits place y)

(l) (*aggravated mother about three-year-old daughter*) "If there were one piece of something STICKY in the universe, AMY would step on it." (Ax = Amy steps on x)

(m) The Abbé's first penitent was a MURDERER. (a = the Abbé, Cxyz = x makes a confession to y at time z, Bxy = x occurs before y)

(n) (*radio ad for novel*) "The world's DEADLIEST$_R$ assassin is a WOMAN and only one MAN can STOP$_R$ her!" (Universe: people; Dxy = x is a more deadly assassin than y, Sxy = x can stop y)

18. (CHALLENGING) This is an exercise in concise symbolization. Symbolize these sentences from the preceding exercise using no more than the indicated number of dyadic-connective occurrences:

    one: (e) and (g)
    two: (f), (h), and (l)
    three: (a), (b), and (d)

The dyadic connectives are: &, v, →, ↔

19. (CHALLENGING) Formulas (a) through (j) were offered by students as symbolizations for S1. Identify the correct symbolizations (if any). For each incorrect symbolization formulate an English sentence that it correctly symbolizes.

(S1) One PHILOSOPHER TAUGHT$_R$ Alexander and he was GREEK.

(a) (∃x)(Px & Txa & Gx)
(b) (x)[(Px & Txa) → Gx]
(c) (∃x)[Px & Txa & (y)(y = x) & Gx]
(d) (x)(y)[(Px & Txa & Gx) → x = y]
(e) (∃x)[Px & Txa & (y)(y = x → Gy)]
(f) (∃x){(y)[(Py & Tya) → y = x] & Gx}
(g) (∃x){(y)[(Py & Tya) ↔ y = x] & Gx}
(h) (x)(∃y){Py & Tya & [(Px & Txa) → y = x] & Gy}
(i) (∃x){Px & Txa & (y)[(Py & Tya) → y = x] & Gx}
(j) (∃x){Px & Txa & Gx & (y)[(Py & Tya & Gy) → y = x]}

(Universe: people)

Instructions for exercises 20 through 22: symbolize each argument. If it is valid, show this by formal proof. If it is invalid, demonstrate this by the method of interpretation.

20. (CHALLENGING) A television ad advances these three propositions (S2 is presented visually):

(S1) INSIDE the truck is every whole WHEAT breakfast cereal that provides all required VITAMINS.
(S2) Total is the only breakfast cereal in the truck.
(S3) Total is the only whole wheat breakfast cereal that provides all required vitamins.

(Universe: breakfast cereals) It seems obvious that an argument is being presented, but it isn't clear which statement is the conclusion. Evaluate the three two-premised arguments that can be composed of these statements.

21. (CHALLENGING) Exercise 3(h) on page 15 concerns this pair of sentences:

(*lyrics*) "Everybody LOVES$_R$ my baby, but my baby don't love nobody but me."
I am my baby.

Prove that the first statement entails the second. (Universe: people)[84]

22. (CHALLENGING) A deeper analysis of the argument in exercise 11:

> **Some BLACK will be PROMOTED to major. Only LIEUTENANTS and those who OUTRANK$_R$ [all] lieutenants can be promoted to major. Smith is the only black lieutenant. He is the highest ranking black officer. If officer a outranks officer b, then b does not outrank a. It follows that Smith will be promoted to major.**

(Universe: City of Miami police personnel) The last premise makes explicit a logical property (asymmetry) of the relation of "outranking" that is crucial to the validity of the argument.

23. (EXTRA CHALLENGING) Breakfast table conversation at the close of the 1995 college football season:

> Howard: *Did Northwestern play Ohio State this year?*
> Carmen: *They couldn't have. Northwestern won all its conference games and Ohio State won all of its conference games except for Michigan.*

On first approach Carmen's argument appears to be simply:

> **Northwestern DEFEATED$_R$ every BIG Ten team it PLAYED$_R$.
> Ohio State defeated every Big Ten team it played except for Michigan.
> Therefore, Northwestern did not play Ohio State.**

In symbols:

(x)[(Pnx & Bx) → Dnx]
(x)[(Pox & Bx) → (Dox ↔ x ≠ m)] & Pom & Bm
⊢ −Pno

The argument symbolized in this way is enthymematic; it is missing several premises. One of those premises is: Ohio State and Northwestern are Big Ten teams. Identify the other suppressed premises that must be added to transform the argument into a valid one, then prove the validity of the augmented argument. Hint for locating missing premises: Begin the proof and see which additional premises are needed to complete the proof.

---

[84]This example is from George Boolos.

§3.7 ANALYSIS BY INSTANTIATION

Many legal arguments apply a general principle (or principles) to a particular situation or case; the principles are expressed in universal statements (statements symbolized with universal quantifications), and facts about the situation are expressed in singular statements. A typical simple form would be:

All A are B.	$(x)(Ax \rightarrow Bx)$
Individual i is A.	Ai
So, i is B.	$\vdash Bi$

If we replace the universal first premise with a statement that is "instantiated" to individual i (as by a step of UO) an argument results that can be treated in propositional logic:

	predicate symbolization	*propositional symbolization*
If individual i is A, then it is B.	$Ai \rightarrow Bi$	$A \rightarrow B$
Individual i is A.	Ai	A
So, i is B.	$\vdash Bi$	$\vdash B$

Now, these two English arguments are not identical, but for many purposes the second simpler one will serve as well as the original.

Consider again the argument contained in the decision in the case of *Yakus v. United States*, U.S. Supreme Court, 321 U.S. 414 (1944) (exercise 2 in Section 3.2).

A statute that grants specified authority to an administrative agency is an unlawful delegation of administrative power only if it fails to contain reasonable standards. However, the <u>Emergency</u> Price Control Act does provide reasonable STANDARDS. Therefore, the Emergency Price Control Act is a VALID delegation of legislative power.

$(x)(-Vx \rightarrow -Sx), Se \vdash Ve$

(Universe: statutes that grant specified authority to an administrative agency) Now, let's give an instantiated analysis to this argument. We replace the general first premise by one that specifically addresses the statute in question. This argument emerges:

The Emergency Price Control Act is an unlawful delegation of administrative power only if it fails to contain reasonable standards. However, that act

provides reasonable STANDARDS. Therefore, it is a VALID delegation of legislative power.

−V → −S, S ⊢ V

While the first formulation of "Yakus" belongs to predicate logic, this instantiated version may be treated in propositional logic. We achieve simplicity by sacrificing generality.

The gain in simplicity was small in our treatment of "Yakus," but when arguments are more complex (and especially when they are relational), the gain in simplicity can be substantial. Consider, for instance, the argument involved in the case of *Asseltyne* v. *Fay Hotel*, 222 Minn. 91, 23 N.W.2d 357 (1946). This case determined whether the proprietor of a hotel destroyed by fire was responsible for a boarder's belongings destroyed in the fire. The argument (paraphrased and symbolized):

An INN is not LIABLE$_R$ for the property of its BOARDERS$_R$ [but only for the property of its guests]. One who makes an inn his or her home for a long period under a SPECIAL contract is a boarder. Mary Asseltyne made the Fay Hotel her home for a long period under a special contract. The Fay Hotel is an inn.[85] **Therefore, the Fay Hotel is not liable for Miss Asseltyne's property.**

(x)[Ix → (y)(Byx → −Lxy)]
(x)[Ix → (y)(Syx → Byx)]
Saf
If
⊢ −Lfa

(Lxy = x is liable for the property of y, Bxy = x is a boarder in y, Sxy = x makes x's home in y for a long period under a special contract) This argument contains two generalizations: a rule concerning innkeeper's liability and a statement of a sufficient condition for being a boarder. We instantiate both generalizations to the Fay Hotel and to Miss Asseltyne.

The Fay Hotel is not LIABLE for the property of Mary Asseltyne if she was a BOARDER there. If Mary Asseltyne made the Fay Hotel her home for a long period under a SPECIAL contract, then she was a boarder. Mary Asseltyne did make the Fay Hotel her home for a long period under a special contract. Therefore, the Fay Hotel is not liable for Miss Asseltyne's property.

B → −L, S → B, S ⊢ −L

---

[85]This premise would normally be taken for granted and not expressed, but without it the argument is not formally valid.

The instantiated argument is more easily symbolized and more easily evaluated than the original version, and that of course is the point of the method.

In both of our examples we began with a predicate argument expressed in English and showed how it could be replaced by a simpler propositional argument (also expressed in English). In each case we provided the predicate symbolization of the original argument, but that step was taken only to make the exposition clear. In practice one would skip the step of predicate symbolization. Part of the point of the procedure is to obviate the need to symbolize in predicate logic.

Some administrative agencies adopt the strategy of instantiating principles in the drafting of regulations. For instance, the instructions on Internal Revenue Form 1040 keep using 'you' instead of the more general 'everybody'.

The method of analysis by instantiation has something in common with the method of propositional analogues explained in Section 3.3. Both employ propositional logic in the treatment of English arguments containing generalizations whose explicit symbolization would be done in predicate logic. That is the extent of the similarity, however. The method of propositional analogues is a technique for evaluating *symbolized* predicate arguments; analysis by instantiation is a technique for bypassing predicate symbolization. Note two other major differences between the methods: (1) analysis by instantiation—but not the method of propositional analogues—may be applied to relational arguments (like "Fay Hotel"). (2) The method of propositional analogues does not apply to arguments containing singular statements, but those are precisely the arguments treated by analysis by instantiation.

EXERCISES

Instructions for exercises 1 through 4: Replace each argument by its propositional instantiation (expressed in English). Symbolize the instantiation using the symbols provided. Evaluate the symbolized argument employing some technique from Chapter Two.

1. In *Regina* v. *Jordan*, 40 Crim. App. 152 (1956), Defendant stabbed the deceased, who was recovering from his stab wounds when a doctor negligently gave him an antibiotic to which he had been found to be allergic, and other hospital employees gave him an excessive amount of intravenous fluids. As a result of these treatments, he died. The court reversed Defendant's murder conviction, saying:

   > *We are disposed to accept it as the law that death resulting from any normal treatment employed to deal with a felonious injury may be regarded as caused by the felonious injury, but . . . this was not normal treatment.*

The court's statement of the law is tantamount to: Death resulting from medical treatment employed to deal with a felonious injury may be regarded as caused by the felonious injury iff the treatment is normal. (R = Deceased's death results from medical treatment employed to deal with a felonious injury, C = Deceased's death may be regarded as caused by the injury, N = The treatment of deceased's injury was normal)

*2. In *Edwards* v. *Aguillard*, 482 U.S. 578 (1987), the Supreme Court considered the constitutionality of Louisiana's Creationism Act (enacted in 1982). Justice Brennan's opinion includes the following statements:

> The Establishment Clause [of the First Amendment to the Constitution] forbids the enactment of any law "respecting an establishment of religion." The Court has applied a three-pronged test to determine whether legislation comports with the Establishment Clause. First, the legislature must have adopted the law with a secular purpose. Second, the statute's principal or primary effect must be one that neither advances nor inhibits religion. Third, the statute must not result in an excessive entanglement of government with religion. . . . State action violates the Establishment Clause if it fails to satisfy any of these prongs. [482 U.S. at 582-3.]
>
> . . .
>
> . . . The preeminent purpose of the Louisiana legislature was clearly to advance the religious viewpoint that a supernatural being created humankind. [Id. at 591.]
>
> . . .
>
> . . . Because the primary purpose of the Creationism Act is to endorse a particular religious doctrine, the Act furthers religion in violation of the Establishment Clause. [Id. at 594.]

(C = The Louisiana Creationism Act is consistent with the Establishment Clause of the First Amendment, S = The Louisiana legislature's preeminent purpose in adopting the law was secular, A = The Act's principal effect is to advance religion, I = The Act's principal effect is to inhibit religion, E = The Act results in an excessive entanglement of government with religion)

3. Nine-year-old Betty Lou Reed was sent to the store by her grandmother to buy whiskey for the grandmother. The state convicted the store owner, Curly Leathers, of selling liquor to a minor. Leathers appealed the conviction in *Leathers* v. *State*, 63 Okla. Crim. 220, 74 P.2d 967, 114 A.L.R. 114 (1937). The court decided as follows:

> . . . If the minor informs the liquor-dealer that the liquor purchased is for the use of another person, who has sent him to buy it, and with whose money he pays for it, such being in truth the case . . . then the sale takes place between the dealer and the adult, the minor is not concerned in it except as the conduit by which the money is conveyed to the one and the liquor

to the other, and consequently the dealer cannot be convicted of selling to the minor.

. . .

Under the facts proven by the State, it is conclusively shown that her grandmother was sick, and gave her the bottle and instructions to go as her messenger and procure the whiskey from the defendant; that she immediately revealed to the defendant that her grandmother had sent her for the whiskey; and the same was immediately returned to her grandmother unmolested. Under the law, this sale was to the grandmother and not to the minor. . . .

The judgment . . . is reversed.

(I = Betty Lou informed Curly Leathers that she was purchasing the whiskey for her grandmother who sent her to buy it, P = Betty Lou purchased the whiskey for her grandmother who sent her to buy it, S = Leathers sold the whiskey to Betty Lou's grandmother, G = Leathers is guilty of selling liquor to a minor)

4. James Bain, a referee, called a foul at the end of a conference tournament game between Iowa and Purdue. The resulting free throws by a Purdue player led to a victory that eliminated Iowa from the Big 10 championship game. Some Iowa fans blamed referee Bain for the loss, claiming that the foul call was mistaken. The owners of a sports paraphernalia store that specialized in Hawkeye merchandise (the Gillispies) sued Bain for damages. The Iowa Court of Appeals heard the case (*Bain v. Gillispie*, 357 N.W.2d 47 (1984)) and ruled on the issue of whether the Gillispies had a cause of action.[86] Judge Snell's opinion (with a few short deletions):

. . . The Gillispies . . . must be direct beneficiaries to maintain a cause of action, and not merely incidental beneficiaries.

A direct beneficiary is either a donee beneficiary or a creditor beneficiary. Gillispies make no claim that they are creditor beneficiaries of Bain, the Big 10 Athletic Conference, or the University of Iowa. The real test [of being a donee beneficiary] is said to be whether the contracting parties intended that a third person should receive benefit that might be enforced in the courts. It is clear that the purpose of any promise that Bain might have made was not to confer a gift on Gillispies. Likewise, the Big 10 did not owe any duty to the Gillispies such that they would have been donee beneficiaries. If a contract did exist between Bain and the Big 10, Gillispies can be considered nothing more than incidental beneficiaries and as such are unable to maintain a cause of action. Consequently, there is no genuine issue for trial that could result in Gillispies obtaining a judgment under a contract theory of recovery.

. . . The judgment of the trial court in favor of Bain on this issue is **affirmed.**

---

[86]The case is discussed in Henry R. Cheeseman, *Business Law* (Upper Saddle River, NJ: Prentice Hall, 1992), pp. 284–85.

(A = The Gillispies are direct beneficiaries of Bain, the Big 10, or the University of Iowa, G = There is a genuine issue for trial, D = The Gillispies are donee beneficiaries of Bain, the Big 10, or the University of Iowa, C = The Gillispies are creditor beneficiaries of Bain, the Big 10, or the University of Iowa, I = Bain, the Big 10, or the University of Iowa intended the Gillispies to receive a benefit enforceable in court)

Instructions for exercises 5 and 6: (a) In each exercise a legal rule is set forth. Symbolize that rule in predicate logic, using the dictionary provided. (b) Suppose the rule to be applied in a case involving the liability of Smith for the death of Jones. Provide an "analysis by instantiation" of the argument for Smith's liability. Use the dictionary provided. (c) Evaluate the symbolized argument employing some technique from Chapter Two.

5. From *Coke, 3rd Institute* 56 (1644):

> If the act be unlawful it is murder. As if A meaning to steale a deer in the park of B, shooteth at the deer, and by the glance of the arrow killeth a boy that is hidden in a bush; this is murder, for that the act was unlawful, although A had no intent to hurt the boy, nor knew not of him. But if B the owner of the park had shot at his own deer, and without any ill intent had killed the boy by the glance of his arrow, this had been homicide by misadventure, and no felony.
>
> So if one shoot at any wild fowle upon a tree, and the arrow killeth any reasonable creature afar off, without any evill intent in him, this is per infortunium: for it was not unlawful to shoot at the wilde fowle: but if he had shot at a cock or hen, or any tame fowle of another mans, and the arrow by mischance had killed a man, this had been murder, for the act was unlawfull.

Dictionary for (a):

Sxy = x is a shot aimed at an animal by y
Kxy = x kills y
Hx = x is a human being
Mxy = x is guilty of murder on account of y
Ux = x is unlawful

Dictionary for (b):

S = Smith shoots at an animal
K = Smith's shot kills Jones
M = Smith is guilty of murder
U = Smith's shot is unlawful

*6. From *Foster's Crown Law* 258-259 (1762):

> *Accidental Homicide: In order to bring the case within this description, the act upon which death ensueth must be lawful: for if the act be unlawful, I mean if it be malum in se, the case will amount to felony, either murder or manslaughter, as circumstances may vary the nature of it. If it be done in prosecution of a felonious intention it will be murder, but if the intent went no farther than to commit a bare trespass, manslaughter: though, I confess, Lord Coke seemeth to think otherwise. . . .*
>
> *[For example,] A shooteth at the poultry of B, and by accident killeth a man; if his intention was to steal the poultry, which must be collected from circumstances, it will be murder by reason of that felonious intent; but if it was done wantonly and without that intention it will be barely manslaughter.*

Dictionary for (a): Use Sxy, Kxy, Hx, Mxy, and Ux from the previous exercise, plus the following: Fx = x is felonious, Axy = x is guilty of manslaughter on account of y. In part (b) symbolize the instantiated argument that shows Smith to be guilty of manslaughter. Abbreviate the legal rule, omitting the irrelevant clause. Dictionary for (b): Use S, K, and U from the previous exercise, plus the following: F = Smith's shot is felonious, A = Smith is guilty of manslaughter for killing Jones.

7. In each of the following paragraphs from *Hale, Pleas of the Crown* 428[87] a legal rule is set forth. (a through d) Symbolize these rules in predicate logic, providing your own dictionaries. (e) Suppose rule (a) applied in an argument aiming to show Smith guilty of homicide. Symbolize the analysis by instantiation of that argument, providing your own dictionary. (f) Similarly, apply rule (b) to show Smith not guilty of homicide. (g) Apply rule (c) to show Smith guilty of homicide in Jones's death resulting from gangrene. (h) Apply rule (d) to show Smith guilty of homicide for hastening Jones's death resulting from AIDS.

> *If a man give another a stroke, which it may be, is not in itself so mortal, but that with good care he might be cured, yet if he die of this wound within the year and day, it is homicide[88] or murder, as the case is, and so it hath been always ruled. 3 Inst. 47.*
>
> *But if the wound or hurt be not mortal, but with ill applications by the party, or those about him, of unwholesome salves or medicines the party dies, if it can clearly appear, that this medicine, and not the wound, was the cause of his death, it seems it is not homicide, but then that must appear clearly and certainly to be so.*

---

[87]The works of Sir Matthew Hale (1609–76) were first published in 1736.

[88]Hale seems to be using the word 'homicide' to include both manslaughter and murder. He seems not to be considering the possibility of justifiable homicide, as in a case of self-defense.

But if a man receives a wound, which is not in itself mortal, but either for want of helpful applications, or neglect thereof, it turns to a gangrene, or a fever, and that gangrene or fever be the immediate cause of his death, yet, this is murder or manslaughter in him that gave the stroke or wound, for that wound, tho it were not the immediate cause of his death, yet, if it were the mediate cause thereof, and the fever or gangrene was the immediate cause of his death, yet the wound was the cause of the gangrene or fever, and so consequently is causa causati.

If a man be sick of some such disease, which possibly by course of nature would end his life in half a year, and another gives him a wound or hurt, which hastens his end by irritating and provoking the disease to operate more violently or speedily, this hastening of his death sooner than it would have been is homicide or murder, as the case happens, in him, that gives the wound or hurt, for he doth not die simply ex visitatione Dei, but the hurt that he receives hastens it, and an offender of such a nature shall not apportion his own wrong, and thus I have heard that learned and wise judge Justice Rolle frequently direct.

4

Disputations

The symbolic machinery developed in the last two chapters can clarify in many ways the concepts, structures, and procedures that lawyers use in presenting their cases. But in law, as in many other aspects of life, it is not enough to present your own case effectively. You have to answer the case presented by an opponent. A formal presentation of opposite sides of a question, with their responses to one another, has traditionally been referred to by logicians as a *disputation*. In this chapter, we will take up disputations and show how they figure in legal cases, especially at the pleading stage.

§4.1 WHAT ARE DISPUTATIONS?

Some years ago, when the Episcopal Church held its General Convention in that very Catholic City, Boston, the following cartoon appeared in the *Boston Herald*: A man in a black suit and a Roman collar was strolling in the park with a large family. Two boys were watching. One said to the other: "He ain't a father. Look at all the children he has." We will refer to the boy who spoke as "Tom" and his pal as "Jerry." Presumably, Jerry had been reasoning this way:

> All men with black suits and Roman collars are Catholic priests.
> All Catholic priests should be called "Father."
> This man has a black suit and a Roman collar.
> So, he should be called "Father."

This is a valid argument. It can be symbolized this way:

$$(x)(Bx \rightarrow Px)$$
$$(x)(Px \rightarrow Cx)$$
$$Bm$$
$$\vdash Cm$$

(Bx = x is a man with a black suit and a Roman collar, Px = x is a Catholic priest, Cx = x should be called "Father," m = this man)

Tom is evidently reasoning:

> Only Catholic priests should be called "Father."
> Catholic priests don't have families.
> This man has a family.
> So, he shouldn't be called "Father."

This, too, is a valid argument:

$$(x)(Cx \rightarrow Px)$$
$$(x)(Px \rightarrow -Fx)$$
$$Fm$$
$$\vdash -Cm$$

(Fx = x has a family)

Obviously, the conclusions of these two arguments can't both be true. One or the other of them has to be false. But, as we have seen, a valid argument can't have all true premises and a false conclusion. Since both arguments are valid and one of them has a false conclusion, one of them must have at least one false premise. So, if one of the two boys is going to show that the other is mistaken in his conclusion, it won't be enough for him to present his own argument. He will have to take issue with at least one of his friend's premises.

What will then take place is an example of a disputation. *The Random House Dictionary* defines this term as:

> an academic exercise consisting of the arguing of a thesis between its maintainer and his opponent.

The thesis of this particular disputation is:

This man should be called "Father."

Jerry is maintaining it, and Tom is opposing it. Lawyers, as well as other non-logicians, may refer to this dialogue as an "argument." But logicians prefer to reserve that term for the marshalling of premises in support of a single conclusion. We will use their terminology, and call the individual presentations on either side "arguments" and the whole set of presentations a "disputation."

§4.2 DENIALS

The first rule for conducting a disputation is that it is not enough for the parties to bring forward their own arguments. They must also answer the arguments brought forward by their opponents. As we noted in Section 1.1, arguments may be assessed in terms of *form* or *content*. In a disputation, one can answer an argument either by claiming it to be invalid or by denying one or more of the premises. (Of course, you could attempt both.) Defects of content are more common than defects of form, and so *denial* is the usual move made in a disputation. In the cartoon, Tom would presumably take this course, denying that all men with black suits and Roman collars are Catholic priests. In this section we concentrate on denials.

In the context of a disputation, *denying* a proposition is refusing to accept it as something agreed upon by both parties. Note that you can deny a proposition without affirming its negation. "Denying" should not be confused with "refuting." To refute a proposition is to disprove it, *i.e.*, demonstrate its falsity. Tom could have *refuted* or *disproved* Jerry's first premise by arguing:

Some men with black suits and Roman collars are Episcopal priests.
No Episcopal priest is a Catholic priest.
So, not all men with black suits and Roman collars are Catholic priests.

$(\exists x)(Bx \,\&\, Ex)$
$(x)(Ex \to -Px)$
$\vdash -(x)(Bx \to Px)$

(Ex = x is an Episcopal priest) While Tom could have refuted Jerry's premise in this way, he might have merely denied that premise. That is, he could have said:

I deny that all men with black suits and Roman collars are Catholic priests.

We will symbolize this denial as:

N: (x)(Bx → Px)

('N' stands for the Latin 'Nego', "I deny.")

Whatever you don't deny, you *admit*. For instance, we may suppose that Tom would admit Jerry's premise that all Catholic priests should be called "Father." We symbolize that admission this way:

C: (x)(Px → Cx)

('C' stands for the Latin 'Concedo', "I admit.")

To admit something is to accept it as something that doesn't have to be proved on this particular occasion. It is not necessarily to claim that it is true. Like denials, admissions do no more than help establish the common ground for a particular disputation. In fact, we know some Catholic priests who prefer to be called "Doctor" or "Monsignor," and others who prefer to be called by their first names. But we may suppose that if the two boys know about Catholic priests who do not wish to be called "Father" they have chosen to leave them out of the discussion.

A disputation need not stop with one argument on each side. After the two arguments displayed in Section 4.1 have been advanced and Tom has refuted one of Jerry's premises, Jerry can come back with a refutation of a premise of Tom's argument. He can point out that some Episcopal priests are called "Father."

N: (x)(Cx → Px)
(∃x)(Ex & Cx)
(x)(Ex → −Px)
⊢ −(x)(Cx → Px)

Or, if he is a follower of the "Anglo-Catholic" position, which holds that Episcopal priests are Catholic (though not *Roman* Catholic) priests, he can deny the second premise of Tom's refutation of his first premise.

N: (x)(Ex → −Px)

EXERCISES

Instructions: In each of the following cases, symbolize the arguments. Do not symbolize bracketed material. Then show how each later argument meets the previous contrary one by identifying the premise in the previous argument that is being denied.

1. *Endicott Johnson Corp. v. Perkins*, 317 U.S. 501 (1943), arose under the Walsh-Healey Act, which required government contractors to meet certain minimum wage requirements. The Secretary of Labor subpoenaed Endicott Johnson's payroll records in order to see if it was paying the required wages. This was a court proceeding to enforce her subpoena.

    (a) The Secretary claimed that she had REASON[1] to believe that Endicott <u>Johnson</u> was a government contractor, and therefore had authority to INVESTIGATE whether it was paying the required wages. She claimed that she could SUBPOENA the payroll records of anyone she could investigate as to this question.

    (b) Endicott Johnson argued that the Secretary could subpoena the payroll records only of employers over whom she had JURISDICTION, and she had jurisdiction only over government CONTRACTORS. Since Endicott Johnson was not such a contractor, she could not subpoena its payroll records.

    (c) The Supreme Court held that if the Secretary had reason to believe an employer was a government contractor, she had jurisdiction over it, and could therefore subpoena its payroll records. [She could decide, subject to judicial review, whether it was a government contractor or not, as well as whether it was paying the required wages or not. Of course, if it was paying the required wages, the question of whether it was a government contractor would be academic.]

*2. In *Spano v. Perini Corp.*, 25 N.Y. 2d 11, 250 N.E.2d 31 (1969), Defendants' blasting in connection with the building of a tunnel caused damage to Plaintiff's garage 125 feet away. There was no showing of negligence. The court said:

> The principal question on this appeal is whether a person who has sustained property damage caused by blasting on nearby property can maintain an action for damages without showing that the blaster was negligent.

The lower court answered this question in the negative, and therefore rendered judgment for Defendants. The Court of Appeals answered

---

[1]Predicate interpretations like 'Rx = the Secretary of Labor has reason to believe x to be a government contractor' are discussed in Chapter Six.

the same question in the affirmative, and therefore reversed the lower court. Symbolize (a) Plaintiff's argument, (b) Defendants' (and the lower court's) response, and (c) the position adopted by the Court of Appeals. Use this dictionary: B = Plaintiff sustains property damage caused by Defendants' blasting on nearby property, D = Plaintiff can maintain an action for damages against Defendants, N = Plaintiff shows Defendants to have been negligent.

3. In *Fuller v. Preis*, 35 N.Y.2d 425, 322 N.E.2d 263 (1974), Plaintiff's decedent was injured in an automobile accident due to Defendant's negligence. As a result of his injuries, he committed suicide, and Plaintiff brought this wrongful death action on behalf of his estate.
   (a) <u>Defendant</u> argued that decedent's <u>suicide</u> constituted an INTERVENING cause, precluding LIABILITY on his part.
   (b) The court held that suicide does not constitute an intervening cause when it is the result of MENTAL derangement, as it appears to have been in this case.

4. In *United States v. Joe Grasso & Son, Inc.*, 380 F.2d 749 (5th Cir. 1967), Grasso sued to recover employment taxes. Grasso owned seven shrimp boats, each operated by a captain, and usually, two crewmen. The taxes were assessed against Grasso on the ground that it was their employer.
   (a) The government sought to bring the captains into the case under F.R.C.P. 14 as third-party defendants on the ground that if Grasso was not the employer, then the captains were.
   (b) But the court held that there was a third possibility, namely, that the crewmen were not anyone's employees. Since the right to introduce a third-party claim under Rule 14 depends on the liability of the defendant and that of the proposed third-party defendant being an "either/or proposition" (i.e., one or the other must be the case), such a possibility precludes the acceptance of the third-party complaint.

   Use this dictionary: G = Grasso is the employer, C = The captains are employers, T = The captains may be made third-party defendants, P = It is possible that the crewmen are not anyone's employees.

5. In *Wong Yang Sung v. McGrath*, 339 U.S. 33 (1950), the question was whether the proper procedure was followed in Wong's deportation case. Section 5 of the Administrative Procedure Act required certain formalities to be followed in any proceeding "required by statute to be determined on the record after opportunity for agency hearing." The decision was, in fact, made on the record after opportunity for agency hearing, but the formalities in question were not followed.
   (a) The government contended that since no statute required a determination on the record in a deportation case, Section 5 of the Administrative Procedure Act did not apply.
   (b) The Supreme Court held that the words "required by statute" in Section 5 included any case where a determination on the record af-

ter hearing was required by the Constitution. Every deportation case is such a case.

Adopt a universe of "administrative determinations" and use this dictionary: Sx = some statute requires x to be made on the record after opportunity for agency hearing, Dx = x is a determination in a deportation case, Ax = Section 5 of the Administrative Procedure Act applies to x, a = the determination in Wong's deportation case, Cx = the Constitution requires x to be made on the record after opportunity for agency hearing.

*6. In *Regina* v. *Chapman*, [1958] 3 All E. R. 143 (C.A.), Chapman was accused of violating a statute that made it an offense to take an unmarried girl under the age of 18 out of her parents' home with the intention that she shall have unlawful sexual intercourse with any man or men. [The question of intention presents complexities discussed in Chapter Six. Here we disregard that aspect of the case.]

   (a) The prosecution argued that Chapman had taken an unmarried girl from her parents' home, that she was under 18, and that he had caused her to have unlawful sexual intercourse.

   (b) Chapman argued that no act of sexual intercourse [however immoral] was unlawful unless forbidden by some statute. [With certain exceptions not relevant here] no statute forbids sexual intercourse between a male and a female over 16, as this girl was. Therefore, the sexual intercourse between Chapman and the girl was not unlawful.

   (c) The court held that the word 'unlawful' as used in this statute included all nonmarital intercourse [whether or not forbidden by statute].

Use these dictionaries: For argument (a): Gx = x is an unmarried girl under 18, Txy = x takes y from y's parents' home, Axy = x is an act by y of sexual intercourse with a man, Ux = x is unlawful, Cxy = x causes act y to occur, Lx = x is criminally liable, g = the girl involved in this case, c = Chapman. For argument (b): Bxyz = x is an act of sexual intercourse between y and z, Sx = x is forbidden by some statute, Ux, Mx = x is male, Fx = x is female, Ox = x is over 16, a = the act of intercourse in question, c, g. For argument (c): Bxyz, Wxy = x is married to y, Ux, a, c, g.

7. (CHALLENGING) In *United States* v. *Florida East Coast Ry.*, 410 U.S. 224 (1973), the Interstate Commerce Commission made a rule for railroads under a statute that required such rules to be made "after hearing." One of the issues in the case was whether the ICC should have allowed the railroads to make their case orally.

   (a) The railroads argued that there cannot be a HEARING unless there is something to LISTEN to, which there cannot be without ORAL presentation. Now if every proceeding involving a hearing involves an oral presentation, then whenever the applicable STATUTE calls for a hearing an oral presentation is REQUIRED. Therefore, in every case where the applicable statute calls for a hearing, it calls for oral presentation.

(b) The court responded by showing that the first sentence of § 553(c) of the Administrative Procedure Act allows rulemaking "with or without opportunity for oral presentation." But § 553(c) applies to all cases in which § 553(b) provides for NOTICE, and § 553(b) provides for notice in some cases where the applicable statute calls for a hearing. So, there are some cases where the applicable statute calls for a hearing, but oral presentation is not required.

(c) The court also pointed out that some rulemaking proceedings covered by § 557(d) of the Act do not require oral presentations. But the only rulemaking proceedings subject to § 557(d) are those made subject to § 557 by the last sentence of § 553(c). And that sentence makes rulemaking proceedings subject to § 557 only if the applicable statute requires a hearing. Therefore, in some cases where the applicable statute requires a hearing, no oral presentation is required.

Adopt "proceedings" as a universe and use this dictionary: Hx = x involves a hearing, Lx = x involves listening to something, Ox = x involves oral presentation, Sx = the applicable statute calls for a hearing in proceeding x, Rx = oral presentation is required in proceeding x, Cx = x is subject to the first sentence of § 553(c) of the Administrative Procedure Act, Nx = x is subject to the notice requirement of § 553(b) of the Administrative Procedure Act, Mx = x is a rulemaking proceeding, Dx = x is subject to § 557(d) of the Administrative Procedure Act, Ax = x is made subject to § 557 of the Administrative Procedure Act by the last sentence of § 553(c).

§4.3 PLEADING

As we have seen (in Section 1.1), a complaint (and any other pleading that isn't just denials) is a form of enthymeme, or argument with missing premises. The missing premises are propositions of law, and the premises actually stated in the pleading are (mostly) facts. In the Anglo-American legal system, the procedures for denying the usually unexpressed legal premises of a pleading are different from the procedures for denying the factual premises. The procedures will be familiar to anyone who studies Civil Procedure, but we need to relate them to the rules of disputation as we are developing them here.

Let us suppose that your client is served with a complaint that reads as follows:

On (date) defendant appeared in a public place frequented by plaintiff wearing a necktie so hideous as to cause plaintiff acute emotional distress.

Wherefore plaintiff demands judgment in the amount of ten thousand dollars and costs.

You will of course realize at once that there is no such tort as the one your client is accused of having committed. Under the Federal Rules of Civil Procedure, you will therefore file a motion under Rule 12(b)(6) to dismiss the complaint for failure to state a claim on which relief can be granted. Under older procedures, you would file a *demurrer*, which would amount to the same thing. Your point is that even if all the allegations of the complaint are true the plaintiff cannot recover anything from your client.

Now put these moves into the form of a logical disputation. The plaintiff has reasoned as follows:

> **Whoever appears in a public place frequented by another person wearing a necktie so hideous as to cause that other person acute emotional distress is liable to that other person.**
> **Defendant appeared in a public place frequented by plaintiff wearing a necktie so hideous as to cause plaintiff acute emotional distress.**
> **Therefore, defendant is liable to plaintiff.**

The first premise, the one that doesn't appear in the pleading, is a proposition of law. You would prove or disprove it by using the standard tools of legal research. Needless to say, you would not have much trouble disproving it. That is just the point you are making with your 12(b)(6) motion. You are denying the unspoken legal proposition that constitutes one of the premises of your opponent's argument.

This brings us to the following principle:

> **A demurrer, 12(b)(6) motion, motion for judgment on the pleadings, or other similar motion constitutes a denial of the legal premises, stated or unstated, on which the opposing pleading is based.**

To apply this principle, we must begin by supplying the missing premises of the pleading to which we want to respond. Then, if they are legal premises, we can use the appropriate procedures for denying them.

Denying factual premises in a pleading is fairly straightforward. There are, however, a few points that need to be looked at. One is the Negative Pregnant Rule—which generally occasions more hilarity than understanding when it comes up in the Civil Procedure course. According to this rule, the literal denial of certain propositions will be taken as admitting all or some part of them. As the old commentators put it, the denial is *pregnant* with an admission. For instance,

> Defendant denies that he executed and delivered the note sued on,

may be taken to admit either that he executed it or that he delivered it. Or

Defendant denies that he shot the plaintiff with a .38 caliber revolver

may be taken to admit that he shot her with some other kind of gun.

Our study of logic helps illuminate the Negative Pregnant Rule. The denial of a complex statement is not, in general, a denial of the individual statements which compose it. Consider, for example, S1 and its components S2 and S3:

(S1) Defendant executed and delivered the note sued on.
(S2) Defendant executed the note sued on.
(S3) Defendant delivered the note sued on.

The denial of S1 will not be a denial of S2 (or of S3). The reason is obvious: *S2 could be true even though S1 is false.* (The defendant might have executed, but not delivered, the note.) Since whatever is not denied is admitted, if S1 but not S2 is denied, S2 will be admitted. The strongest denial that can be made in this case is the denial that either S2 or S3 is true, that is:

N: (S2 v S3)

To see what's wrong with the revolver example, we will have to go behind the denial and look at the claim the plaintiff is trying to make with the allegation that is being denied. The plaintiff would have just as good a case if the defendant had used some other kind of gun (or a bow and arrow for that matter). She has put in a description of the weapon simply for the sake of clarity or rhetorical effect. The defendant will not be allowed to divert the case into an inquiry about the nonessential question of what kind of weapon was used. The relevant question is whether he shot the plaintiff at all, and unless he puts that question in issue he will lose. Therefore, the only acceptable denial will be a denial that he shot the plaintiff at all. The correct response to the allegation

Defendant shot plaintiff with .38 caliber revolver

is

Defendant denies that he shot plaintiff with a .38 caliber revolver or any other weapon.

In this way, the specific instance brought forward by the plaintiff is specifically denied, and it is also denied that any other specific instance could be brought forward.

Putting these points into symbolic form, we come up with the following rules:

1. **An allegation in conjunctive form (A & B) must be denied in disjunctive form (N: (A v B)).**

2. **A singular allegation (Fa) where a different singular allegation in the same form (Fb) would support the same conclusion (as in the argument (x)(Fx → P), Fa ⊢ P) must be denied both specifically and in the form of an existential quantification (N: [Fa v (∃x)Fx] rather than N: Fa).**

Note that these are not general rules of logic or of legal disputation. They are rules of law that govern this particular form of disputation.

Another such rule is the rule against argumentative denials. For instance, the allegation in Federal Rule Form 9

> *On June 1, 1936, in a public highway called Boylston Street in Boston, Massachusetts, defendant negligently drove a motor vehicle against plaintiff who was then crossing said highway*

cannot properly be denied by pleading

> Defendant has spent his entire life in and around Independence, Missouri, where he was born, and has never been in Boston, Massachusetts.

An allegation must be expressly denied. It will not do to make a statement from which the negation of the allegation can be inferred. That is why the second rule above requires 'N: Fa' to be specifically pleaded even though it follows from 'N: (∃x)Fx', which must also be pleaded.

There are also rules of law governing what happens if you say nothing in response to an allegation. In criminal pleading, saying nothing is taken for a denial, while in civil pleading it is taken for an admission.

As far as logic is concerned, you can deny anything you don't care to admit, but in litigation there are both legal and ethical rules that apply. For instance, Federal Rules of Civil Procedure Rule 8(b) provides that a pleader who has no information about a matter pleaded may say so, and the statement will have the effect of a denial. But Rule 11 subjects all denials to the requirement that there must be good ground to support them, and that they must not be interposed for delay. It would be a violation of Rule 11 to deny something you know your opponent can prove, because the only purpose in the denial would be to cause delay.

Students of legal ethics argue about whether it is acceptable in a civil case to deny something that you know is true, but that you think your opponent won't be able to prove. On the other hand, there seems to be no ethical objection to pleading not guilty in a criminal case, even if you know you are guilty. This is because you have no obligation to help the state convict you.[2]

Turning to the proceedings on a motion for summary judgment in a civil case, another kind of disputation, we find that the Federal Rules of Civil Procedure do not permit a party to deny an opponent's premises without refuting them. Rule 56(e) says:

> *When a motion for summary judgment is made and supported as provided in this rule, an adverse party may not rest upon the mere allegations or denials of the adverse party's pleading, but the adverse party's response . . . must set forth specific facts showing that there is a genuine issue for trial.*

To sum up what we have said about pleading, the rules of disputation say only:

1. **If your opponent makes an argument, you must answer it.**

2. **If you can't show that the argument is invalid, you must deny (or refute) one or more of the premises.**

3. **If you don't deny a premise, you admit it.**

Beyond the rules of disputation, there are legal and ethical rules that cover such questions as whether it is legitimate to deny something, whether if you do deny it you must go further and refute it, and what will be the significance of saying nothing at all.

EXERCISES

1. Deny each of the following allegations. Symbolize the allegation and the denial.
   (a) On June 10 in the parking lot of Eastgate Mall Defendant stole a Rolex watch from Plaintiff.
   *(b) Defendant negligently drove 65 miles per hour in a 30-mile-per-hour zone.
   (c) Defendant owes Plaintiff five hundred dollars, being the sum which Defendant won from Plaintiff in an unlawful game of seven card stud.

[2]Model Rules of Professional Conduct, Rule 3.1.

(d) As Plaintiff was born on April 3, 1980, Plaintiff was less than 16 years of age when he entered into the contract on February 17, 1996. (Assume that the correctness of the date of the contract is not in dispute.)

(e) Defendant, with another person whose name is unknown to Plaintiff, beat Plaintiff about the face, arms, and head, causing serious injury. (Note that the claim of beating and injuring has to be split to be effectively symbolized.)

*(f) Plaintiff was injured while playing on a slide negligently manufactured by Defendant. (Presumably Defendant will not know whether or not Plaintiff was injured while playing on a slide. Accordingly, Defendant can disclaim information. Under F.R.C.P. 8(b), the disclaimer will be treated as a denial.)

Instructions: In each of the following cases, symbolize the argument you are to respond to, symbolize your response in accordance with the principles presented in this section, and state your response in English.

2. You represent a corporation that was formed for the purpose of running a marina. The directors have decided also to run a theater, and have leased a building for that purpose. The lessor has filed a complaint seeking to avoid the lease on the ground that your client's leasing the building is ultra vires (outside the powers granted the corporation in its charter). You propose to rely on a statute that says that no corporate action shall be invalidated on the ground that it is ultra vires. See *711 Kings Highway Corp.* v. *F.I.M.'s Marine R.S. Inc.*, 273 N.Y.S.2d 299 (Sup. 1966).

*3. Your client is sued under a contract that provides for the payment of a reasonable attorney's fee if plaintiff has to sue. The complaint alleges that a reasonable attorney's fee would be $2,000. You claim that not more than $500 is justified. See *Wingfoot California Homes Co.* v. *Valley Nat. Bank*, 80 Ariz. 133, 294 P.2d 370 (1956).

4. Your client is being sued for removing timber from a certain parcel of land, Blackacre, which, according to the complaint, belongs to the plaintiff. Your client claims title to the land in question by adverse possession. See *Denham* v. *Cuddeback*, 210 Or. 485, 311 P.2d 1014 (1957).

5. You represent an insurance company being sued on a fire policy. The complaint alleges that all the property covered by the policy was destroyed by fire. Your client's investigators believe that not all the property was destroyed, and that some of the more valuable pieces have been hidden by the plaintiff. See *Curnow* v. *Phoenix Ins. Co.*, 46 S.C. 79, 24 S.E. 74 (1896).

6. Your client is being sued for $1,050, which the plaintiff alleges to be the value of 300 bushel baskets sold and delivered to your client. Your client claims that they only held 7/8 of a bushel each, and were worth no more

than two dollars each. See *Marshall Mfg. Co. v. Dickerson*, 55 Okla. 188, 155 Pac. 224 (1916).

*7. You represent the defendant in a suit on a promissory note. The complaint alleges that on September 3 of last year at Fairfield, Iowa, your client executed and delivered to plaintiff a note for $5,000. Your client claims it is a forgery. See *Spencer v. Turney & Co.*, 5 Okla. 683, 49 Pac. 1012 (1897). One solution uses this dictionary:

> Nx = x is a note, Exyzw = x executes y in favor of z at place w, Dxyzw = x delivers y to z at place w, Oxyz = x owes y the amount stated in z, i = this instrument, f = Fairfield, Iowa

8. You represent a hospital being sued by the estate of a person who died on the operating table. The complaint states that the plaintiff's decedent died on account of negligent treatment by one or more of the hospital's employees, acting in the scope of their employment. Several nurses and interns employed by the hospital participated in the operation, but you claim that if anyone was negligent it was the anesthesiologist or his assistant, neither of whom is a hospital employee. Assume you have sufficient evidentiary basis to plead that this was the case, and also to deny that anyone involved was negligent. See *Willinger v. Mercy Catholic Medical Center*, 241 Pa. Super. 456, 362 A.2d 280 (1976). One solution uses this dictionary:

> Exy = x is an employee of y acting in the scope of his or her employment, Nxy = x is a negligent act or omission on the part of y, Dxy = x caused y's death, Lxy = x is liable to y or y's estate, h = the hospital, d = plaintiff's decedent

§4.4 DISTINGUO

There is a man in the business of fitting and installing Venetian blinds and other similar window coverings who drives around in a truck with the sign:

BLIND MAN DRIVING THIS TRUCK

He presumably hopes to get attention for his business from people who are startled by the idea of a blind man driving a truck, and who then realize that although the driver is a blind man (just as a man in the insurance business is an insurance man), he is not the kind of blind man who can't see. The first, and startling, line of reasoning of the passerby is:

(S1) **When a blind man drives a truck, he acts dangerously.**
(S2) **This man is a blind man and is driving a truck.**
(S3) **Therefore, he is acting dangerously.**

This looks like a valid argument. Furthermore, it would seem that the premises are true while the conclusion (assuming that the "blind man" is driving normally) is false. But a logician will point out that it is impossible for a valid argument to have true premises and a false conclusion. This one is no exception. It trades on the ambiguity of the expression 'blind man'. If the expression means "sightless man" in both premises, then the second premise (S2) is false. If, however, it means "man in the blind business" in both premises, then the first premise (S1) is false. In each case we have a valid argument with a false premise, and we are not surprised by the falsity of the conclusion. There is a third possibility, that the expression means "sightless man" in the first premise and "man in the blind business" in the second. In this case both premises are true, but the argument is invalid. The two premises do not link together in a way that leads to the conclusion. There is no way of understanding the argument in which it has both valid form and two true premises.

This kind of analysis is referred to in traditional logic as *distinguishing* the premises (or a premise and the conclusion) of an argument. The logician will put it this way:

Distinguo (I distinguish) S1:
   (S1A) When a blind man, *i.e.*, a man who can't see, drives a truck, he acts dangerously, Concedo (I admit).
   (S1B) When a blind man, *i.e.*, a man in the blind business, drives a truck, he acts dangerously, Nego (I deny).

Distinguo S2:
   (S2A) This man is a blind man, *i.e.*, a man in the blind business, and is driving a truck, Concedo.
   (S2B) This man is a blind man, *i.e.*, a man who can't see, and is driving a truck, Nego.

Note that the interpretation of the expression 'blind man' that makes S1 true (S1A) makes S2 false (S2B), and vice versa. The reasons why a person would probably accept S1 in fact support only S1A, and the reasons why a person would accept S2 support only S2A. On the other hand, the conclusion (S3) cannot be established without accepting either S1B or S2B. Therefore, the distinguishing strategy has provided an effective response to the argument.

In most cases, it will be sufficient to distinguish the ambiguous term in just one of the propositions containing it (for example, S1 above); in some cases, however, we will distinguish both statements for the sake of clarity. Another shortcut: on occasion we may express only the version of the statement that is being denied (and not the version being conceded).

Now, let's put this process into symbolic form. The original argument is symbolized as follows:

$(x)[(Bx \& Tx) \rightarrow Dx]$
$Bm \& Tm$
$\vdash Dm$

(Bx = x is a blind man, Tx = x drives a truck, Dx = x acts dangerously, m = this man) Our distinguishing strategy is based on the fact that there are two possible meanings for the words 'blind man'. We can represent these two meanings by attaching subscripts to the predicate letter 'B':

B_1x = x is a blind man, *i.e.*, one who can't see
B_2x = x is a blind man, *i.e.*, one in the blind business

Then using 'D' for Distinguo in the same way we have been using 'C' for Concedo and 'N' for Nego, we can express the whole procedure as follows:

D: $(x)[(Bx \& Tx) \rightarrow Dx]$
C: $(x)[(B_1x \& Tx) \rightarrow Dx]$
N: $(x)[(B_2x \& Tx) \rightarrow Dx]$

Now let's try this technique on another argument. Here is a classic puzzle that goes back to the ancient Greeks:

We buy raw meat in the market.
Whatever we buy in the market, we eat.
Therefore, we eat raw meat.

This argument may be symbolized as follows:

$(\exists x)(Mx \& Bx \& Rx)$
$(x)(Bx \rightarrow Ex)$
$\vdash (\exists x)(Mx \& Rx \& Ex)$

(Mx = x is meat, Bx = we buy x in the market, Rx = x is raw, Ex = we eat x)

It's obvious what is wrong with this argument. Even if the meat is raw when we buy it, it doesn't stay that way. We cook it before we eat it. To put this point in symbolic terms, we have to distinguish the first premise of the argument above and its conclusion by introducing a reference to time when we interpret the predicate 'Rx':

R_1x = x is raw when purchased
R_2x = x is raw when eaten

D: (∃x)(Mx & Bx & Rx)
C: (∃x)(Mx & Bx & R_1x)
N: (∃x)(Mx & Bx & R_2x)

D: ⊢ (∃x)(Mx & Rx & Ex)
C: ⊢ (∃x)(Mx & R_1x & Ex)
N: ⊢ (∃x)(Mx & R_2x & Ex)

In both of the above examples, we have distinguished statements instead of denying them because we can't deny them. Each of them can be interpreted in such a way as to make it true. But sometimes we distinguish a statement instead of denying it not because it is capable of being true but because, although false, it is not obviously so. Our distinguishing strategy can further the disputation by showing how the statement in question is false. Consider this conversation:

> Sam: *Charlie has a remarkable car. He drove it 800 miles without stopping once for gas.*
> Susie: *Oh, it's not all that remarkable. It's a diesel car.*

If Charlie drives a diesel car, it's not remarkable that he didn't stop for gas, because a diesel car doesn't use gas. We don't know how often Charlie stopped for diesel fuel on his 800-mile trip.

Now, in the above conversation, Sam is reasoning:

(x)[(Mx & −Gx) → Rx]
Mc & −Gc
⊢ Rc

(Universe: cars; Mx = x is driven 800 miles, Gx = x stops for gas, Rx = x is remarkable, c = Charlie's car)

We know that Sam's first premise is false. If a car doesn't burn gasoline, there is nothing remarkable about driving it 800 miles without stopping for gas. Susie could simply deny Sam's premise:

N: (x)[(Mx & −Gx) → Rx]

But if she did this she would not be furthering the discussion, because Sam would not understand why she is denying the seemingly obvious statement that a car that can go 800 miles without stopping for gas is remarkable. So she distinguishes the statement instead, admitting that it is true of gasoline-fueled cars, but denying that it is true of other cars:

D: (x)[(Mx & −Gx) → Rx]
C: (x)[(Mx & Bx & −Gx) → Rx]
N: (x)[(Mx & −Bx & −Gx) → Rx]

(Bx = x burns gasoline)

Note that in both the "Blind Man" and the "Raw Meat" examples, the distinction was made by assigning alternative meanings, designated by subscripts, to one of the terms in the premise being distinguished. But in the "Diesel Car" example, we make the distinction by introducing an additional term into one alternative version of the premise being distinguished, and a negation of the same term into the other version.

An example of this second style of distinguishing, using the traditional terminology, appears in the comedy *The Imaginary Invalid*, written in 1673 by the famous French playwright Molière.[3] Angélique is trying to avoid marrying Thomas, the dull and pedantic young medical student picked out by her hypochondriac father, who hopes to have a physician son-in-law. She is asking Thomas for time to think about his offer, hoping that some means of escape will occur to her. Thomas, on the other hand, wants an immediate commitment:

> Angélique: *When a marriage pleases us, we know very well how to go into it without being dragged. Be patient. If you love me, Monsieur, you should wish for everything I wish for.*
> Thomas: *True, Mademoiselle, except as regards my interest as a lover.*
> Angélique: *But the great mark of love is to submit to the will of the beloved.*
> Thomas: *Distinguo, Mademoiselle. In whatever does not pertain to the possession of the beloved,* concedo, *but in whatever does pertain to it,* nego.

Often, important questions of public policy are addressed by means of the distinguishing process. Here is an example: in *N.L.R.B. v. Hearst Publications, Inc.*, 322 U.S. 111 (1944), the question was whether certain news vendors were "employees" of the newspaper they sold, and therefore entitled to collective bargaining under the National Labor Relations Act. The newspaper contended that they were not employees but "independent contractors," because the newspaper exercised very little control over the way they did

---

[3]*Le Malade Imaginaire*, Act II, Scene 6 (our translation).

their work. But the court pointed out that they were just the kind of people who needed the protection provided by this and other federal laws for the protection of workers. The common law cases relied on by the newspaper dealt mainly with the application of rules making an employer liable for an employee's negligence. The court found them not to be helpful in interpreting the statute involved in this case:

> *Unless the common law tests are to be imported and regarded as exclusively controlling, without regard to the statute's purposes, it cannot be irrelevant that the particular workers in these cases are subject, as a matter of economic fact, to the evils the statute was designed to eradicate and that the remedies it affords are appropriate for preventing them or curing their harmful effects in the special situation.* (322 U.S. at 127.)

We can formalize the newspaper's argument this way:

$$(x)(Bx \rightarrow Ex), (x)(Vx \rightarrow -Ex) \vdash (x)(Vx \rightarrow -Bx)$$

(Bx = x has a right of collective bargaining, Ex = x is an employee, Vx = x is a news vendor) And the court's refutation:

D: $(x)(Bx \rightarrow Ex)$
C: $(x)(Bx \rightarrow E_1 x)$
N: $(x)(Bx \rightarrow E_2 x)$

D: $(x)(Vx \rightarrow -Ex)$
C: $(x)(Vx \rightarrow -E_2 x)$
N: $(x)(Vx \rightarrow -E_1 x)$

($E_1 x$ = x is an employee within the meaning of the statute involved in this case, $E_2 x$ = x is an employee under the common law tests) The court's distinction refutes the newspaper's argument because on each interpretation of the term 'employees', one of the premises is rejected. That this distinction involves a major policy judgment is shown by the fact that Congress, a few years later, amended the National Labor Relations Act to exclude any person "having the status of an independent contractor" from the definition of an "employee." See 29 U.S.C. § 152 (3). Obviously, the analytical technique offered here will not solve any policy questions involved in a particular case. But it may help us see just what the questions are.

EXERCISES

Instructions for exercises 1 through 6: Symbolize the initial argument (a) and the response (b). In symbolizing each response, employ the notation 'D:', 'C:', and 'N:'.

1. (a) The man at the telephone in the Charles Addams cartoon seems to be reasoning that since he is allowed one phone call and is making one phone call, he is within his rights. (b) The policeman at the desk is denying that he is allowed an obscene phone call.

"When I said you were allowed one phone call, I did not mean <u>another</u> obscene one."

Drawing by Charles Addams, ©1974/The New Yorker Magazine, Inc.

*2. *O'Callahan* v. *Parker*, 395 U.S. 258 (1969) involved a court martial conviction of a serviceman for a rape committed while on leave. (a) The government argued that court martial jurisdiction extended to all cases involving military personnel. (b) But the court held that cases had to be service-connected to be subject to court martial.

3. In the famous *Palsgraf* case, *Palsgraf* v. *Long Island R.R.*, 248 N.Y. 339, 162 N.E. 99 (1928), an employee of the defendant, by pushing a passenger onto a train, negligently caused the passenger to drop a package. The package contained fireworks, which exploded, knocking over a scale at the other end of the station platform. The scale injured plaintiff when it fell. (a) Plaintiff argued that since she was injured as a result of a negligent act of defendant's employee in the scope of his employment, defendant was liable to her. (b) But the court held that a negligent act does not create liability unless it is negligent in relation to the person seeking recovery. One solution uses this dictionary:

> $Axy = x$ is an act of y's employee in the scope of his or her employment, $Ixy = x$ results in injury to y, $Nx = x$ is negligent, $Lxy = x$ is liable to y, $r =$ the railroad, $p =$ plaintiff, $Txy = x$'s negligence relates to y

4. *Parker* v. *Motor Boat Sales, Inc.*, 314 U.S. 244 (1941) arose under the federal Longshoremen's and Harbor Workers' Compensation Act, which compensated employees for injuries on the navigable waters of the United States in cases where compensation could not constitutionally be provided under state law. Under *Southern Pac. Co.* v. *Jensen*, 244 U.S. 205 (1917), there were many such cases. (a) The employer argued that *Jensen* should be overruled, and that there are, in fact, no cases of injuries on the navigable waters of the United States in which compensation cannot constitutionally be provided under state law. Therefore, the Longshoremen's Act applies to nothing. (b) But the court held that a statutory reference to the Constitution means the Constitution as interpreted at the time the statute was enacted. Therefore, the constitutional reference in the Longshoremen's Act must be interpreted in accordance with the *Jensen* case, even though that case is contrary to the current understanding of the Constitution.

5. *Nashville & K.R.R.* v. *Davis*, 78 S.W. 1050 (Tenn. 1902):

> WILKES, J. *This is an action for damages against the railroad company for running over and killing three geese of the value of $1.50. The owner of the geese lived about one mile from the railroad, but permitted them to run at large, and they went upon the railroad track near a public crossing. The engineer blew the whistle and rang the bell for the crossing, but there is no proof that he rang the bell or sounded the alarm for the geese. Whether the geese knew of this failure to whistle for them does not appear. We think there is no evidence of recklessness or common-law negligence shown in the case, and the only question is whether a goose is an ani-*

mal or obstruction in the sense of the statute (section 1574, subsec. 4, Shannon's Compilation), which requires the alarm whistle to be sounded, and brakes put down, and every possible means employed to stop the train and prevent an accident when an animal or obstruction appears on the track. It is evident that this provision is designed, not only to protect animals on the track, but also the passengers and employees upon the train from accidents and injury. It would not seem that a goose was such an obstruction as would cause the derailment of a train, if run over. It is true, a goose has animal life, and, in the broadest sense is an animal; but we think the statute does not require the stopping of trains to prevent running over birds, such as geese, chickens, ducks, pigeons, canaries, or other birds that may be kept for pleasure or profit. Birds have wings to move them quickly from places of danger, and it is presumed that they will use them (a violent presumption, perhaps, in the case of a goose, an animal which appears to be loath to stoop from its dignity to even escape a passing train). But the line must be drawn somewhere, and we are of the opinion that the goose is a proper bird to draw it at. We do not mean to say that in the case of recklessness and common-law negligence there might not be a recovery for killing geese, chickens, ducks, or other fowls, for that case is not presented. Snakes, frogs, and fishing worms, when upon railroad tracks, are, to some extent, obstructions; but it was not contemplated by the statute that for such obstructions as these trains should be stopped, and passengers delayed.

We are of the opinion that there is error in the court below giving judgment for the plaintiff, and the judgment is reversed, and the case having been heard without a jury, the suit is dismissed, at the plaintiff's cost.

One solution uses this dictionary for the plaintiff's argument:

A = An animal appears on the track, S = The crew attempts to stop the train, V = The railroad violates the statute, L = The railroad is liable

For the court's distinction, it adds A_1 and A_2 to the foregoing dictionary.

*6. Symbolize the above disputation between Angélique and Thomas (see page 224). Assume that the unspoken conclusion to Angélique's argument is that if Thomas loves her, he will submit to her wish in the matter of granting more time to give her answer.

7. (CHALLENGING) The following is an excerpt from the book *Chutzpah*, by Professor Alan Dershowitz of the Harvard Law School.[4] He refers to the argument—which he does not fully accept—that all race-specific affirmative action plans are objectionable because they call for discrimination on grounds of race. (a) Symbolize that argument and show how Jamin Dershowitz refutes it. (b) On the basis of Jamin's position as

---

[4](Boston: Little Brown, 1991), pp. 78–79. We are grateful to Professor Dershowitz for permission to use this excerpt.

stated, symbolize an argument to the effect that race-specific affirmative action plans are necessary in view of the effects of racism on American society. (c) Professor Dershowitz continues to insist that some plans that are necessary according to the above argument are still indefensible. Symbolize his response to Jamin's argument.

> *The issue of affirmative action has contributed significantly to racial tensions between the Jewish and black communities, despite the fact that Jews tend to support affirmative action programs—of every type—to a significantly greater degree than do whites from other ethnic groups. There are, of course, many Jews who—recalling the way "quotas," "diversity," and "discretion" were used against them—are wary of specific types of affirmative action programs, especially those that discriminate against Jews today. This is entirely understandable, in light of our history, and should be taken into account by those who jump too quickly to single out Jews for criticism on the affirmative action issue. Indeed, singling out Jews—from among other white ethnic groups—for special criticism over affirmative action is, itself, a subtle manifestation of anti-Semitism, especially in light of Jews' greater support for even race-specific affirmative action.*
>
> *Over the past several years, I have been engaged in an ongoing debate with my younger son, Jamin, about these concerns. My son, who recently graduated from Yale Law School, is a strong supporter of affirmative action programs of all kinds. Like me, he prefers those that do not focus exclusively on race and do not give undue advantage to elite blacks from wealthy families. But he is prepared to give substantial weight to the race of an applicant, since race alone is often a factor in discrimination, subtle and overt.*
>
> *His argument—which has persuaded me to a degree—is that in light of America's unique history of racism and its continuing impact today, it is better to err on the side of giving some blacks who have not suffered from discrimination a measure of undeserved benefit than to risk denying benefits to some blacks who have suffered discrimination. He has also persuaded me that there is an enormous difference between the white majority imposing racially motivated discrimination on a black minority, and the white majority imposing discrimination against other whites (on an equal basis) in a positive effort to undo the effects of past discrimination. The person discriminated against—whether black or white—is undoubtedly hurt, but there is a real difference between the institutional impact and intensity of the hurt suffered as part of an invidious pattern of racial subordination and as part of a benevolent pattern of racial equalization. As Oliver Wendell Holmes once put it: "Even a dog understands the difference between being stumbled over and being kicked."*
>
> *I have learned a great deal from my debates with Jamin, but I still insist that it is indefensible for any program of affirmative action to do what Harvard's continues to do: namely, hold relatively constant the white, Anglo-Saxon, Protestant "quota" of admittees through the discriminatory vehicles of "geographic distribution" and "alumni preference" (i.e., giving preference to descendants and relatives of alumni), while adjusting for the black affirmative action quota largely by reducing the number of Jewish, Catholic, and now Asian-American admittees. Jamin and I are in full*

agreement that the "burdens" of any fair affirmative action program must be spread equitably throughout the entire applicant pool and not imposed most disproportionately on groups that were relatively recent victims of Harvard discrimination. Indeed, if the burdens must fall disproportionately on some group, it would be fairer if the groups historically favored by discrimination were to bear a heavier burden of current efforts to achieve equality.

§4.5 AFFIRMATIVE DEFENSES

In the case of *Barber* v. *Vincent*, Freeman 531, 89 Eng. Rep. 397 (C.P. 1680), the plaintiff brought a quasi-contractual action for the price of a horse that he had sold to the defendant. The defendant pleaded that he was under age at the time of the transaction. The plaintiff replied that he had "sold him the horse for his conveniency to carry him about his necessary affairs." The defendant then demurred, arguing "that an infant was chargeable only for necessaries, as meat, drink, clothes, lodging, and education." The court rejected the defendant's contention and gave judgment for the plaintiff.

We will put the whole disputation involved in this case into logical form, with the symbolizations opposite the English propositions. The unstated premises are in brackets.

ARGUMENT I—PLAINTIFF

1. [Anyone who buys something must pay the seller the price.]	$(x)(y)(z)(Bxyz \rightarrow Pxzy)$
2. Defendant bought a certain horse from plaintiff.	$Bdhp$
3. Therefore, defendant must pay plaintiff the price of the horse.	$\vdash Pdph$

($Bxyz$ = x buys y from z, $Pxyz$ = x must pay y for z)

ARGUMENT II—DEFENDANT

4. [Minors need not pay for what they buy.]	$(x)(y)(z)[(Bxyz \,\&\, Mx) \rightarrow -Pxzy]$
5. Defendant was a minor when he bought this horse from plaintiff.	$Bdhp \,\&\, Md$
6. Therefore, he need not pay plaintiff for the horse.	$\vdash -Pdph$

§4.5 Affirmative Defenses

RESPONSE TO ARGUMENT I

7. Distinguish line 1:
   D:(x)(y)(z)(Bxyz → Pxzy)
   C:(x)(y)(z)[(Bxyz & −Mx) → Pxzy]

   7A. Anyone who is not a minor and buys something must pay the seller the price—*Concedo*.

   7B. Any minor who buys something must pay the seller the price —*Nego*.
   N:(x)(y)(z)[(Bxyz & Mx) → Pxzy]

ARGUMENT III—PLAINTIFF

8. [A minor who buys a necessary must pay the seller the price.]
   (x)(y)(z)[(Bxyz & Mx & Ny) → Pxzy]

9. Defendant, although he was a minor, bought this horse, which was a necessary, from plaintiff.
   Bdhp & Md & Nh

10. Therefore, defendant must pay plaintiff the price.
    ⊢ Pdph

RESPONSE TO ARGUMENT II

11. Distinguish line 4:
    D:(x)(y)(z)[(Bxyz & Mx) → −Pxzy]
    C:(x)(y)(z)[(Bxyz & Mx & −Ny) → −Pxzy]

    11A. Minors need not pay for things that they buy that are not necessaries, *Concedo*.

    11B. Minors need not pay for things that they buy that are necessaries, *Nego*.
    N:(x)(y)(z)[(Bxyz & Mx & Ny) → −Pxzy]

RESPONSE TO ARGUMENT III—DEFENDANT

(Note: Defendant's demurrer is not a new argument. It merely attacks the unspoken legal premise (line 8) of plaintiff's last argument (Argument III).)

12. Distinguish line 8:
    D:(x)(y)(z)[(Bxyz & Mx & Ny) → Pxzy]
    C:(x)(y)(z)[(Bxyz & Mx & N_1y) → Pxzy]

    12A. As to necessaries such as meat, drink, clothes, lodging, and education, *Concedo*.

    12B. As to necessaries such as transportation, *Nego*.
    N:(x)(y)(z)[(Bxyz & Mx & N_2y) → Pxzy]

(N_1x = x is a necessary such as meat, drink, clothes, lodging, or education, N_2x = x is a necessary such as transportation)

Note that the distinction offered in line 12, like those in the "Blind Man" and "Raw Meat" examples, is based on an interpretation, rather than a denial, of the premise being distinguished. The defendant is not denying that minors must pay for necessaries; he is denying that the term 'necessary' in line 8 includes things he needs for purposes of transportation. We have shown this by symbolizing the distinction with subscripts instead of with additional predicates. On the other hand, the distinctions offered in lines 7 and 11, like that in the "Diesel Car" example, are tantamount to denials of the premises being distinguished. If minors do not have to pay for what they buy, then it is not the case that anyone who buys something must pay for it. Similarly, if minors must pay for necessaries, then it is not the case that they need not pay for anything they buy. Why, then, does the defendant distinguish line 1 instead of denying it, or why does the plaintiff distinguish line 4 instead of denying it? Because that is what the law requires.

As we have seen, the way to deny a legal premise like lines 1 and 4 is to file a demurrer or corresponding motion. The claim of such a motion is that there is no such principle of law as the one to which the motion is addressed. But the law will not allow you to prevail on a claim that there is no such principle as the one stated in line 1 or the one stated in line 4. Admittedly, there are exceptions to both principles; but if you claim the benefit of such an exception, you must plead and prove that you come within it. Federal Rules of Civil Procedure Rule 8(c) lists some claims that have to be made in this way: it calls them "affirmative defenses." Other such claims are determined by history and precedent, or even by common sense. The general idea is that some exceptions are so obscure that it would be misleading to invoke them without calling attention to them.

We can now summarize our rules on the distinguishing of statements:

1. **A statement should be distinguished rather than simply denied if any of the following holds:**

    (a) It is capable of different interpretations, and under at least one interpretation is true (the "Blind Man" example and the defendant's claim on line 12 of the *Barber* case);

    (b) Although true, it will be misleading unless additional circumstances are brought forward (the "Raw Meat" example); or

    (c) It is subject to an exception which law or common sense requires to be specifically brought forward (the "Diesel Car" example and the claims on lines 7 and 11 of the *Barber* case).

2. **The steps in responding to an argument by distinguishing one or more statements are as follows:**

    (a) Distinguish the statements in the argument either by uncovering an ambiguity in some term they contain ('blind man' for

§4.5 Affirmative Defenses 233

example) or by specifying some additional condition ('things they need' for example).

(b1) **If two premises are distinguished, deny one premise under one interpretation and the other premise under the other interpretation. You have undermined the argument for the conclusion.**

(b2) **If a premise and the conclusion are distinguished, deny them both under one of these interpretations.**

EXERCISES

Instructions: Each of the following cases involves an affirmative defense, the avoidance of an affirmative defense, or both. As in Barber v. Vincent, *above, the party introducing an affirmative defense or avoidance must support it by a complete argument with premises and conclusions. To introduce the legal premise of that argument, the party must also distinguish the unspoken legal premise of the previous argument offered by the other side. Therefore, these cases differ from the ones dealt with in the exercises following Section 4.4 above. In those cases, we distinguished premises in order to show that the party relying on the premises had failed to prove a case. It was not necessary for us to go on and prove a case of our own. The difference arises from rules of law such as F.R.C.P. 8(c) rather than from rules of logic. In each of the following cases, state and symbolize the argument supporting the affirmative defense or avoidance, and show how you distinguish the legal premise of the preceding argument to support the legal premise of that argument.*

1. Plaintiff sued defendant for false imprisonment because he had detained her against her will. Defendant was a security guard at a department store. He observed plaintiff and reasonably suspected her of shoplifting. He detained her for no longer than was reasonably necessary to establish that she had paid for everything she took from the store. See *Bonkowski* v. *Arlan's Dept. Store*, 12 Mich. App. 88, 162 N.W.2d 347 (1968).

*2. Plaintiff tied his sailboat up to defendant's dock during a storm, and defendant cut it loose so that it was wrecked and plaintiff was injured. See *Ploof* v. *Putnam*, 81 Vt. 471, 71 Atl. 188 (1908).[5]

---

[5]For the benefit of students who have not yet taken Torts, we state the applicable principles of law: (1) Anyone who causes harm by an intentional act is liable for that harm. (2) A property owner may use reasonable force to prevent another from entering or remaining on the property. (3) A person may enter or remain on another's property when necessary to avoid danger. (2) is an exception to (1), and (3) is an exception to (2). Both must be affirmatively shown.

3. Plaintiff was a spectator at a baseball game, and was hit by a foul ball and injured. He sued the defendant, owner of the field, for negligence in not fencing in the stands. The defendant claimed that plaintiff assumed the risk of injury from foul balls because he understood the game of baseball, and voluntarily sat in an unscreened part of the stands. See *Kavafian* v. *Seattle Baseball Club Ass'n*, 105 Wash. 215, 181 Pac. 679 (1919).

4. Plaintiff was a witness at a trial. Defendant, one of the lawyers, attempted to discredit her testimony by making false statements in closing argument regarding her character. She sued for slander, and he claimed an absolute privilege because his remarks were relevant to the matter in issue in the trial. See *Irwin* v. *Ashurst*, 158 Ore. 61, 74 P.2d 1127 (1938).

5. Plaintiff sues on a promissory note executed by defendant for the purchase price of an automobile. Defendant claims that he was induced to buy the automobile by fraudulent misrepresentations made by the seller. Plaintiff claims to be a holder[6] in due course of the note.

*6. Plaintiff claims to have been negligently run over by defendant. He has executed a release, but claims that it was procured by fraud.

7. Plaintiff sues for injuries sustained from a front-end loader negligently designed by defendant. He seeks discovery as to changes made in the design after the accident. He claims that the information is discoverable under F.R.C.P. 26(b)(1) because it will constitute admissible evidence or else it may lead to the discovery of admissible evidence. Defendant claims such information is privileged. See *Lindberger* v. *General Motors Corp.*, 56 F.R.D. 433 (W.D. Wis. 1972).

---

[6]A holder in due course is a person who takes a note or other instrument in good faith and for value. He or she is not subject to defenses like fraud.

5

Pitfalls and Paradoxes

In this chapter we discuss a number of problems that may arise in the application of propositional and predicate logic to legal materials. Some of the problems considered (the pitfalls) result from complexities in the material analyzed, while others (the paradoxes) are the product of limitations within standard logic. In the first section we consider a problem that may occur when one symbolizes statutes and other legal rules.

§5.1 NECESSARY AND SUFFICIENT CONDITIONS

We noted in Section 2.2 (page 33) that the antecedent of a conditional statement expresses a *sufficient condition* for the consequent and that the consequent expresses a *necessary condition* for the antecedent. We can extend these points to universally quantified conditionals as well. In S1, for example, being a misdemeanor is declared to be a sufficient condition for being a crime and the latter is declared to be a necessary condition for the former.

(S1) All misdemeanors are crimes.
(F1) (x)(Mx → Cx)

You should be aware that when an antecedent is disjunctive (as in S2) *each* disjunct (as well as the entire disjunction) is presented as a sufficient condition; and when a consequent is a conjunction (as in S3), *each* conjunct (as well as the conjunction) is presented as a necessary condition.

(S2) If Smith is a congresswoman or a senator, then she is OVER 24.
(F2) (C v S) → O

(S3) All U.S. presidents are at LEAST 35 and NATIVE-born.
(F3) (x)[Px → (Lx & Nx)]

Smith's being a congresswoman is declared by S2 to be a sufficient condition for her being over 24. According to S3 being at least 35 is a necessary condition for being a U.S. president. On the other hand, when an antecedent is a conjunction, each conjunct is just a part of the sufficient condition and not presented by itself as a sufficient condition. Likewise, the individual disjuncts in a disjunctive consequent are not presented as necessary conditions.

In biconditionals and quantified biconditionals, each of the two constituents is set forth as both a necessary and a sufficient condition for the other. In S4, for instance, being a lawyer is asserted to be a necessary and sufficient condition for eligibility to represent clients in court (and vice versa).

(S4) All and only LAWYERS are eligible to REPRESENT clients in court.
(F4) (x)(Lx ↔ Rx)

When we symbolize universal legal statements, we must consider whether the conditions we represent are to be treated as sufficient, necessary, or both.

Here is an example that will help clarify this matter. 42 United States Code § 402, part of the Social Security Act, reads in part as follows:

Every individual who—
(1) is a fully insured individual (as defined in section 414(a) of this title),
(2) has attained age 62, and
(3) has filed application for old-age insurance benefits, or was entitled to disability insurance for the month preceding the month in which he attained the retirement age (as defined in section 416(l) of this title),
shall be entitled to an old-age insurance benefit. . . .

The symbolization of this provision looks pretty simple:

(x){[Ix & Ax & (Fx v Dx)] → Bx}

> Ix = x is a fully insured individual
> Ax = x has attained age 62
> Fx = x has filed an application for old-age insurance benefits
> Dx = x was entitled to disability insurance for the month preceding the month in which x attained the retirement age
> Bx = x is entitled to an old-age insurance benefit

But let us suppose we have a client, Charlie, who is 61 years old, is a fully insured individual, and wants to know whether he will be eligible for an old-age benefit if he applies. Any competent lawyer would tell him that he would not be eligible because he has not yet reached the required age of 62 years.

Our advice to Charlie can be symbolized (using the above notation) as:

> $-Ac \vdash -Bc$

(c = Charlie) This argument is obviously enthymematic. If we attempt to complete it by adding to the premise set our symbolization of § 402, we find that the argument remains invalid. It will involve a mistake related to "denying the antecedent," a fallacy that we described in Section 2.3.

The mistake arises from our different ways of interpreting § 402. When we set out, as logicians, to symbolize it, we treat it as stating a *sufficient* condition for entitlement to an old-age benefit. That is how we interpret it because that is how it reads. But when we go as lawyers to apply it, we treat it as also stating *necessary* conditions for the benefit in question. We know that public officials are not going to pay money unless the person asking for the money comes within some law calling for the money to be paid. Because a person wishing to collect public funds must demonstrate a legal right to them, whatever sufficient condition a claimant chooses to invoke is, for purposes of that claim, a necessary condition. So unless Charlie can find an alternative ground for collecting an old-age benefit (there isn't any), we must evaluate his claim by replacing the arrow in the above symbolization of § 402 with a double arrow.

A logical principle underlies our procedure here:

If A is the sole sufficient condition for B, then A is also a necessary condition for B.[1]

(This principle is validated in exercise 14 at the end of the section.) So, if in a given symbolization you realize that the condition expressed in the antecedent of the sentence is the sole sufficient condition for the consequent, you may

---

[1] In fact there will always be additional sufficient conditions for B, for example, any sufficient condition for A. So, a more adequate statement of the principle is: If A is a sufficient condition for B and every sufficient condition for B is also a sufficient condition for A, then A is a necessary condition for B.

symbolize with a double arrow instead of an arrow. Admittedly your biconditional formula will assert more than the sentence claims, but what it asserts will be true (if you are right that the condition is the sole sufficient condition).

There are many other situations in which legal principles may be treated as stating both necessary and sufficient conditions for some result when they seem from their language to state only sufficient conditions. For instance, the criminal laws of the state set forth a multitude of alternative sufficient conditions for criminal liability. But, because the accused is entitled to know exactly what he or she is accused of, when the prosecutor chooses one of the sufficient conditions to be the basis of an indictment or information, it is the sole sufficient condition for conviction in that particular case and so becomes also a necessary condition for conviction in that case.[2] A similar transformation takes place with other legal rules that people use as a basis for claiming relief before a court or government agency. Since the claimant has the burden of establishing the claim, the sufficient conditions brought forth for the purpose become necessary conditions for prevailing.

Another principle that enables us to find necessary conditions where the language states only sufficient conditions is the maxim *inclusio unius est exclusio alterius*—the inclusion of one thing is the exclusion of another. We can apply this principle, for instance, to F.R.C.P. 37(d), which imposes sanctions if "a party or the officer, director, or managing agent of a party" fails to respond to certain discovery requests. It is generally interpreted to mean that an officer, director, or managing agent of a party, like a party, must respond to these requests without a subpoena, whereas any employee of a party who is not an officer, director, or managing agent is in the same position as any other nonparty, and does not have to respond unless served with a subpoena. The inclusion of the mentioned groups implies the exclusion of those who are not mentioned. So the rule of subdivision 37(d) as written is:

$(x)[(Px \lor Ox \lor Dx \lor Mx) \to Rx]$

Px = x is a party
Ox = x is an officer of a party
Dx = x is a director of a party
Mx = x is a managing agent of a party
Rx = x must respond to certain discovery requests without a subpoena

But it is interpreted as:

$(x)[(Px \lor Ox \lor Dx \lor Mx) \leftrightarrow Rx]$

---

[2]Of course, the accused doesn't have to establish the absence of a necessary condition in order to be acquitted. Since the prosecution has the burden of proof, all that is required for acquittal is to undermine the claim of a sufficient condition. This can be done with the techniques discussed in Chapter Four.

Of course, *inclusio unius* isn't a logical principle. It is a principle of legal interpretation based on the recognition that most enumerations in legal material are meant to be exhaustive. When they are, they invoke the following principles of logic:

> **If each of a set of conditions is sufficient for X and the set is exhaustive, then *their disjunction* is a necessary condition for X.**[3]
>
> **If each of a set of conditions is necessary for X and the set is exhaustive, then *their conjunction* is a sufficient condition for X.**[4]

Inclusio unius doesn't apply to all enumerations because not all enumerations are intended to be exhaustive. A court often has to decide as a matter of policy or legislative intent whether the principle applies in a particular case. For instance, in *Humphrey's Executor v. United States*, 295 U.S. 602 (1935), an issue presented to the court for decision was:

> Do the provisions of Section 1 of the Federal Trade Commission Act stating that "any commissioner may be removed by the President for inefficiency, neglect of duty, or malfeasance in office" restrict or limit the power of the President to remove a commissioner except upon one or more of the causes named?

The court determined on the basis of legislative history and policy that the answer was "yes."

EXERCISES

1. Symbolize each of these schematic statements in propositional logic.
   (a) A and B are each sufficient conditions for C.
   *(b) A and B together are a sufficient condition for C.
   (c) A and B are each necessary conditions for C.
   (d) A and B together are a necessary condition for C.

---

[3] A more adequate statement of the principle: If each of a set of conditions is sufficient for X and every sufficient condition for X is also a sufficient condition for the disjunction of the conditions in the set, then the disjunction of the conditions in the set is a necessary condition for X.

[4] A more adequate statement of the principle: If each of a set of conditions is necessary for X and every necessary condition for X is also a necessary condition for the conjunction of the conditions in the set, then the conjunction of the conditions in the set is a sufficient condition for X.

Examples of this principle are harder to find than examples of the other, but there are a few. For instance, Rule 1.8(f) of the Model Rules of Professional Conduct provides that a lawyer shall not accept compensation for representing a client from anyone other than the client unless three conditions are met. This is generally interpreted as allowing the payment if the conditions are met.

(e) C is a sufficient condition for A and also for B.
*(f) C is a necessary condition for A and also for B.

Which of these statements (if any) are logically equivalent?

2. Identify the false statements in this set. (Hint: symbolize the material.)
   (a) If A is a sufficient condition for C, then the conjunction of A and B must be also.
   *(b) If A is a sufficient condition for C, then the disjunction of A and B must be also.
   (c) If A is a necessary condition for C, then the conjunction of A and B must be also.
   (d) If A is a necessary condition for C, then the disjunction of A and B must be also.

3. Revisit the exchange in the House Judiciary Committee quoted on pages 9–10 and treated in exercise 5 in Section 2.7. Congressman McClory's motion (stated as S1 in that exercise) provides a sufficient condition for postponement. Assume that it provides *the sole* sufficient condition for postponement. Treat the proposed redraftings of the motion (S4 and S5) according to the same assumption. Would that assumption change any of the answers to questions (a) through (d) in exercise 5?

*4. An abbreviated newspaper story:[5]

> ... On Monday, the Supreme Court agreed for the first time to determine the constitutionality of one method of restricting the number of times a member of Congress may be re-elected.
> The test case comes from Arkansas, where a 1992 amendment to the state constitution forbids printing on ballots the names of U.S. senators who have served 12 years and House members in office six years.
> ...
> The Supreme Court case, expected to be resolved next year, will focus on whether the Arkansas term-limits amendment violates the very first article of the U.S. Constitution.
> Article I lists qualifications for members of Congress. A member of the House of Representatives must be at least 25 years of age and a U.S. citizen for seven years. A senator must be at least 30 and a U.S. citizen for nine years. Both senators and House members must be residents of the states that elected them.
> ...
> What Arkansas did, according to its highest court, was add a new limitation to the age, nationality and residency requirements prescribed by the Constitution.
> "Term limitations for congressional representation may well have come of age," wrote Arkansas Associate Justice Robert L. Brown.

---

[5]"High Court to Determine if Term Limits Are Legal," *Asheville (NC) Citizen-Times* (June 21, 1994), p. 3A. The article was prepared by Knight-Ridder. The decision of the Arkansas Supreme Court was affirmed in *U.S. Term Limits, Inc. v. Thornton*, 115 S.Ct. 1842 (U.S. 1995).

> "But to institute such a change, an amendment to the U.S. Constitution is required, ratified by three-fourths of the states."
>
> But dissenting state judges argued that Article I was intended to prescribe the minimum qualifications, not the only ones. . . .

The qualifications for representatives are set out in Section Two of Article One of the Constitution and the qualifications for senators in Section Three. The passages are formally alike, so we focus on the first:

> *No Person shall be a Representative who shall not have attained to the age of twenty five Years, and been seven Years a Citizen of the United States, and who shall not, when elected, be an Inhabitant of that State in which he shall be chosen.*

(a) Symbolize this passage as it is understood in the majority opinion of the Arkansas Supreme Court. Adopt "persons" as the universe and use this dictionary: Ex = x is eligible to be a Representative, Ax = x has attained the age of 25, Cx = x has been a citizen for at least seven years, Ix = x is an inhabitant of the state in which x is chosen. (b) Symbolize it as it is understood by the dissenting justices. (c) Use the concepts of "necessary condition" and "sufficient condition" to clarify the disagreement between the two positions.

Exercises 5 and 6 concern Rule 23 of the Federal Rules of Civil Procedure. Here are subdivision (a) and part of subdivision (b) of that rule:

Rule 23. Class Actions

> **(a) Prerequisites to a Class Action.** *One or more members of a class may sue or be sued as representative parties on behalf of all only if (1) the class is so numerous that joinder of all members is impracticable, (2) there are questions of law or fact common to the class, (3) the claims or defenses of the representative parties are typical of the claims or defenses of the class, and (4) the representative parties will fairly and adequately protect the interests of the class.*
>
> **(b) Class Actions Maintainable.** *An action may be maintained as a class action if the prerequisites of subdivision (a) are satisfied, and in addition:*
>
> *(1) the prosecution of separate actions by or against individual members of the class would create a risk of*
>
> *(A) inconsistent or varying adjudications with respect to individual members of the class which would establish incompatible standards of conduct for the party opposing the class, or*

(B) adjudications with respect to individual members of the class which would as a practical matter be dispositive of the interests of the other members not parties to the adjudications or substantially impair or impede their ability to protect their interests; or

(2) the party opposing the class has acted or refused to act on grounds generally applicable to the class, thereby making appropriate final injunctive relief or corresponding declaratory relief with respect to the class as a whole; or

(3) the court finds that the questions of law or fact common to the members of the class predominate over any questions affecting only individual members, and that a class action is superior to other available methods for the fair and efficient adjudication of the controversy.

5. Symbolize this material using the dictionary displayed below. Provide two symbolizations of subdivision (b): First provide the literal symbolization; then provide the stronger symbolization that is warranted by the principle of *inclusio unius*.

   Mx = x may be maintained as a class action
   $A_1 x$ = x complies with condition (1) of Rule 23(a)
   $A_2 x$ = x complies with condition (2) of Rule 23(a), etc.
   $B_1 x$ = x complies with condition (1) of Rule 23(b), etc.

6. Indicate whether fulfilling each combination (i) through (vii) below constitutes a necessary condition to maintaining a class action, a sufficient condition, or both, or neither. Use the strong interpretation of subdivision (b) based on the principle of *inclusio unius*.

   i. (a)(3) and (b)(3)
   ii. (a)(1) or (a)(2)
   iii. (a)(1) or (a)(2) or (b)(3)
   iv. (a)(1) and (a)(2) and (b)(1) and (b)(2)
   v. (b)(1) or (b)(2) or (b)(3)
   vi. (a)(1) and (a)(2) and (a)(3) and (a)(4) and (b)(2)
   vii. (a)(1) and (a)(2) and (a)(3) and (a)(4), and either (b)(1) or (b)(2) or (b)(3)

Instructions for exercises 7 through 12: Each exercise contains an argument that lawyers would find valid or at least plausible. Symbolize the argument. (The device of analysis by instantiation explained in Section 3.7 may be applied to some of the exercises.) Demonstrate the invalidity of the symbolized argument using a technique described in Chapters Two or Three. Then, on the basis of one of the principles displayed below, either add a premise or strengthen an existing premise so as to produce a valid symbolized argument. Demonstrate the formal validity of this augmented argument. Principles:

>P1. *Inclusio unius est exclusio alterius.*
>
>P2. A person seeking to restrain the liberty of another must point to a specific principle of law authorizing him or her to do so.

7. In *Sommersett's Case*, 20 Howell's State Trials 1 (1772), Lord Mansfield held that Sommersett, the slave of a Jamaica planter, could not be confined in England by his master because there was no specific principle of law authorizing slavery in England.

*8. Rule 25(a) of the Federal Rules of Civil Procedure says "If a party dies and the claim is not thereby extinguished, the court may order substitution of the proper parties." It follows that if the claim is extinguished by the death of a party, no substitution is proper.

9. *Regina* v. *Jackson*, [1891] 1 Q.B. 671 (C.A.) involved a married woman who had left her husband for no good reason. He took her back to his house by force, and kept her there until the court released her by a writ of habeas corpus. The court held that although she was legally obliged to live with her husband, the law does not allow him to enforce the obligation by confining her.

10. In *Canfield* v. *Tobias*, 21 Cal. 349 (1863), the complaint alleged a debt, and added that a purported release was procured by fraud. The answer claimed the benefit of the release but did not respond to the allegation of fraud. The plaintiff claimed that the allegation of fraud was admitted under a statute providing that any material allegation of the complaint was admitted unless denied in the answer. But the court held that the allegation of fraud was not material (because it belonged in a later stage of the pleadings), and therefore was not admitted.

11. Federal Rule of Civil Procedure 26(b)(1) says: "It is not ground for objection [to a discovery request] that the information sought will be inadmissible at the trial if the information sought appears reasonably calculated to lead to the discovery of admissible evidence." It follows that information that is neither admissible nor calculated to lead to the discovery of admissible evidence is not discoverable. Add as a premise the obvious point that if information is inadmissible and its inadmissibility is ground for objection to the discovery request, then it is not discoverable.

*12. 5 U.S.C. § 553(c), part of the federal Administrative Procedure Act, says: "When rules are required by statute to be made on the record after opportunity for an agency hearing, sections 556 and 557 of this title apply. . . ." In *United States* v. *Florida East Coast Railway*, 410 U.S. 224

(1973), it was held that § 556 does not apply to a rulemaking proceeding in the absence of a statutory requirement that it be on the record.

13. (CHALLENGING) A Federal statute, 42 U.S.C. § 1981, provides as follows:

> *All persons . . . shall have the same right in every State and Territory to make and enforce contracts, to sue, be parties, give evidence, and to the full and equal benefit of all laws and proceedings for the security of persons and property as is enjoyed by white citizens, and shall be subject to like punishment, pains, penalties, taxes, licenses, and exactions of every kind, and to no other.*

In *McDonald v. Santa Fe Trail Transp. Co.*, 427 U.S. 273 (1976), the Supreme Court recognized that if this language were to be read literally, it might not give white people any relief against being treated worse than blacks, but that in view of our constitutional and historical commitment to racial equality, it would be given a broader interpretation so as to protect whites also. Symbolize both interpretations. Hint: The statutory categories of "all persons" and "white citizens" overlap. But insofar as the statute gives white citizens the rights enjoyed by white citizens, it is tautologous. In symbolizing, use the actual nontautologous effect of the statute by changing "all persons" to "all persons other than white citizens."

14. (CHALLENGING) We assert this principle on page 237:

> If A is the sole sufficient condition for B, then A is also a necessary condition for B.

Prove the truth of the principle by demonstrating the validity of this argument:

Sab & (x)(Sxb → Sxa)	A is the sole sufficient condition for B.[6]
(x)(y)[Sxy ↔ (Ox → Oy)]	Definition of 'sufficient condition'.
(x)(y)[Nxy ↔ (Oy → Ox)]	Definition of 'necessary condition'.
⊢ Nab	A is a necessary condition for B.

(Universe: possible states; Sxy = x is a sufficient condition for y, a = possible state A, b = possible state B, Ox = x obtains, Nxy = x is a necessary condition for y)

---

[6]A more adequate statement: A is a sufficient condition for B and every sufficient condition for B is also a sufficient condition for A.

§5.2 PROBLEMATIC UNIVERSAL STATEMENTS

Statements of the following five forms may be perplexing:

Only A are B.
None but A are B.
The only A are B.
All A except B are C.
All A and B are C.

ONLY A ARE B. Consider this example:

(S1) Only males play in the National Football League.

This is clearly a universal statement and is therefore equivalent to either S2 or S3, but which one?

(S2) All males play in the NFL. $(x)(Mx \rightarrow Nx)$
(S3) All who play in the NFL are males. $(x)(Nx \rightarrow Mx)$

As S1 is true and S2 false, those two statements can hardly be equivalent. It follows that the statement equivalent to S1 is S3. Note that both S1 and S3 have the same content as S4:

(S4) No nonmales play in the NFL. $(x)(-Mx \rightarrow -Nx)$

It makes a logical difference whether the subject of an 'only' sentence is singular or plural. Consider these examples:

Only MALES took the EXAM. $(x)(Ex \rightarrow Mx)$

Only <u>Oliver</u> took the exam. $Eo\ \&\ (x)(Ex \rightarrow x = o)$ *or*
                              $(x)(Ex \leftrightarrow x = o)$

The first sentence expresses one claim; the second expresses two.

NONE BUT A ARE B. The following sentence means the same as S1 and is therefore equivalent to S3.

None but males play in the National Football League.

'Only' and 'none but' are synonymous. When either expression is replaced in a statement by the standard quantifier 'all' the subject and predicate terms of the statement must be interchanged.

THE ONLY A ARE B. One must be careful to distinguish 'only' from 'the only'. S5 and S6 do not make the same claim:

(S5) The only cars I own are Japanese sports cars.
(S6) Only cars I own are Japanese sports cars.

S5 (but not S6) is equivalent to:

All the cars I own are Japanese sports cars. $\quad (x)(Ox \rightarrow Jx)$

Even though the terms 'only' and 'all' are not equivalent, the expressions 'the only' and 'all the' are often interchangeable. From the account of 'Only A are B' statements given above, we can see that S6 is equivalent to the outrageous falsehood:

All Japanese sports cars are cars I own. $\quad (x)(Jx \rightarrow Ox)$

ALL A EXCEPT B ARE C. Consider the "exceptive" claim S7:

(S7) All faculty except instructors enjoy retirement benefits.

S7 is ambiguous. It certainly embodies the claim made in S8. It is often intended also to embody the claim made in S9.

(S8) All faculty who aren't instructors enjoy retirement benefits. $\quad (x)[(Fx \,\&\, -Ix) \rightarrow Bx]$

(S9) No faculty who are instructors enjoy retirement benefits. $\quad (x)[(Fx \,\&\, Ix) \rightarrow -Bx]$

If it embodies both claims, then it is equivalent to S10.

(S10) Faculty enjoy retirement (x)[Fx → (Bx ↔ −Ix)]
benefits iff they are
not instructors.

Sentences containing 'except' must be clarified before being symbolized. Note that when the term following 'except' is singular, the sentence makes two claims:

All faculty except <u>Smith</u> (x)[(Fx & x ≠ s) → Tx] & −Ts *or*
are tenured. (x)[Fx → (Tx ↔ x ≠ s)]

ALL A AND B ARE C. Consider S11.

(S11) All Toyotas and Hondas are Japanese vehicles.

This sentence may be symbolized by F11, but not by F12.

(F11) (x)[(Tx v Hx) → Jx]
(F12) (x)[(Tx & Hx) → Jx]

(Universe: vehicles) F12 symbolizes the quite different claim S12.

(S12) Each vehicle that is both a Toyota and a Honda is Japanese.

Unless and until the Toyota and Honda corporations enter into a joint venture, the class corresponding to the subject of S12 will remain empty. In contrast, the class about which S11 makes a claim is a very large class—one that embraces all Toyotas as well as all Hondas. S11 is equivalent to S11A, and, as you would expect, F11 and F11A are logically equivalent.

(S11A) All Toyotas are Japanese vehicles and all Hondas are Japanese vehicles.

(F11A) (x)(Tx → Jx) & (x)(Hx → Jx)

It is surprising that formula F11, with its wedge connective, should represent S11 with its 'and', but notice that if we rephrase S11 in a way that makes explicit the predicates involved ('is a Toyota', 'is a Honda', and 'is a Japanese vehicle') the 'and' gives way to an 'or':

Anything that is *either* a Toyota *or* a Honda is a Japanese vehicle.

We summarize the information presented to this point:

Only A are B None but A are B	= All B are A	(x)(Bx → Ax)
Only individual a is B		(x)(Bx ↔ x = a)
The only A are B	= All A are B	(x)(Ax → Bx)
All A except B are C		(x)[(Ax & −Cx) → Bx)] *or* (x)[Ax → (Cx ↔ −Bx)]
All A except individual b are C		(x)[Ax → (Cx ↔ x ≠ b)]
All A and B are C		(x)[(Ax v Bx) → Cx]

ALL A ARE NOT B. S13 is ambiguous; it may mean S14 or S15.

(S13) All events are not caused.

(S14) No events are caused. (x)(Ex → −Cx)
(S15) Not all events are caused. −(x)(Ex → Cx)

Clues about which meaning S13 has in a particular case may be provided by the context in which it appears or the intonation used (if uttered). While S13 is ambiguous, S14 and S15 are not.

DEFINITIONS: Definitions combine two claims. Definition S16, for example, entails both S17 and S18.

(S16) 'Father' means "male parent."
(S17) All fathers are male parents. (x)[Fx → (Mx & Px)]
(S18) All male parents are fathers. (x)[(Mx & Px) → Fx]

Definitions may be symbolized concisely by quantified biconditionals; F16 symbolizes S16.

(F16) (x)[Fx ↔ (Mx & Px)]

Speaking strictly F16 represents S19 which follows from S16 but is not quite equivalent to S16.

(S19) All and only fathers are male parents.

The difference between S16 and S19 is that the former but not the latter makes a claim (or a stipulation) about the meaning of a word. For our purposes, this difference may be ignored.

DEFINITIONAL TRUTHS: Some statements are true by definition and others are true because they make an assertion about the world that happens to be the case. Consider S20 and S21.

(S20) All fathers are males.   (x)(Fx → Mx)
(S21) All humans are less than 200 years old.   (x)(Hx → Lx)

S20 is not a definition (if it were it would be a terrible one!), but it is a definitional truth because it follows from definition S16. S21 is not a definitional truth but a descriptive truth. The truth of S20 may be determined by consulting the dictionary; knowing that S21 is true requires more than just looking at a dictionary—one must have knowledge of the extra-linguistic world. S20 is necessarily true, while S21 is only contingently true. In spite of these important differences the two statements are symbolized in the same manner.

UNIVERSALS IN SINGULAR FORM: Sometimes we use a sentence with an indefinite article (*a* or *an*) and a singular subject to express a general claim:

A whale is a mammal.   (x)(Wx → Mx)
A growing boy needs spinach.   (x)[(Gx & Bx) → Nx]

Such sentences are not always universal, however:

An actor won the presidency.   (∃x)(Ax & Px)

Putting universals in singular form is a common device in legislative drafting:

Federal Rules of Evidence, Rule 201(c)
A court may take judicial notice,   (x)(Cx → Nx)
whether requested or not.

Model Penal Code, Section 1.06

A prosecution for murder may be commenced at any time.   (x)(Mx → Cx) *or*
(x)[Mx → (y)Cxy]

(Mx = x is a prosecution for murder, Cx = x may begin at any time, Cxy = x may begin at time y)

STATEMENTS WITHOUT QUANTIFIERS: Many statements with plural subjects lack quantifier terms (such as 'all', 'most', or 'some'); here are a few examples:

> Mockingbirds are rowdy.
> Mockingbirds are in the bottlebrush tree.
> Tourists buy postcards.
> Cigarettes cause cancer.

Statements like these are at least potentially equivocal. The first sentence, for instance, might be used to make any of these claims:

All mockingbirds are rowdy.	(x)(Mx → Rx)
Typically, mockingbirds are rowdy.	A
Most mockingbirds are rowdy.	B
Many mockingbirds are rowdy.	C
Some mockingbirds are rowdy.	(∃x)(Mx & Rx)

In deciding which claim was intended, we rely on factors such as the context in which the sentence was uttered or written, as well as our knowledge of the beliefs and attitudes of the person making the assertion (if we know them). Note that only the first and last of the interpretations displayed above can be handled adequately in predicate logic. If we need to symbolize one of the other statements, we must either fall back on propositional logic or go with the weaker symbolization '(∃x)(Mx & Rx)'.

When treating the sentence 'Mockingbirds are in the bottlebrush tree', we will naturally choose S22 rather than a false interpretation like S23.

> (S22) Some mockingbirds are in the bottlebrush tree.
> (S23) All mockingbirds are in the bottlebrush tree.

Unless there is reason to hold otherwise, we assume that the speaker (writer) intended to express a truth.

How shall we treat the ambiguous 'Tourists buy postcards'? For example, which of the following is best?

All tourists buy postcards.	(x)(Tx → Px)
Tourists usually buy postcards.	A
Most tourists buy postcards.	B
Some tourists buy postcards.	(∃x)(Tx & Px)

If we assume the person who asserted 'Tourists buy postcards' was expressing a truth, then the falsity of the first of these four possible reformulations excludes it. The remaining statements are true, but the last statement is weaker than the second or the third. By P is "weaker than" Q we mean that Q entails P but not vice versa. As a general rule we should interpret someone as making the strongest of the plausible interpretations that the person would take to be true. (More precisely, we should adopt this strategy when dealing with someone's premises so that we do justice to the argument being advanced.) Notice that if we employ the second or third interpretation we are precluded from applying predicate logic. In a given argumentative context the weaker (but symbolizable) version may be adequate. This argument provides such a context:

Tourists buy postcards.	(∃x)(Tx & Px)
Therefore, somebody buys postcards.	⊢ (∃x)Px

Universe: people
Tx = x is a tourist
Px = x buys postcards

The argumentative setting in which a statement resides will often influence the symbolization we choose.
    May we interpret S24 as S25?

(S24) Cigarettes cause cancer.
(S25) All cigarettes cause cancer.   (x)(Ax → Bx)

S25 means that each and every cigarette causes cancer; we know this to be false. For example, a cigarette that is never smoked does not cause cancer. Furthermore, it is the cumulative effect of smoking that causes cancer. Since S24 is true, the false S25 is not a satisfactory restatement of it. Here are two plausible reformulations of S24:

    All cigarettes are carcinogenic.
    All people who regularly smoke cigarettes increase their probability of
        contracting cancer.

We understand 'carcinogenic' to mean "having the capacity to induce cancer."

NORMATIVE STATEMENTS: A "normative" statement is one that says what *ought* to be the case rather than what *is* the case. A normative statement may concern morality or legality, or it may be merely prudential ("You ought to replace your engine coolant once a year"). The symbolization of such statements may pose problems. If we take S26 to be universal in scope, then it should probably be regarded as a normative statement and rephrased as S27 rather than being treated as a descriptive statement and reworded as S28. We prefer S27 because S28 is obviously false.

(S26) Catholics attend Mass on Sundays.
(S27) All Catholics ought to attend   (x)(Cx → Ox)
      Mass on Sundays.
(S28) All Catholics attend Mass on Sundays.

(Ox = x ought to attend Mass on Sundays) Another reasonable normative treatment of S26 would be this:

All Catholics are required by a rule of   (x)(Cx → Rx)
their church to attend Mass on Sundays.

(Rx = x is required by a rule of x's church to attend Mass on Sundays) There is further discussion of the treatment of normative statements in Chapter Six and Appendix One.

LEGISLATIVE STATEMENTS: Statements like S29 and S30 that occur in legislation *embody* legal enactments rather than *describe* them; in some ways they are more like normative statements and definitions than descriptive statements. Nevertheless, they can be symbolized in our system; more precisely, statements expressing their legal content can be symbolized.

(S29) All persons born or naturalized in the United States and subject to the jurisdiction thereof, are citizens of the United States and of the State wherein they reside (U.S. Const., Amendment XIV).

S29, for instance, may be symbolized in this way:

(x){[(Bx v Nx) & Jx] → (Ux & Sx)}

Universe: people
Bx = x was born in the U.S.
Nx = x was naturalized in the U.S.
Jx = x is subject to the jurisdiction of the U.S.

Ux = x is a citizen of the U.S.
Sx = x is a citizen of the state in which x resides

Care must be taken in symbolizing legislative statements to avoid distorting their meaning. This symbolization of S30 is a case in point.

(S30) Persons guilty of class B   (x)(Gx → Ix)
       misdemeanors are subject to
       imprisonment for not more
       than six months.[7]

(Universe: people; Gx = x is found guilty of a class B misdemeanor, Ix = x is subject to imprisonment for not more than six months) Consider Sam, who has been found guilty of both a class B misdemeanor (say, disorderly conduct) and robbery; he will be imprisoned for four months for the misdemeanor conviction and five years for the robbery. Sam's situation falsifies the proposed symbolization since the predicate 'G' applies to him but the predicate 'I' does not. While Sam's case falsifies the symbolization, it does not conflict with S30—so much the worse for the symbolization. The real subject matter of S30 is the sentences that may be imposed for certain types of offense; accordingly, the quantifier in our symbolization should range over *offenses* rather than *people*.

(S30A) All class B misdemeanors   (x)(Bx → Sx)
       are punishable by imprisonment
       for not more than six months.

(Universe: offenses; Bx = x is a class B misdemeanor, Sx = x is punishable by imprisonment for not more than six months)

EXERCISES

1. Symbolize the following statements.
   (a) (*proverb*) "He who HESITATES is LOST." (Universe: people)
   *(b) (*Leona Helmsley*) "Only the little [i.e., middle-class] people pay TAXES." (Universe: people; Mx = x is middle class)
   (c) (*Bobby Bowden, speaking about his team following a loss*) "The MEEK shall inherit the BENCH." (Universe: FSU football players)

---

[7]The Model Penal Code uses the language "may be sentenced to imprisonment."

(d) (*logic manuscript*) "[Unfortunately,] all REASONING is not VALID; people sometimes reason invalidly]." (Rx = x is a piece of reasoning. Do not symbolize bracketed material.)

(e) (*Luke 12:48, King James edition*) ". . . Unto whomsoever much is GIVEN, of him shall be much REQUIRED. . . ." (Universe: people)

*(f) (*Samuel Johnson*) "Nobody but mean [*i.e.*, poor] people WALK in Paris." (Universe: Parisians; Px = x is poor)

(g) (*philosophy exam*) "Suicide can only be JUSTIFIED for people in COMAS." (Universe: people; Jx = x is justified in committing suicide)

(h) (*Steve Spurrier*) "I have not made a STAR player out of every quarterback I've COACHED, and every quarterback that has been under my tutelage has not PANNED out."

(i) (*Supreme Court ruling*) INVOCATIONS and BENEDICTIONS at public school graduation ceremonies VIOLATE the constitution. (Ix = x is an invocation given at a public school graduation ceremony, Bx = x is a benediction given at a public school graduation ceremony)

*(j) CORONERS are the only DOCTORS who make HOUSE calls.

(k) (*Lutie Dugger's tombstone, colonial Williamsburg*) "None KNEW her but to LOVE her." (Universe: people; Kx = x knew Lutie, Lx = x loved Lutie)

(l) (*dictionary*) "SPERM: a MALE GAMETE."

(m) JUDGES are HUMAN.

*(n) JUDGES ATTENDED the lobbyists' breakfast.

(o) A U.S. MARSHAL is ENTITLED to statutory fees.

(p) A U.S. MARSHAL was WOUNDED during the robbery.

(q) Students are RESPONSIBLE for the ODD-numbered exercises except for the ones marked "CHALLENGING." (Universe: exercises; Rx = students are responsible for exercise x)

*(r) (*George Allen*) "The only TEAM that gets BOOED$_R$ is the HOME$_R$ team." (Tx = x is a football team, Bxy = x is booed at time y, Hxy = x is playing at home at time y)

(s) (*Julius Caesar, speaking of Gauls*) "No one is allowed to DISCUSS$_R$ politics except in a public ASSEMBLY$_R$." (Gx = x is a Gaul, Dxy = x is allowed to discuss politics at time y, Axy = x is in a public assembly at time y)

(t) (*Noel Coward*) "[Only] mad DOGS and ENGLISHMEN go OUT in the noonday sun."

(u) (*Hume*) ". . . None but a FOOL or MADMAN will ever PRETEND to dispute the authority of experience. . . ."

*(v) (*sign*) "No ANIMALS On BEACH Except SEEING-eye Dogs." (Bx = x is permitted on the beach)

(w) (*tax instructions*) ". . . Someone who PREPARES$_R$ your return for you but does not CHARGE$_R$ you should not SIGN$_R$ your return." (Universe: people; Pxy = x prepares y's tax return, Cxy = x charges y for preparing y's tax return, Sxy = x is permitted to sign y's tax return)

(x) (*Samuel Johnson*) "No man but a BLOCKHEAD ever WROTE$_R$ except for MONEY." (Px = x is a person, Wxyz = x writes y for purpose z, Mx = x is the earning of money)

(y) (*book ad*) "SALES territory is worldwide except traditional BRITISH market (but including Canada)." (Universe: nations; Sx = the book may be sold in x, Bx = x is a traditional British market, c = Canada).

2. Symbolize the following rules and statutes.

(a) (*Alice in Wonderland*) "[... The King ... read out from his book 'Rule Forty-two.] All persons more than a mile high to leave the court.'" (Universe: people; Mx = x is more than a mile high, Rx = x is required to leave the court)

*(b) (*Iowa regulation*) It is illegal to stop on an Interstate except for an emergency. (Sx = x is an act of stopping on an Iowa Interstate highway, Ex = x is done for an emergency, Ix = x is illegal)

(c) (*sign on New York beach, state ordinance*) "Swimming is permitted only where a lifeguard is on duty." (Universe: state-owned beaches in New York; Sx = swimming is permitted at x, Lx = a lifeguard is on duty at x)

(d) (*sign on copier*) "Do not copy unless all staples and paper clips are removed." (Sx = x has staples, Px = x has paper clips, Cx = it is permissible to copy x)

(e) (*Federal Rules of Evidence, Rule 607*) "The credibility of a WITNESS may be attacked by any PARTY, including the party calling the witness." (Axy = x may attack the credibility of y)

*(f) (*Code of Judicial Conduct, Canon 3 (B)(3)*) "A JUDGE shall require order and decorum in proceedings before the judge." (Oxy = x shall require order in y, Dxy = x shall require decorum in y, Pxy = x is a proceeding before y)

(g) (*Administrative Procedure Act, 5 U.S.C. §704*) "Agency action made reviewable by STATUTE and FINAL agency action for which there is no other ADEQUATE remedy in a court are subject to judicial REVIEW." (Universe: agency actions; Ax = there is adequate remedy in a court (besides judicial review) for x)

(h) (*Id., §556(b)*) "There shall preside at the taking of evidence—

(1) the agency;

(2) one or more members of the body which comprises the agency; or

(3) one or more administrative law judges appointed under section 3105 of this title."

(Pxy = x shall preside at y, Ex = x is a procedure of evidence taking, a = the agency, Mxy = x is a member of the body which comprises y, Ax = x is an administrative law judge appointed under section 3105 of the Administrative Procedure Act)

(i) (*Code of Canon Law, can. 1090, §1*) "One who, with a view to entering marriage with a particular person, has killed that person's spouse, or his or her own spouse, invalidly attempts this marriage." (Universe: people; Kxyz = x kills y with a view to marrying z, Sxy = x was a spouse of y, Vxy = x may validly marry y)

*(j) (*Id., §2*) "They also invalidly attempt marriage with each other who, by mutual physical or moral action, brought about the death of either's spouse." (Universe: people; Vxy, Pxyz = x and y together bring about the death of z by physical action, Mxyz = x and y together bring about the death of z by moral action, Sxy)

*3. A newspaper story from a less liberated era:[8]

> *Miami's Tiger Bay Political Club voted 180-96 today to continue their policy of prohibiting women members. Only female reporters are allowed to attend the club's luncheons.* . . .

The reporter wrote S1, but intended to express S2. S3 is another meaning S1 could have. Symbolize these statements.

(S1) Only **FEMALE REPORTERS** are **ALLOWED** to attend the club's luncheons.

(S2) Among females only reporters are allowed to attend the club's luncheons.

(S3) Among reporters only females are allowed to attend the club's luncheons.

4. A Florida driver's license exam contained a multiple-choice question about the meaning of this sign:

---

[8]"Tiger Bay Votes against Women," *Miami News* (January 4, 1972), p. 1A.

The exam key identifies the following answer as correct:

> Left turn from left lane only and traffic in adjoining lane may turn left or continue straight ahead.[9]

This answer is contradictory. The test constructor confused S1 with S2.

(S1) Left turns permitted from left lane only.
(S2) Only left turns permitted from left lane.

Symbolize S1 and S2. (Universe: turns at the signed intersection; Lx = x is a left turn, Px = x is permitted, Mx = x is made from the left lane)

5. Consider the familiar aphorism, "God helps those who help themselves." This maxim can be understood in three ways:

(S1) God helps all those who help themselves.
(S2) God helps only those who help themselves.
(S3) God helps all and only those who help themselves.

Symbolize each interpretation, employing "people" as the universe and using these symbols: Gx = God helps x, Sx = x is a self-helper.

6. A newspaper box displays these instructions: 'Use any coin combination (totalling 50¢). Do not use pennies.' (a) Formulate an improved version of these instructions with one sentence that can be symbolized with this dictionary: Universe: coin combinations; Tx = x totals 50¢, Px = x contains pennies, Ux = x may be used. (b) Symbolize your sentence.

*7. Signs bearing S1 are posted in rest areas along highways in Arkansas:

(S1) Soliciting, selling *or* LOITERING in the rest area is PROHIBITED.

The signs would issue the same warning had they carried S2:

(S2) Soliciting, selling *and* LOITERING in the rest area are PROHIBITED.

---

[9]John Keasler, "Driver's License Exam: No Passing without Failure," *Miami News* (November 23, 1971), p. 8B. Keasler spotted the error.

We can use our symbolic machinery to explain why S1 and S2 are equivalent even though they employ different connectives. Symbolize S1 using the wedge and S2 using the ampersand. If these symbolizations are correct they will be logically equivalent. (Universe: activities taking place in the rest area; Ox = x is a solicitation, Ex = x is an act of selling)

8. (SEMI-CHALLENGING) In the example about Toyotas and Hondas we claimed that F11 and F11A are logically equivalent formulas. Prove this by constructing two formal proofs.

$$(F11) \quad (x)[(Tx \vee Hx) \rightarrow Jx]$$
$$(F11A) \quad (x)(Tx \rightarrow Jx) \mathbin{\&} (x)(Hx \rightarrow Jx)$$

9. (SEMI-CHALLENGING) Consider this statement by Spence Carlson:

If human animals have rights, . . . so do non-human animals.[10]

(a) This is a multiply amphibolous statement because it lacks quantifier terms ('all' and 'some') governing 'human animals' and 'non-human animals'. State and symbolize the four distinct statements that emerge when these quantifier terms are provided in different combinations. Use this dictionary: Hx = x is human, Ax = x is an animal, Pxy = x possesses y, Rx = x is a right. (b) The following related statement is also amphibolous, and for the same reason:

Any right possessed by human animals is also possessed by non-human animals.

State and symbolize the four distinct statements that emerge when quantifier terms are provided. Use the same dictionary.

10. (CHALLENGING) The Fifth Amendment to the United States Constitution reads in part:

No person shall be held to answer for a capital, or otherwise infamous crime, unless on a presentment or indictment of a Grand Jury, except in cases arising in the land or naval forces, or in the Militia, when in actual service in time of War or public danger. . . .

As we saw in Chapter One, this is amphibolous. Does the expression 'when in actual service in time of war or public danger' attach just to 'in

---

[10] "Animals Are Victims of Vast Human-Regulated System of Slavery," *Los Angeles Times* (April 22, 1987).

the militia' or to all of 'in the land or naval forces, or in the militia'? (a) Symbolize both interpretations using this dictionary:

> Px = x is a person
> Axy = x shall be held to answer for y
> Cx = x is a capital or otherwise infamous crime
> Ixy = x is presented[11] or indicted by a grand jury for y
> Fx = x is a member of the land or naval forces
> Mx = x is a member of the militia
> Wx = x occurs in time of war or public danger

(b) From which interpretation of the amendment coupled with the following facts may we deduce '−Aab'?

> Pa
> Cb
> −Wb
> −Iab

§5.3 THE PARADOX OF MATERIAL IMPLICATION

When John McEnroe was playing in a tennis tournament in Australia, he became dissatisfied with the conditions of the court. After exchanging some heated words with the officials, he turned to a particularly bald official and said, "Listen, if that court is in good condition, you've got hair on your head."[12] This statement is evidently intended to constitute one of the premises in the following argument—the hearers being expected to supply the other premise and the conclusion:

> **If this court is in good condition, the person addressed has hair.**
> **The person addressed has no hair.**
> **Therefore, this court is not in good condition.**

Of course, McEnroe is not employing this argument as a means of *proving* the unsatisfactoriness of the court; rather, he is using the argument to emphasize the court's bad condition in an insulting manner. This is a valid argument in the form of *modus tollens*:

> C → H, −H ⊢ −C

---

[11]Grand jury presentments were made by the jurors from personal knowledge rather than from evidence presented by the prosecutor. They are no longer used in the federal system.
[12]*Chicago Tribune* (November 30, 1985), p. 2C.

McEnroe's use of the word 'if' corresponds exactly to our definition of the corresponding symbol, the arrow. For McEnroe (at least on this particular occasion), the claim that if A is true then B is true is not a claim that there is any causal relation or even any evidentiary relation between them; it is simply the claim that A cannot be true while B is false. (In addition to '$A \rightarrow B$', that claim can be symbolized by '$-(A \& -B)$'; of course, both of these formulas are equivalent to '$-A \vee B$'.) Accordingly, a conditional statement is true if the antecedent is false or if the consequent is true—no matter how unrelated the two halves of the statement are, or how peculiar they look in each other's company.

Using 'if' in this way makes the following arguments valid:

Jean Chrétien is Prime Minister of Canada.
So, if the moon is made of green cheese, Jean Chrétien is Prime Minister of Canada.

The moon is not made of green cheese.
So, if the moon is made of green cheese, Jean Chrétien is President of the United States.

The first of these arguments is in the form "$A \vdash B \rightarrow A$," the second in the form "$-A \vdash A \rightarrow B$." Both of these forms are valid according to our definition of the arrow.

But if we define the word 'if' in the same way, as McEnroe impliedly did in complaining about the court, we will sometimes get bizarre results:

Jean Chrétien is Prime Minister of Canada.
Therefore, if he is not Prime Minister of Canada, he is Prime Minister of Canada.

Jean Chrétien is not President of the United States.
Therefore, if he is President of the United States, he is not President of the United States.

Al Gore does not have two wives.
Therefore, if he has two wives, he is a monogamist.

Al Gore is a monogamist.
Therefore, if he has two wives, he is a monogamist.

In common understanding, these arguments would be considered invalid; that is, they would be regarded as having true premises and false conclusions. The reason is that most of the time when people say, "If A, then B," they mean more than that it is not the case that A is true and B false. They mean that there is some causal or explanatory connection between the antecedent event (or state) and the consequent event. Typically, the relation between statements asserted by a conditional is closer to an evidentiary relationship than a truth-functional one.

So the common understanding of the word 'if' is not quite the same as our definition of the corresponding symbol, the arrow. Standard logic provides a truth-functional definition of the arrow because that justifies useful rules of inference such as Arrow In, Arrow, and Chain Argument, and it allows us to use truth tables to evaluate arguments using the symbol. But we pay a price for this convenience, and the price is that we make it possible for an English conditional sentence generally regarded as false to have a true symbolization.

Logicians refer to this problem as the *paradox of material implication*.[13] Much of the time, they are able to disregard it. But in applying logic to legal discourse, we have to pay attention to it. Many legal statements can be most easily formalized as conditionals with conditional antecedents or as negated conditionals. Both forms are particularly likely to generate the paradox. Many other statements involve an "as if" construction or express conditions contrary to fact. In these the paradox is inherent.

1. Conditionals with Conditional Antecedents—Many legal statements take the form "$(A \rightarrow B) \rightarrow C$." For instance:

> (S1) If Judge Smith will decide the case in my favor if I will give him a thousand dollars, then he is a corrupt judge.
>
> (S2) If Harold will behave as a law-abiding citizen if he is let out of jail, he should be let out.
>
> (S3) If this utility will serve the public convenience and necessity if licensed, it should be licensed.

Similar statements can be made that take the form of universal quantifications "$(x)[(Ax \rightarrow Bx) \rightarrow Cx]$" from which statements in the above form can be derived by UO.

> (S1A) Any judge who will decide a case in my favor if I give him or her a thousand dollars is a corrupt judge.

Similar statements can be made corresponding to S2 and S3.

The trouble is that under our definition of the arrow, the following arguments are valid:

$(A \rightarrow B) \rightarrow C, -A \vdash C$
$(A \rightarrow B) \rightarrow C, B \vdash C$

---

[13]Some logicians distinguish two kinds of implication: *material* and *formal*. We have drawn the same distinction with the terms 'implication' and 'entailment' (see note 3 on page 28).

Each of these symbolized arguments is valid because the second premise entails the antecedent of the first premise. Although these symbolized arguments are valid, they have absurd English interpretations:

> **If Judge Smith will decide the case in my favor if I will give him a thousand dollars, then he is a corrupt judge.**
> **I will not give him a thousand dollars.**
> **Therefore, he is a corrupt judge.**

You can also "establish" that Judge Smith is corrupt whether or not I bribe him if the persuasiveness of my case will lead him to decide in my favor. Similarly, from S2 and the fact that Harold remains in jail you can "establish" that he should be let out, and from S3 and the fact that the utility doesn't have a license you can "prove" that it should be given one. The corresponding universals may be used in "proofs" that all judges are corrupt, that everyone in jail should be let out, and that all unlicensed utilities should be licensed.

2. *Negations of Conditionals*—Statements exhibiting the form "$-(A \rightarrow B)$," and the corresponding quantifications, raise similar difficulties. Section 41 of the Model Business Corporation Act[14] provides that a transaction with an interested director, if properly ratified, or if fair to the corporation, will not be rendered void or voidable by the participation of the interested director in the voting for the transaction. This legislation sets aside the rule previously enforced in some jurisdictions that made a transaction void or voidable any time an interested director participated in the voting. A natural way to state and symbolize the old rule is:

(S4) Every transaction is void or voidable if an interested director participated in the voting on it.

(F4) $(x)(Px \rightarrow Vx)$

(Universe: transactions; Px = an interested director participated in the voting on x, Vx = x is void or voidable) It would seem to follow that the section of the Model Act abolishing this rule in certain cases could be stated and symbolized as:

(S5) As to every transaction that is either properly ratified or fair, it is not the case that if an interested director participated in the voting on it, it is void or voidable.

(F5) $(x)[(Rx \lor Fx) \rightarrow -(Px \rightarrow Vx)]$

---

[14]Predecessor to the *Revised* Model Business Corporation Act, whose provisions on this subject are too complicated to be used here.

(Rx = x is properly ratified, Fx = x is fair) The trouble is that while F4 creates no problems, F5, because of our definition of the arrow, leads to absurd results. As we have seen, 'A → B' is equivalent to '−(A & −B)'. It follows that '−(A → B)' is equivalent to 'A & −B'. Therefore, you can derive from F5:

(F6) (x)[(Rx v Fx) → (Px & −Vx)],

which translates into

(S6) As to every transaction that is properly ratified or fair, an interested director participated in the voting on it, and it is neither void nor voidable.

That statement is so nonsensical that there is not much point in worrying about it. But other rules derivable by standard logical inferences from F5, although they are not good law, are not obviously absurd; students are sometimes taken in by them:

(F7) (x)(Fx → −Vx)
(S7) No transaction is void or voidable if it is fair.

(F8) (x)(Rx → −Vx)
(S8) No transaction is void or voidable if it is properly ratified.

In the famous *Globe Woolen* case (*Globe Woolen Co. v. Utica Gas & Electric Co.*, 224 N.Y. 483, 121 N.E. 378 (1918)), decided before the Model Act was adopted, the court voided a contract to supply electricity to the plaintiff at ruinously low rates. The plaintiff's chief stockholder and chief executive officer was a member of the defendant's board of directors at the time the contract was adopted. He dominated the proceedings of the board, and the other members relied on him for predictions of how the contract would work out. However, the other board members were all disinterested, and all voted for the transaction. If the Model Act had been in force, the ratification required by Section 41 would have occurred.

Students of corporation law, when asked if the Model Act would have changed the result of the *Globe Woolen* case, tend to answer that it would. If F8 and S8 followed from the rule of Section 41, then, of course, the students would be right. And if the rule were properly stated in F5 and S5, then F8 and S8 would follow from it. In fact, though, the students are wrong. The Model Act would not change the result of the case, because Section 41 says nothing about domination. It may no longer be the case that mere participation by an interested director makes the transaction voidable, but domination still does so.

It follows, of course, that S5 and F5 cannot properly be used to state and symbolize Section 41. The same problem arises in other cases where a rule would normally be stated in the form of a conditional. We encounter no difficulties when we state the proposed rule as a quantified conditional, but we get in trouble if we try to state a denial of the same rule in the form of a quantification of the negation of that same conditional. For instance, F.R.C.P. Rule 21 says, "Misjoinder of parties is not ground for dismissal of an action." This is a repeal of the common law rule by which misjoinder of parties *was* ground for dismissal of an action. The common law rule could easily be stated as:

(S9) Every action is subject to dismissal if the parties are misjoined.

(F9) $(x)(Mx \rightarrow Dx)$

(Universe: actions; Mx = the parties to x are misjoined, Dx = x is subject to dismissal) But Rule 21 cannot be stated as:

(S10) As to any action, it is not the case that if the parties are misjoined, it is subject to dismissal.

(F10) $(x)-(Mx \rightarrow Dx)$[15]

because from F10 we can derive that all actions have misjoined parties, and that no action is subject to dismissal.

3. Conditions Contrary to Fact—Many legal dispositions—wills, contracts, statutes, and judge-made rules—call for treating a situation as if something *were* the case that is not in fact the case. Take, for instance:

(S11) My executor shall have the same power to invest the assets of my estate that I myself *would have* if alive and acting.

It would seem that S11 could be symbolized as:

(F11) $(x)[(Ai \rightarrow Pxi) \rightarrow Pxe]$

(Ax = x is alive and acting, Pxy = x is a power of investment possessed by y, i = I, e = the executor) Other legal dispositions, although they are not ex-

---

[15]Note that Rule 21 would not be adequately expressed as a mere denial of S9 (i.e., as "It is not the case that every action is subject to dismissal if the parties are misjoined") because that is equivalent to "Some actions in which the parties are misjoined are not subject to dismissal." The common law rule was applicable to every action, and Rule 21 is applicable to every action also.

pressly stated in "as if" form are still best understood in that form. For instance:

(S12) Blacks have the same rights as whites.

and

(S13) A properly adopted administrative regulation has the effect of a statute.

really mean

(S12A) Every black person is entitled to every right he or she would have under the circumstances if he or she were white.

(F12A) $(x)(y)(z)[(Wx \rightarrow Rxyz) \rightarrow (Bx \rightarrow Rxyz)]$

(Wx = x is white, Rxyz = x has the right to y under set of circumstances z, Bx = x is black) and

(S13A) Every properly adopted administrative regulation has every effect it would have if it were a statute.

(F13A) $(x)(y)[(Sx \rightarrow Exy) \rightarrow (Ax \rightarrow Exy)]$

(Ax = x is a properly adopted administrative regulation, Sx = x is a statute, Exy = x has effect y)

Here again, our definition of the arrow produces absurd results. From F11 and '−Ai' (which will have to be true before my will takes effect) you can derive '(x)Pxe'—that is, that my executor has unlimited powers. Similarly, from F12A and the fact that Bill Cosby is black (and therefore not white) you can derive a formula that would give him the right to do anything under any circumstances, and from F13A and the trivial truth that no administrative regulation is a statute, you can derive a formula that would give any effect to any given administrative regulation.

You may have noticed a shift in the form of our examples since we started this discussion of conditions contrary to fact. All the conditional statements treated elsewhere in the book have had the form "If such-and-such *IS* the case, then so-and-so *IS* the case," while here we are examining conditionals having the form "If such-and-such *WERE* the case, then so-and-so *WOULD BE* the case"; that is, our examples have shifted from conditional statements expressed in the indicative mood to those expressed in the subjunctive. Logicians call these "subjunctive" or "counterfactual" conditionals and recognize that they are especially prone to the paradox of material implication because they signal the presupposition that their antecedents are false.[16] For this reason most logicians recommend against symbolizing counterfactual conditionals with the arrow, and we will follow their advice.

Some logicians have developed revisions of standard logic intended to eliminate the paradox of material implication, but no revision has gained general acceptance.[17] Our strategy will be to remain with standard logic (and its truth-functional definition of the arrow) but to offer ways of rephrasing troublesome conditionals before we symbolize them. We will avoid formulas that contain conditionals with conditional antecedents or negations of conditionals when at all possible, and we will refuse to use the arrow to represent counterfactual conditionals. In the process, we will learn more about what we mean by the legal arguments we use. One of the things we learn from the study of logic is that the word 'if' doesn't have as precise a meaning as we may have thought. Here, then, are three strategies for avoiding the paradox of material implication.

1. *Replacement of Conditional by Relational Predicate*—In many cases we use 'if' when we mean something much more specific. For instance, when we say, "If the sheep are growing thick coats, it will be a hard winter," we mean that the thick coats on the sheep constitute *evidence* that the winter will be hard. When we say "If you go out without a scarf you will catch cold," we mean that going out without a scarf will *cause* you to catch cold. When we say, "If two of the sides of a triangle are equal, it is an isosceles triangle," we mean that a triangle with two equal sides *is by definition* an isosceles triangle.

It is possible, therefore, to get rid of a number of problems with material implication by simply saying what we really mean instead of saying 'if'. For instance, S1 could be restated as:

(S14) If my giving Judge Smith a thousand dollars will cause him to decide the case in my favor, then he is a corrupt judge.[18]

(F14) $(\exists x)(Gx \ \& \ Dxs) \rightarrow Cs$

(Gx = x is a gift from me of \$1,000, Dxy = x causes y to decide in my favor, s = Judge Smith, Cx = x is corrupt) Similarly, S1A could be stated as:

---

[16]More precisely, such a sentence signals the presupposition that the condition expressed by its antecedent does not obtain, and a statement in the indicative mood expressing that condition would be false.

[17]The two best known systems are those propounded by Robert Stalnaker and David Lewis. See Stalnaker's "A Theory of Conditionals" and Lewis's "Counterfactuals and Comparative Possibility," which are reprinted in William Harper et al (eds.), *Ifs* (Dordrecht: D. Reidel Publishing Company, 1981).

[18]There is still an 'if' in S14, but it does not present a problem. S1 presented a problem because its antecedent was itself a conditional. The antecedent of S14, on the other hand, is a simple statement.

§5.3 The Paradox of Material Implication

(S14A) Any judge whom my gift of a thousand dollars will cause to decide a case in my favor is a corrupt judge.

(F14A) (x)[(∃y)(Gy & Jx & Dyx) → Cx]

(Jx = x is a judge) This strategy will cope with many conditional antecedent cases, but not all. It is harder, for instance, to replace the 'if's in S2 and S3 with relational predicates than it was in S1.

The strategy is particularly effective on negations of conditionals. In fact, the framers of Section 41 of the Model Business Corporation Act used it. Instead of S5, which contains the negation of a conditional, they said that a properly ratified or fair transaction would not be void or voidable *by reason of* the participation of the interested director.

(F15) (x)(y){[Tx & (Rx v Fx) & Py] → −Vyx}

(Tx = x is a transaction, Px = x is an act of participation by an interested director, Vxy = x renders y void or voidable) Similarly, the framers of F.R.C.P. Rule 21 did not enact S10. Rather, they provided that misjoinder of parties is not *ground* for dismissal.

(F16) (x)(y)[(Ax & My) → −Gyx]

(Ax = x is an action, Mx = x is a misjoinder of parties, Gxy = x is ground for dismissing y)

Sometimes, however, we may need a more elaborate symbolization than is provided in F15 and F16. To illustrate, let us take the opposite of F16—the rule that was in force before F.R.C.P. Rule 21 was adopted. That rule would be that misjoinder *is* a ground for dismissal of an action. Under that rule, it would be possible to argue:

(S17) **Misjoinder of parties is ground for dismissal of an action.**
(S18) **The parties to this action are misjoined.**
Therefore, there is ground for dismissal of this action.

You will find that if you try to symbolize S17 using the same symbols we used for F16, you won't be able to symbolize the rest of the argument. There will not be any predicate in the symbolization of S17 that can be used to symbolize S18. For S18, you will have to use something like 'Ma' (Mx = the parties to x are misjoined, a = this action). So there will be no predicate common to both premises, and therefore no way of reaching a conclusion.

The problem is that 'misjoinder of parties' in S17 is a nominative expression symbolized by an *individual constant* while 'misjoined' in S18 is an adjectival expression symbolized by a *predicate*. The connection between the two expressions is lost in the symbolization. To symbolize the argument successfully, we will have to restate both premises so that there will be some predicate involving misjoinder appearing in both of them. The following will do:

(S17A) **If there is a misjoinder of the parties to an action, that misjoinder is ground for dismissing the action.**
(S18A) **There is misjoinder of the parties to this action.**
Therefore, there is ground for dismissing this action.

Stated in this way, the argument yields the following valid symbolization:

(F17A) (x)(y)(Mxy → Gxy)
(F18A) (∃x)Mxa
⊢ (∃x)Gxa

(Mxy = x is a misjoinder of the parties to action y, a = this action) Since we may need the symbols of F17A to express adequately the rule abolished by F.R.C.P. Rule 21, we may want to express Rule 21 in the same symbols:

(x)(y)(Mxy → −Gxy)

In the same way, we may prefer the following to F15 as a symbolization of Section 41 of the Model Business Corporation Act:

(F15A) (x)(y){[Tx & (Rx v Fx)] → (Pyx → −Vyx)}

(Tx = x is a transaction, Rx = x is ratified, Fx = x is fair, Pxy = x is an act of participation by an interested director in the adoption of y, Vxy = x renders y void or voidable)

2. *Burying Conditional in Predicate* —Often, "if" clauses in legal dispositions do not have to be unpacked to draw conclusions from them. Take for instance this argument based on S11:

(S11) **My executor shall have the same power to invest the assets of my estate that I myself would have if alive and acting.**
If alive and acting, I would have the power to invest the assets of the estate in Consolidated Kiddiecar stock.
Therefore, my executor has the power to invest the assets of the estate in Consolidated Kiddiecar stock.

This argument can be symbolized as follows:

(F11A) (x)(Px → Ex)
Pc
⊢ Ec

(Px = x is a power I would have if alive and acting, Ex = x is a power my executor has, c = the power to invest in Consolidated Kiddiecar stock) By beginning with F11A instead of F11, we have avoided the necessity of using an arrow to symbolize a condition contrary to fact, and therefore have avoided introducing the paradox of material implication.

This strategy will work for many conditions contrary to fact. It will also work tolerably well in other cases. For instance, one way of dealing with Section 41 of the Model Business Corporation Act would be to symbolize it as:

(x)[(Rx v Fx) → −Vx]

(Universe: transactions; Rx = x is ratified, Fx = x is fair, Vx = x is rendered void or voidable if an interested director participated in approving it) While this is an acceptable symbolization, we prefer the ones we offered earlier (F15 and F15A), because they make it clearer why Section 41 does not change the result of the *Globe Woolen* case. By stating the rule of *Globe Woolen* as '(x)Vdx' (d = domination by an interested director), and setting it alongside F15, we can make the difference obvious.

The strategy does not work as well for all conditions contrary to fact as it does for S11. Take the rule that blacks have the same rights as whites. Assume that Susie is black, and that if white she would be entitled to attend the first grade in the James Madison School. Then, we will have no trouble with the following argument:

**If any right is one Susie would have if white, Susie has that right.
The right to attend first grade in the James Madison School is one Susie would have if white.
Therefore, Susie has the right to attend the first grade in the James Madison School.**

This can be effectively symbolized by burying the counterfactual in one of the predicates:

(x)(Wx → Rx)
Wf
⊢ Rf

(Wx = x is a right Susie would have if white, Rx = x is a right Susie has, f = the right to attend first grade in the James Madison School) Well and good. But how do we know that Susie would have the right to attend first grade in the James Madison School if she were white? Because there are a batch of other qualifications for admission to that grade and school, and Susie possesses all those other qualifications. To make this point, we will have to find some way of unpacking the counterfactual symbolized by 'W' in the above argument:

> **If any right is one that a white person possessed of certain nonracial qualifications would have, then, if Susie has those qualifications, she has that right.**
> **There is a set of nonracial qualifications such that every white person possessing them has the right to attend first grade in the James Madison School, and Susie has those qualifications.**
> **Therefore, Susie has the right to attend first grade in the James Madison School.**

3. *Universal Quantifications*—In the argument just stated, the claim that if Susie were white she would have the right to attend first grade in the James Madison School has been turned into a claim that any white person with Susie's nonracial qualifications would have that right.[19] That way, it can be symbolized with a universal quantification (within an existential quantification), and the whole argument can be stated this way:

(x)(y){(z)[(Wz & Pzx) → Rzy] → (Psx → Rsy)}
(∃x){(y)[(Wy & Pyx) → Ryf] & Psx}
⊢ Rsf

(Wx = x is white, Pxy = x possesses set of nonracial qualifications y, Rxy = x has right y, s = Susie, f = the right to attend first grade at the James Madison School)

Our final example is drawn from *Federal Crop Insurance Corp.* v. *Merrill*, 332 U.S. 380 (1947), plaintiff, a farmer, bought crop insurance from defendant, a government corporation. Unbeknownst to the plaintiff, or to the agent who sold him the insurance, his crop was of a kind that properly adopted administrative regulations forbade defendant to cover. When his crop was destroyed he sued for the loss, and the defendant set up the administrative regulations as a defense. The lower court adopted the analogy of a private insurance company whose agent sold a policy in violation of instructions from the home office. Since the private company could not escape lia-

---

[19]If there were no such white persons, there would be further difficulties, as we will see in §5.5.

bility on such a ground, the court held the defendant liable. But the Supreme Court reversed. It pointed out that defendant was not a private insurance company but a government agency, and that government agencies are always subject to legal restrictions on what they can do. If the statute setting up the defendant had forbidden it to insure this crop, there would be no question. Since a properly adopted administrative regulation has the effect of a statute, there should still be no question.

To symbolize this argument, we have to begin with the first premise—that a properly adopted administrative regulation has the effect of a statute. This premise is one we stated earlier on in two forms, S13 and S13A. S13, the simpler form, seems the easier of the two to symbolize by a universal quantification. But before we do so, we will have to figure out what it means. What *do* we mean when we say that a properly adopted administrative regulation has the effect of a statute? We can't very well mean that every regulation has every effect that any statute has: the Manual of Public Assistance Administration doesn't have all the effects of the Uniform Commercial Code. Nor can we mean that some administrative regulation has the effect of some statute: even if that were true, it wouldn't be very informative. We could mean that all regulations have every effect that all statutes have. That might work in the *Merrill* case, because all statutes (at least, all the ones we can think of) do have the effect of binding people whether or not they know about them. (*Ignorantia legis neminem excusat*.) In most cases, though, that interpretation won't work because we will be considering effects that some, but not all, statutes have.

So the only meaning that will effectively cover all the ground is that every properly adopted administrative regulation has the effect that *it would itself have* if it were a statute—the form we have stated as S13A. But in that form, we can't symbolize it by a universal quantification in such a way as to avoid the paradox of material implication. The best we can do is bury the conditional in a predicate:

(x)(y)[(Ax & Sxy) → Hxy]

(Ax = x is an administrative regulation, Sxy = x, if it were a statute, would have effect y, Hxy = x has effect y)

EXERCISES

Instructions for exercises 1 through 9: Symbolize each of the following statements in a way that sidesteps the paradox of material implication. That is, avoid symbolizations that involve the negations of conditionals and conditionals with conditional antecedents, and do not symbolize counterfactual conditionals with arrows. Provide dictionaries.

1. An employer who won't hire blacks violates the Civil Rights Act.

*2. It is not true that sexual intercourse with a girl over the age of sixteen is not unlawful. (See Section 4.2, exercise 6.)

3. The corporate veil may be pierced[20] if recognition of corporate existence would lead to injustice.

4. If the requirements of a job violate a person's religious commitments, then the person will not lose unemployment benefits if he or she turns down the job. (See *Sherbert v. Verner*, 374 U.S. 399 (1963).)

5. The fact that a job requires violation of union rules doesn't prevent turning it down from being a bar to receiving unemployment compensation benefits. (See *Norman v. Emp. Sec. Agency*, 83 Idaho 1, 356 P.2d 913 (1960).)

*6. The innocent party to a putative marriage[21] has the same rights in the property of the other party as he or she would have if the marriage were valid.

7. (CHALLENGING) Here is subdivision (b) of Rule 15 of the Federal Rules of Civil Procedure. Symbolize each of the underlined passages.

> ***(b) Amendments to Conform to the Evidence.*** <u>When issues not raised by the pleadings are tried by express or implied consent of the parties, they shall be treated in all respects as if they had been raised in the pleadings.</u> Such amendment of the pleadings as may be necessary to cause them to conform to the evidence and to raise these issues may be made upon motion of any party at any time, even after judgment; but failure so to amend does not affect the result of the trial of these issues. <u>If evidence is objected to at the trial on the ground that it is not within the issues made by the pleadings, the court may allow the pleadings to be amended and shall do so freely[22] when the presentation of the merits of the action will be subserved thereby and the objecting party fails to satisfy the court that the admission of such evidence would prejudice the party in maintaining the party's action or defense upon the merits.</u> The court may grant a continuance to enable the objecting party to meet such evidence.

8. (CHALLENGING) Symbolize the following passage from Rule 19(a) of the Federal Rules of Civil Procedure:

> ***(a) Persons to Be Joined if Feasible.*** *A person who is subject to service of process and whose joinder will not deprive the court of juris-*

---

[20] I.e., liability for corporate transactions may be imposed directly on the individuals involved.

[21] A putative marriage is an invalid marriage where at least one of the parties is unaware of the invalidity—as where one of them, unbeknownst to the other, is already married. The following statement will need some qualification if there is a genuine spouse also in the picture.

[22] The word 'freely' may be disregarded in your symbolization.

diction over the subject matter of the action shall be joined as a party in the action if (1) in the person's absence complete relief cannot be accorded among those already parties, or (2) the person claims an interest relating to the subject of the action and is so situated that the disposition of the action in the person's absence may (i) as a practical matter impair or impede the person's ability to protect that interest or (ii) leave any of the persons already parties subject to a substantial risk of incurring double, multiple, or otherwise inconsistent obligations by reason of the claimed interest.

9. (CHALLENGING) Symbolize the following provision from a model unemployment compensation act:

> (5) Compensation shall not be denied to any otherwise eligible individual for refusing to accept new work under any of the following conditions:
>
> (A) if the position offered is vacant due directly to a strike, lockout, or other labor dispute;
> (B) if the wages, hours, or other conditions of the work offered are substantially less favorable to the individual than those prevailing for similar work in the locality;
> (C) if as a condition of being employed, the individual would be required to join a company union or to resign from or refrain from joining any bona fide labor organization. . . .

§5.4 QUANTIFIER SCOPE

In Section 3.5 we symbolized S1 with F1A. Is either F1B or F2 also a correct symbolization?

(S1) Whoever KILLS$_R$ a person BREAKS the fifth commandment.

(F1A) (x)[(∃y)Kxy → Bx]
(F1B) (x)(y)(Kxy → Bx)
(F2) (x)(∃y)(Kxy → Bx)

(Universe: people) The (possibly surprising) answer is that F1B is a correct symbolization of S1, while F2 is incorrect. In this section we explain why this is so; understanding this matter should enhance your symbolization skills.

We begin the discussion by noting the following equivalences:

(A) (∃x)Fx → P   *is equivalent to*   (B) (x)(Fx → P)
(C) (x)Fx → P   *is equivalent to*   (D) (∃x)(Fx → P)

Formulas A and B differ in two ways: (1) one formula has an existential quantifier while the other employs a universal, and (2) the scope[23] of one of those quantifiers is the antecedent of a conditional while the scope of the other includes the entire conditional, 'Fx → P'.[24] Note that C differs from D in the same two ways.

The equivalence of A and B is established by these two proofs:

(1)	(∃x)Fx → P	A	1	(1)	(x)(Fx → P)	A	
(2)	−(x)(Fx → P)	PA	2	(2)	(∃x)Fx	PA	
(3)	(∃x)−(Fx → P)	2 QE	2	(3)	Fa	2 EO	
(4)	−(Fa → P)	3 EO	1	(4)	Fa → P	1 UO	
(5)	Fa & −P	4 AR	1,3	(5)	P	4,3 →O	
(6)	Fa	5 &O	1	(6)	(∃x)Fx → P	2-5 →I	
(7)	(∃x)Fx	6 EI					
(8)	P	1,7 →O					
(9)	−P	5 &O					
(10)	P & −P	8,9 &I					
(11)	(x)(Fx → P)	2-10 −O					

(The equivalence of C and D is the subject of exercises 3 and 8 at the end of the section.)

Note also that neither A nor B is equivalent to either C or D. We can show the nonequivalence of A and C with this interpretation:

(A) (∃x)Fx → P,   (C) (x)Fx → P

Universe: people

Fx = x is unmarried
P = There are no husbands

(A) If someone is unmarried, then there are no husbands.   (F)

(C) If everyone is unmarried, then there are no husbands.   (T)

---

[23]Recall that the scope of a quantifier consists of the quantifier, the quantifier-scope groupers, and all the symbols enclosed within those groupers. Remember, too, that in cases where no ambiguity results from the omission of quantifier-scope groupers, they are usually left out. They are omitted when only one predicate falls within the scope of a quantifier; formulas A and C serve as examples. And when two quantifiers are contiguous, scope groupers for the first are omitted; formulas F1B and F2 illustrate this.

[24]In this section it will be convenient to refer to formula fragments such as 'Fx → P' as *conditionals*, even though they are not, strictly speaking, statements at all, since they contain unquantified variables.

If we refer to an existential and a universal quantifier involving the same variable as "counterparts," the patterns exemplified above may be summarized as follows:

When the scope of a quantifier is expanded from the antecedent of a conditional to include the entire conditional, the quantifier must be replaced by its counterpart to preserve equivalence.

We can apply this principle to the formulas considered at the beginning of the section:

(F1A) (x)[(∃y)Kxy → Bx]
(F1B) (x)(y)(Kxy → Bx)
(F2) (x)(∃y)(Kxy → Bx)

The scope of the *y*-quantifier in F1A ((∃y)Kxy) is expanded in F1B and F2 to include the entire conditional. (We underline the quantifiers in question for emphasis.) F1B is equivalent to F1A because the existential quantifier in F1A has been replaced by its counterpart (a universal *y*-quantifier) in F1B. F2 is not equivalent to F1A because the *y*-quantifier remains existential in F2. As F1A is a correct symbolization of S1, the equivalent F1B is also correct, but the nonequivalent F2 is unacceptable. The equivalence of F1A and F1B is evident in these readings of the two formulas:

(S1A) Any person who kills *some*one breaks the fifth commandment.

(S1B) Any person who kills *any*one breaks the fifth commandment.

Formula F2 symbolizes this statement:

(S2) For every person, there is some person whom the first person would break the fifth commandment by killing.

Note that S2 follows from S1A (and from S1B), but not the other way around. If the fifth commandment forbade only suicide, S2 would be true, but S1A and S1B would be false.

The symbolization of S3 presents a related problem. S3 may be symbolized by either F3A or the logically equivalent F3B, but not by the logically distinct F4. The *y*-quantifier has an expanded scope in both F3B and F4, but only in F3B is the quantifier replaced by its counterpart.

(S3) Anyone who LOVES$_R$ everyone is a SAINT.

(F3A) $(x)[(y)Lxy \rightarrow Sx]$
(F3B) $(x)(\exists y)(Lxy \rightarrow Sx)$
(F4) $(x)(y)(Lxy \rightarrow Sx)$

(Universe: people)
S3B gives a literal reading of F3B.

(S3B) For each person there is a person such that if the first loves the second, the first is a saint.

Since F3A and F3B are equivalent, it seems that the statements they represent, S3 and S3B, will also be equivalent; but are they? If we understand the 'if' in S3B to have the minimal meaning discussed in Section 5.3 (so that 'if p then q' means no more than "either not p or q"), then S3B and S3 are equivalent—as this series of statements shows:

(S3B) For each person there is a person such that if the first loves the second, the first is a saint.

For each person there is a person such that either the first doesn't love the second or the first is a saint.

For each person either there is a person the first doesn't love or the first is a saint.

Each person either doesn't love everyone or else is a saint.

(S3) Anyone who loves everyone is a saint.

(Establishing the equivalence of these five sentences is the subject matter of exercises 6 and 7 following this section.) On the other hand, if we attach a stronger meaning to the 'if' in S3B, the equivalence with S3 may be lost.
Be sure to distinguish S3B from the very different S5, which is symbolized with F5:

(S5) For each person, if there is a second person that the first loves, the first is a saint.

(F5) $(x)[(\exists y)Lxy \rightarrow Sx]$

S5 confers sainthood on everyone who loves anyone at all.[25] The condition for sainthood proposed by S3 and S3B is much more demanding.

Here are some additional examples of the importance of quantifier scope for symbolization. In each case the two formulas printed below the sentence exhibit different quantifier scopes, but are logically equivalent and equally correct symbolizations of the sentence. Should you replace any quantifier in any of these formulas with its counterpart, you will produce a mistaken symbolization.

(*television newscast*) "If the jury CONVICTS any of the officers it will convict Laurence Powell." (Universe: officers on trial)
  $(\exists x)Cx \to Cp$
  $(x)(Cx \to Cp)$

(*Tsar Nicholas II*) "I will never make PEACE so long as a single ENEMY remains on RUSSIAN soil." (P = Nicholas makes peace, Ex = x is an enemy of Nicholas, Rx = x is on Russian soil)
  $(\exists x)(Ex \& Rx) \to -P$
  $(x)[(Ex \& Rx) \to -P]$

Anyone HOLDING$_R$ an INFANT will be SEATED. (Universe: people)
  $(x)[(\exists y)(Iy \& Hxy) \to Sx]$
  $(x)(y)[(Iy \& Hxy) \to Sx]$

If everyone HELPS, we'll get the job DONE. (Universe: people; D = We'll get the job done)
  $(x)Hx \to D$
  $(\exists x)(Hx \to D)$

(*poster slogan*) "No one is FREE when others are OPPRESSED." (Universe: people)
  $(x)[(\exists y)(y \neq x \& Oy) \to -Fx]$
  $(x)(y)[(y \neq x \& Oy) \to -Fx]$

(*folklore*) Any seventh son of a seventh son can HEAL. (Sxy = x is the seventh son of y)
  $(x)[(\exists y)(\exists z)(Sxy \& Syz) \to Hx]$
  $(x)(y)(z)[(Sxy \& Syz) \to Hx]$

Curiously, the patterns of equivalence we have discussed do not apply when the scope of a quantifier is expanded from the *consequent* (rather than

---

[25] As would any statement symbolized by F4.

the antecedent) of a conditional to include the entire conditional.[26] Consider formulas E through H.

(E) $P \to (\exists x)Fx$ *is equivalent to*   (F) $(\exists x)(P \to Fx)$
(G) $P \to (x)Fx$ *is equivalent to*   (H) $(x)(P \to Fx)$

In summary:

When the scope of a quantifier is expanded from the consequent of a conditional to include the entire conditional, the quantifier must remain the same to preserve equivalence.

The box on page 279 contains examples of equivalent correct symbolizations that illustrate this principle.

When a formula has a quantified antecedent *and* a quantified consequent, there will be several equivalent symbolizations. We illustrate by symbolizing a statement made by Justice Frankfurter:

If one man is allowed to DETERMINE for himself what is law, then every man can.

$(\exists x)Dx \to (y)Dy$
$(x)[Dx \to (y)Dy]$ *(scope of antecedent quantifier extended)*
$(y)[(\exists x)Dx \to Dy]$ *(scope of consequent quantifier extended)*
$(x)(y)(Dx \to Dy)$ *(scope of both quantifiers extended)*

(Universe: people; Dx = x is allowed to determine for x what is law)

We should also note that the scope of a quantifier may be expanded from a conjunct (disjunct) to an entire conjunction (disjunction) without changing the quantifier. The formulas in each horizontal pair below are logically equivalent:

---

[26]How is this asymmetry between antecedent scope and consequent scope to be explained? If a quantifier has a disjunct as its scope, the scope of that quantifier may be extended across the disjunction without changing the content of the formula. Examine the formulas below. Each formula in each list is equivalent to the formulas above it; hence the top and bottom formulas in each list are equivalent. The left-hand list concerns the expansion of the scope of an antecedent quantifier; the other list treats the expansion of the scope of a consequent quantifier. Note that a quantifier transformation occurs as you move from the second to the third formula in the left-hand list, while no such transformation occurs in the other list.

$(x)Ax \to (y)By$ $(x)Ax \to (y)By$
$-(x)Ax \vee (y)By$ $-(x)Ax \vee (y)By$
$(\exists x)-Ax \vee (y)By$
$(\exists x)[-Ax \vee (y)By]$ $(y)[-(x)Ax \vee By]$
$(\exists x)[Ax \to (y)By]$ $(y)[(x)Ax \to By]$

(x)Fx & P,	(x)(Fx & P)	P & (x)Fx,	(x)(P & Fx)
(∃x)Fx & P,	(∃x)(Fx & P)	P & (∃x)Fx,	(∃x)(P & Fx)
(x)Fx v P,	(x)(Fx v P)	P v (x)Fx,	(x)(P v Fx)
(∃x)Fx v P,	(∃x)(Fx v P)	P v (∃x)Fx,	(∃x)(P v Fx)

In view of the patterns studied in this section you can see that it will ordinarily be possible to rewrite a formula containing embedded quantifiers in a way that brings those quantifiers to the front of the formula. Such a transformation can facilitate proof construction.

If we SURRENDER, we all PERISH.
(Universe: members of the party; S = We surrender)
    S → (x)Px
    (x)(S → Px)

Whoever commits suicide KILLS$_R$ someone. (Universe: people)
    (x)[Kxx → (∃y)Kxy]
    (x)(∃y)(Kxx → Kxy)

Every POLITICIAN KNOWS$_R$ some LOBBYIST or other.
    (x)[Px → (∃y)(Ly & Kxy)]
    (x)(∃y)[Px → (Ly & Kxy)]

Every POLITICIAN KNOWS$_R$ every LOBBYIST.
    (x)[Px → (y)(Ly → Kxy)]
    (x)(y)[Px → (Ly → Kxy)]

A SEARCH is LEGITIMATE only if it is APPROVED$_R$ by some MAGISTRATE.
  (Axy = x approves y)
    (x){Sx → [Lx → (∃y)(My & Ayx)]}
    (x){Sx → (∃y)[Lx → (My & Ayx)]}
    (x)(∃y){Sx → [Lx → (My & Ayx)]}

As you work the exercises following this section, remember the requirement that no variable may fall within the scope of two quantifiers employing that variable. F6 through F6C satisfy this requirement; F7 through F9 do not. The latter three formulas are not well formed. In accord with the patterns we have been discussing, F6 through F6C are equivalent formulas.

(F6) (x)Fx → (x)Gx

well-formed
(F6A) (∃x)[Fx → (y)Gy]
(F6B) (x)[(y)Fy → Gx]
(F6C) (∃x)(y)(Fx → Gy)

ill-formed
(F7) (∃x)[Fx → (x)Gx]
(F8) (x)[(x)Fx → Gx]
(F9) (∃x)(x)(Fx → Gx)

Keep in mind also the requirement that every variable in a formula must fall within the scope of some quantifier employing that variable. If you contract the scope of the *y*-quantifier in F10 you reach formula F11 or F12, which are improperly formed because they have an occurrence of the variable *y* falling outside the scope of the *y*-quantifier.

(F10) (x)(y)(Fy → Gxy)
(F11) (x)[Fy → (y)Gxy]   *(ill-formed)*
(F12) (x)[(∃y)Fy → Gxy]   *(ill-formed)*

In Section 3.1 we offered the following "rule of thumb" for symbolizing: the universal quantifier ordinarily pairs with the arrow and the existential quantifier with the ampersand. In Section 3.5 we refined this observation for relational formulas: when a formula begins with several quantifiers followed by a quantifier-scope grouper, expect this pairing to obtain between the connective and the quantifier nearest it. Exceptions to this rule abound when the scope of quantifiers is expanded. Formula F3B (on page 276) is one example; there are several others in this section.

When you bring a quantifier to the front of a formula you should locate it to the right of any quantifiers already at the front. Violating this warning may lead to changes in the content of the formula. The "Suicide" sentence will serve as an example:

sentence: Whoever commits suicide KILLS$_R$ someone.

initial
symbolization: (x)[Kxx → (∃y)Kxy]

correct
transformation: (x)(∃y)(Kxx → Kxy)

incorrect
transformation: (∃y)(x)(Kxx → Kxy)

The incorrect transformation is not equivalent to the initial symbolization; instead it represents the outlandish statement, "There is one person who is killed by all who commit suicide."

When you change the order of quantifiers you change their scope. For example, in the correct transformation displayed in the preceding paragraph, the universal quantifier is outside the scope of the existential quantifier; in the mistaken transformation the universal quantifier falls within the scope of the existential quantifier. In some cases this difference in scope makes a difference in the content of the formulas, in other cases it does not. There are, in fact, three situations in which the order of two contiguous quantifiers is immaterial:

(1) when both quantifiers are universal,

EXAMPLES:
(x)(y)Rxy *and* (y)(x)Rxy *are equivalent.*

(2) when both are existential, and

(∃x)(∃y)Rxy *and* (∃y)(∃x)Rxy *are equivalent.*

(3) when the two quantifier-variables do not appear in the same relational predicate.

(x)(∃y)(Fx → Gy) *and* (∃y)(x)(Fx → Gy) *are equivalent.*

Conversely, when one quantifier is universal and the other existential and both quantifier-variables occur in at least one relational predicate, quantifier order becomes crucial. In that case, if you should reverse the order of the quantifiers you will almost certainly change the content of the formula. Consider the following as an example:

(S13) Each ANTITRUST case is HEARD$_R$ by some federal JUDGE (or other).

(F13A) (x)[Ax → (∃y)(Jy & Hyx)]
(F13B) (x)(∃y)[Ax → (Jy & Hyx)]   *(correct transformation)*
(F14) (∃y)(x)[Ax → (Jy & Hyx)]   *(incorrect transformation)*

(Hxy = x hears y) F13A is the intuitive symbolization of S13, and F13B is equivalent to it. F14, which differs from F13B only in the order of quantifiers, is not equivalent to F13A or F13B, and does not represent S13. Consider another example:

(S15) There is a federal judge who hears all antitrust cases.

(F15A) (∃x)[Jx & (y)(Ay → Hxy)]
(F15B) (∃x)(y)[Jx & (Ay → Hxy)]   *(correct transformation)*
(F16) (y)(∃x)[Jx & (Ay → Hxy)]   *(incorrect transformation)*

F15A and F15B, but not F16, are proper symbolizations of S15.

You might suppose that while F14 does not represent S13, it does represent S15; but it does not. F14 does not entail the existence of any federal judge, while S15 (and F15A and F15B) do have that entailment. F14 corresponds to this statement:

(S14) If there are any antitrust cases, then there is a federal judge who hears them all.

S14 differs in content from both S13 and S15. In a similar way F16 does not represent S13 because they have different existential entailments. The following section examines more closely this matter of existential entailments.

EXERCISES

1. For each of the following sentences provide four equivalent symbolizations with different quantifier scopes.
    (a) If somebody AGREES with Lucy, somebody is RIGHT.
   *(b) If somebody agrees with Lucy, everybody is right.
    (c) If everybody agrees with Lucy, somebody is right.
    (d) If everybody agrees with Lucy, everybody is right.

    (Universe: people; Ax = x agrees with Lucy)

2. For each of the following sentences from earlier exercises provide two equivalent symbolizations with different quantifier scopes.
    (a) If anyone can TURN the Georgia Tech football program around, Pepper <u>Rogers</u> can do it. (Universe: people; Tx = x can turn around the Georgia Tech football program)
   *(b) Every VARIABLE falls within the SCOPE$_R$ of some QUANTIFIER. (Sxy = x falls within the scope of y)
    (c) LIFE SPRINGS$_R$ only from life.
    (d) Anyone RAPED$_R$ by a person who has GONORRHEA will also have the disease. (Universe: people; Rxy = x rapes y)
    (e) (<i>James Herriot</i>) "No MAN is as STRONG$_R$ as a COW." (Sxy = x is stronger than y)
   *(f) Any PERSON who is HOOKED$_R$ on <u>alcohol</u> is hooked on a DRUG.

3. Show by formal proof that formula D entails formula C.

    (C) (x)Fx → P   (D) (∃x)(Fx → P)

4. (a) Prove the equivalence of formulas E and F by constructing two formal proofs.

    (E) P → (∃x)Fx   (F) (∃x)(P → Fx)

   *(b) Prove the equivalence of G and H by constructing two formal proofs.

    (G) P → (x)Fx   (H) (x)(P → Fx)

(c) Prove the nonequivalence of E and G by the method of interpretation.

*5. Jesse McCrary's aphorism, "Injustice anywhere is injustice everywhere," is correctly symbolized by each of these formulas:

    (F1) ($\exists$x)Ix → (x)Ix
    (F2) (x)[($\exists$y)Iy → Ix]
    (F3) (x)[Ix → (y)Iy]
    (F4) (x)(y)(Ix → Iy)

(Ix = injustice exists in area x) Establish the logical equivalence of these formulas by showing (by proof) that each of the first three formulas entails the next formula, and that F4 entails F1.[27]

6. On page 276 we claimed the equivalence of the five sentences in this list (assuming that the 'if' in S3B carries the minimal meaning):

  (S3B) For each person there is a person such that if the first LOVES$_R$ the second, the first is a SAINT.

  (S3C) For each person there is a person such that either the first doesn't love the second or the first is a saint.

  (S3D) For each person either there is a person the first doesn't love or the first is a saint.

  (S3E) Each person either doesn't love everyone or else is a saint.

  (S3A) Anyone who loves everyone is a saint.

Support the claim of equivalence by symbolizing the sentences, and showing by proof that each of the first four sentences entails the next. In each case employ the literal symbolization.

7. (SEMI-CHALLENGING) Show by proof that S3A entails S3B (see previous exercise). This completes the demonstration of the logical equivalence of the five statements displayed in exercise 6.

8. (CHALLENGING) Prove by formal proof that formula C entails formula D. Can you do it in 11 (or fewer) lines?

    (C) (x)Fx → P   (D) ($\exists$x)(Fx → P)

---

[27]Since entailment is transitive, these four proofs show that every sentence in the set entails every other, and therefore that every sentence in the set is logically equivalent to every other.

9. (CHALLENGING) For each of the following sentences (which appeared in previous sections as examples or exercises) provide a symbolization in which all quantifiers are at the front of the formula.
   (a) Everybody LOVES$_R$ a lover. (Universe: people; Lxy)
   (b) (*editorial*) "<u>CBS</u> spends part of every corporate DAY KNUCKLING$_R$ under to somebody about something." (Dx = x is a business day, Kxyzw = x knuckles under to y about z on day w)
   (c) (*Teddy Roosevelt*) "This country will not be a good place for any of us to live in unless we make it a good place for all of us to live in." (Universe: Americans; Gx = America is a good place for x to live in)
   (d) (*adage*) "Nothing VENTURED$_R$, nothing GAINED$_R$." (Vxy = x ventures y, Gxy = x gains y)
   (e) (*Kipling*) "Those who KILL$_R$ SNAKES get killed by snakes."
   (f) Any person who has a child who doesn't aggravate that person doesn't have children. (Universe: people; Cxy = x is a child of y, Axy = x aggravates y)
   (g) (*legal decision*) Conflict of INTEREST$_R$ exists if client and attorney are being PROSECUTED$_R$ by the same OFFICE. (Ixy = x's representation of y constitutes conflict of interest, Cxy = x is legal client of y)

10. (CHALLENGING) Support the following two claims made in text.
    (a) On page 281 we claimed the equivalence of these two formulas:

    (x)(∃y)(Fx → Gy)
    (∃y)(x)(Fx → Gy)

    Don't bother proving that the second entails the first; there's no challenge there. Instead, prove that the first entails the second.
    (b) On page 281 we also asserted that F14 represents S14. Vindicate this contention by proving that F14 and F14A (S14's literal symbolization) are equivalent.

    (F14)   (∃y)(x)[Ax → (Jy & Hyx)]
    (S14)   If there are any antitrust cases, then there is a federal judge who hears them all.
    (F14A) (∃x)Ax → (∃y)[Jy & (x)(Ax → Hyx)]

11. (SUPER-CHALLENGING) S1 is correctly symbolized by F1. Is F2 (which differs from F1 only in the scope of the second quantifier) also correct? Find out by determining whether these two formulas are logically equivalent.

    (S1) The INDICTED students are ELIGIBLE to enter a pretrial diversion program iff they REPAY$_R$ all the MONEY they obtained FRAUDULENTLY$_R$.

(F1) (x)(Ix → {Ex ↔ (y)[(My & Fxy) → Rxy]})

(F2) (x)(y)(Ix → {Ex ↔ [(My & Fxy) → Rxy]})

(Ix = x is an indicted student, Fxy = x obtains y fraudulently)

12. (SUPER-CHALLENGING) Symbolize and evaluate this argument:[28]

> **Any STATE that PERMITS_R someone to run a GAME_R of a particular kind MUST permit any INDIAN tribe to run a game of that kind without any LIMIT. <u>Florida</u> is a state and it permits some people to run <u>poker</u> games. The <u>Seminole</u> Tribe is an Indian tribe. Therefore, Florida must permit the Seminole Indian Tribe to run a poker game without limit.**

(Gxyz = x is a game of kind y run by z, Pxy = x does permit y, Mxy = x must permit y, Lx = there are betting limits on x)

§5.5 EXISTENTIAL IMPORT

Consider this argument:

We have no effective gun-control legislation.   −(∃x)Gx
Therefore, any effective gun-control legislation   ⊢ (x)(Gx → Ix)
interferes with the liberties of American citizens.

(Gx = x is effective gun-control legislation, Ix = x interferes with the liberties of American citizens) The English argument seems very poor indeed, yet its symbolization is demonstrably valid:

(1)	−(∃x)Gx	A
(2)	−(x)(Gx → Ix)	PA
(3)	(x)−Gx	1 QE
(4)	(∃x)−(Gx → Ix)	2 QE
(5)	−(Ga → Ia)	4 EO
(6)	Ga & −Ia	5 AR
(7)	Ga	6 &O
(8)	−Ga	3 UO
(9)	Ga & −Ga	7,8 &I
(10)	(x)(Gx → Ix)	2-9 −O

[28]See "Seminoles May Offer Poker Games," *Miami Herald* (August 7, 1993), p. 5B.

The situation grows more curious, for the following argument also has a valid symbolization:

> **We have no effective gun-control legislation.** $-(\exists x)Gx$
> **Therefore, any effective gun-control legislation** $\vdash (x)(Gx \rightarrow Sx)$
> **strengthens the liberties of American citizens.**

(Sx = x strengthens the liberties of American citizens) We have before us two (seemingly) poor English arguments with the same premise and (apparently) conflicting conclusions and their symbolizations are both valid; this is a paradox. The situation is reminiscent of the so-called paradox of material implication discussed in Section 5.3. Indeed that paradox is lurking in the shadows; the derivation of line 7 (in two steps) from line 5 in the proof above amounts to one form of that paradox $(-G \vdash G \rightarrow I)$ "stood on its head."[29]

$$
\begin{align}
(5) \quad & -(Ga \rightarrow Ia) \quad && 4 \text{ EO} \\
& \ldots\ldots \\
(7) \quad & Ga \quad && 6 \text{ \&O}
\end{align}
$$

However, something more than the paradox of material implication is at work here and that something involves the notion of "existential import."

We define this technical term as follows:

> **A general subject-predicate sentence has *existential import* iff it entails that its subject class has membership.**

The subject class is simply the class corresponding to the sentence's grammatical subject. Here are some examples:

(S1) Some Lutherans are Republicans.
(S2) No Lutherans are Jews.

In standard predicate logic those sentences are understood to mean exactly the following:

(S1A) There exists at least one Lutheran who is a Republican.
(S2A) There does not exist a Lutheran who is a Jew.

---

[29] '−A' entails 'B' iff '−B' entails 'A'.

Quite clearly statement S1A, but not statement S2A, entails S3.

> (S3) There exists at least one Lutheran.

Therefore, S1A has existential import and S2A lacks it. As S1 is understood in standard logic to mean the same as S1A, it is held to possess existential import; since S2 is equated with S2A, S2 is regarded as lacking existential import.

The ascription of existential import to S1 and the denial of it to S2 has some intuitive support. Imagine a world in which (due to plague or massive changes in religious affiliations) there are no Lutherans. In that situation we would regard S1B as false and S2B as true.

> (S1B) Some Republicans are Lutherans.
> (S2B) No Jews are Lutherans.

Surely S1B and S1 make the same claim and so do S2B and S2. So, in such a world we take S1 to be false and S2 true. That we take S2 to be true in that instance proves that we do not ascribe existential import to it. That we take S1 to be false in that hypothetical world indicates that we do ascribe existential import to it.

Many people find counterintuitive the logician's view that universal statements lack existential import, yet there are universal statements with empty subject classes we readily accept as true. 'Every moving body subject to no outside forces continues to move in a straight line' is an example.

Note that a statement that lacks existential import can never entail a statement that has it. That means, for example, that as logicians interpret statements S4 and S5, S4 does not entail S5.

> (S4) All Lutherans are Protestants.
> (S5) Some Lutherans are Protestants.
> (S6) There are Lutherans.

This argument shows that S4 does not entail S5:

If one statement entails a second, then it must also entail any statement the second entails. S5 entails S6. S4 does not entail S6. Thus, S4 does not entail S5.

The same point can be made more informally. The conclusion of a valid argument cannot go beyond what is contained in the premises. S5 does go beyond S4 since it entails the existence of Lutherans. So, the argument with S4 as premise and S5 as conclusion is not valid.

It may be instructive to attempt a proof of validity for this invalid argument.

All Lutherans are Protestants. (x)(Lx → Px)
So, some Lutherans are Protestants. ⊦ (∃x)(Lx & Px)

Plainly, a proof deriving the conclusion by EI will not be possible because the premise for the EI step will be a conjunction ('La & Pa', for example) and you cannot derive a conjunction from a conditional (such as 'La → Pa'). A proof employing the Dash Out Strategy fares no better.

We can transform the argument into a valid one by adding a simple existential claim: "There are Lutherans." A proof of validity for the augmented argument is easily constructed.

All Lutherans are Protestants. (x)(Lx → Px)
There are Lutherans. (∃x)Lx
So, some Lutherans are Protestants. ⊦ (∃x)(Lx & Px)

(1)	(x)(Lx → Px)	A	
(2)	(∃x)Lx	A	{*existence assumption*}
(3)	La	2 EO	
(4)	La → Pa	1 UO	
(5)	Pa	4,3 →O	
(6)	La & Pa	3,5 &I	
(7)	(∃x)(Lx & Px)	6 EI	

That standard logic views S4 as not having existential import is clear from the symbolization it provides. The formula '(x)(Lx → Px)' means "for any individual, *if* it is a Lutheran then it is a Protestant." We say that standard logic gives a "hypothetical" interpretation rather than a "categorical" interpretation to universal statements.

Any formula that is the existential quantification of a conjunction will have existential import and any one that is the universal quantification of a conditional will lack existential import. If the conclusion of an argument has existential import and all of the premises lack it, the argument will be invalid. Here is an example:

No insular possessions of the United States are represented in the Senate.
All insular possessions of the United States are places subject to American law.
Hence, some places subject to American law are not represented in the Senate.

We can transform this argument into a valid one by adding this existential assumption: "There are insular possessions of the United States."

Not only are all universal statements viewed by logicians as lacking existential import, the truth of any such statement is entailed by any statement claiming emptiness of the subject class. Why this should be so is clear from the following proof sketch.

(S7) There are no ghosts.	$-(\exists x)Gx$	is equivalent to:
	$(x)-Gx$	which entails:
	$(x)(-Gx \lor Mx)$	which is equivalent to:
(S8) All ghosts are mathematicians.	$(x)(Gx \rightarrow Mx)$	

It ceases to be a surprise that S7 entails S8 when you realize that S8 may be restated as follows:

There are no ghosts who $-(\exists x)(Gx \,\&\, -Mx)$
fail to be mathematicians.

We call a universal statement whose subject class is empty "vacuously true."

It would be a mistake to conclude that the entailment of S8 by S7 is nothing more than a paradoxical byproduct of a symbolic mechanism. The following argument establishes the same point without reliance on symbolization:

Consider these four statements:

(S4) All Lutherans are Protestants.
(S9) Some Lutherans are not Protestants.
(S6) There are Lutherans.
(S10) There are no Lutherans.

The first two statements in this list are exact opposites of one another and so are the last two statements in the list. If one statement entails a second, then the opposite of the second entails the opposite of the first. S9 entails S6. So, S10 entails S4. Q.E.D.

We are now in a position to explain the paradoxical argument displayed at the beginning of the section. If the conclusion of that argument is interpreted hypothetically (as logicians understand it), as meaning that there exists no effective gun-control legislation that does not interfere with the liberties of American citizens, then it clearly follows from its premise, which claims that the subject class of the conclusion is empty. If, on the other hand, the conclusion is interpreted categorically, the symbolization provided is not an adequate representation of the argument. In either case the paradox is resolved.

But the resolution leaves us with an important question. How can we make arguments about effective gun-control legislation if everything favorable or unfavorable that we can say about it is true until we enact it?[30] The answer is that we can use our logic in hypothetical worlds as well as in the real world.[31] Consider these two statements:

(S11) All unicorns have horns in their foreheads.
(S12) All clergymen of Barchester have horns in their foreheads.

Unicorns are mythical horses with horns in their foreheads. Barchester is a fictitious cathedral city that figures in Trollope's novels. In the real world, there are no unicorns, so S11 is true if interpreted hypothetically. Equally, in the real world, there is no diocese of Barchester, and therefore there are no clergymen of that diocese. It follows that in the real world, S12 (interpreted hypothetically) is just as true as S11.

On the other hand, in the world of mythological literature, there are unicorns with horns in their foreheads. And in the world of Trollope's fiction, there is also a diocese of Barchester with clergymen who look like any other Victorian clergymen. It follows that in the world of mythological literature, S11 is true (even when interpreted categorically), while in the realm of Trollope's fiction, S12 is false (because in that realm there exist Barchester clergymen without horns). In the same way, there can be a world of possible legislation in which statements about this or that law will be either true or false. It is this world that we need to enter into when we argue about effective gun-control legislation—or any other legislation that we are thinking about enacting.

So far, we have been discussing the existential import of *general* subject-predicate statements, both particular and universal. We have said that such a statement has existential import iff it entails the existence of some member of the subject class, and we have shown why, in standard logic, particular statements have such import and universal statements do not. It remains for us to deal with *singular* subject-predicate statements (sentences like 'Bill Clinton is a Democrat'). For such statements, we can define "existential import" as follows:

A singular subject-predicate statement has *existential import* iff it entails the existence of the individual named by its grammatical subject.

---

[30]Everything we can say about it by means of universal statements, that is. Everything we can say about it by means of existential statements is, of course, false until we enact it.

[31]Or, to put it more technically, we can adopt as our universe of discourse the class of objects existing in some world other than the real world.

Standard predicate logic accords existential import to all singular statements. For example, since in standard logic F13 entails F14, this logic adopts the stance that S13 entails S14.[32]

(S13) Bill <u>Clinton</u> is a DEMOCRAT. (F13) Dc
(S14) Bill Clinton exists. (F14) $(\exists x)(x = c)$

In fact, according to standard logic, *every* formula entails F14 because F14 is logically true, as this proof shows.

[ ] (1) $c = c$ =I
[ ] (2) $(\exists x)(x = c)$ 1 EI

The empty brackets in the assumption-dependence column call attention to the fact that each line in the proof is free of assumptions, a sign of logical truth.

We have a paradox here because S14, which is obviously not a logical truth, has a logically true symbolization. The paradox deepens when we consider S15.

(S15) <u>Santa</u> Claus does not exist. (F15) $-(\exists x)(x = s)$

If S15 is true, then its apparent symbolization, F15, is not even a legitimate formula. And if S15 is false, F15 is legitimate, but logically contradictory. This peculiar result demands further explanation.

We should bear in mind that standard logic makes two (and only two[33]) existential presuppositions:

(1) The universe of discourse chosen has at least one member.

(2) Every individual constant (that is, lower-case letter) employed refers to some member of the universe of discourse.[34]

---

[32] In a famous exchange with Bertrand Russell, P. F. Strawson argued that singular statements do *not* have existential import. His view, put simply, is that the relationship between S13 and S14 is not "entailment," but the weaker relation of "presupposition." Saying that S13 presupposes S14 is claiming that S13 can possess a truth value only if S14 is true. See Strawson's "On Referring," *Mind*, LIX (1950), 320–44.

[33] Note in particular that standard logic does not presuppose that every predicate employed applies to some member of the universe of discourse.

[34] There are nonstandard logics (known as "free logics") which do not make these existential presuppositions.

(Both of these presuppositions were noted in Section 3.4 in the discussion of the method of interpretation.) The first of these two presuppositions is revealed in the logical truth of F16 and F17, and the second in the logical truth of F14 and F18.

(F16) (∃x)(Fx v −Fx)
(F17) (∃x)(x = x)

(F14) (∃x)(x = c)
(F18) Dc → (∃x)Dx

If we adopt as our universe of discourse "earth residents," then a logician who embraces standard logic and denies Santa's existence must hold that F15 is not a legitimate formula. Does this stricture prevent the logician from applying standard logic to reasoning about nonexistent individuals? For instance, are we unable to assess the following argument using the system of logic presented in this book?

**Santa likes everyone.
So, someone likes Santa.**

Two avenues are available to us. The first is the approach taken above with arguments about effective gun-control legislation. We can place ourselves in a world (such as the world of Christmas legends) where Santa is one of the inhabitants.

The second avenue is to adopt a device introduced in Section 3.3:[35] we may absorb the name 'Santa' into a predicate ('Sx = x is [identical to] Santa') and then treat singular statements about Santa as universal generalizations. Using this device the argument displayed above may be symbolized:

(x)[Sx → (y)Lxy] ⊢ (x)[Sx → (∃y)Lyx]

This device also overcomes the problem mentioned above that the contingent S14 has a logically true symbolization. F14' is suitably contingent.

(S14)   Bill Clinton exists.
(F14') (∃x)Cx

(Cx = x is Bill Clinton)

---

[35]This device was described by W. V. Quine in "On What There Is," *Review of Metaphysics*, 2 (1948).

EXERCISES

Instructions for exercises 1 through 4: Each of the following arguments fails to be valid only because its premise set lacks existential import while its conclusion has such import. (a) Establish invalidity by interpretation. (b) Supply a single, one-predicate existential premise that will transform the argument into a valid one. (c) Establish the validity of the augmented argument by proof or propositional analogue.

1. **All motorcycle HELMET laws are PATERNALISTIC yet JUSTIFIABLE. Hence, not all paternalistic laws are unjustifiable.**

   (Universe: laws)

*2. Ornithologist Knut Schmidt-Nielsen:[36]

   The best argument against considering . . . [pneumatized bones] as being necessary for flight is provided by bats. They . . . do not . . . have pneumatized bones, and yet they are excellent fliers.

   His argument:

   No BATS have PNEUMATIZED bones. All bats are FLIERS. This shows that pneumatized bones are not necessary for flight.

   (Universe: vertebrates)

3. An argument considered by Bertrand Russell:[37]

   All HUMANS are EVIL. Nothing evil is PART of God. Thus, the thesis [of pantheism] that everything is part of God is false.

4. A philosopher visiting the University of Miami employed this argument in a lecture:

   All PHYSICAL events can be explained in purely physical TERMS. No event of EXTRA-sensory perception is explicable in purely physical terms. Every ESP event is a MENTAL event. This proves that some mental events are not physical events.

---

[36] "How Birds Breathe," *Scientific American* (December, 1971), p. 74.
[37] *Religion and Science* (New York: Henry Holt and Company, Inc., 1935), pp. 193–94.

(Universe: events; Tx = x can be explained in purely physical terms)
The crucial existential assumption is controversial, to say the least.

5. We offered above two symbolizations for this argument:

> **Santa LIKES$_R$ everyone.**
> **So, someone likes Santa.**
>
> (a) (x)Lsx ⊢ (∃x)Lxs
> (b) (x)[Sx → (y)Lxy] ⊢ (x)[Sx → (∃y)Lyx]

(Universe: characters in the Santa legend; s = Santa, Sx = x is Santa) Establish the validity of each symbolization.

*6. Exercise 6 in Section 5.2 concerned these instructions: 'Use any coin combination (totalling 50¢). Do not use pennies.' (a) Symbolize these statements using "coin combinations" as the universe and this dictionary: Tx = x totals 50¢, Px = x contains pennies, Ux = x may be used. (b) Prove that the two formulas are not logically contradictory by giving an interpretation that makes them both true. (c) Supply a true existential statement which makes the set of formulas contradictory. (d) Derive a standard contradiction from the augmented set of formulas.

7. (SEMI-CHALLENGING) We claimed on page 292 that in our system of logic each of the following is a logical truth. Back this claim up by deriving each formula free of assumptions.

> (F16) (∃x)(Fx v −Fx)
> (F17) (∃x)(x = x)
> (F18) Dc → (∃x)Dx

Can you do the proof of F16 in four (or fewer) lines? The proof of F17 in two lines?

8. (CHALLENGING) In Section 5.2 we pointed out that 'All events are not caused' can mean either S14 or S15. Obviously S15 does not entail S14. Does S14 entail S15?

> (S14) No events are caused.
> (S15) Not all events are caused.

We can symbolize these statements as F14A and F15A or, using "events" as a universe, as F14B and F15B.

> (F14A) (x)(Ex → −Cx)

(F15A) $-(x)(Ex \to Cx)$

(F14B) $(x)-Cx$
(F15B) $-(x)Cx$

(a) Use a proof or an interpretation to determine whether F14A entails F15A. (b) Do the same for F14B and F15B. (c) If you get discrepant results, can you account for them?

9. (CHALLENGING) F.R.C.P. 8(d) provides:

> *Averments in a pleading to which a responsive pleading is required . . . are admitted when not denied in the responsive pleading.*

Amanda, an associate in your law firm, has served a complaint on behalf of a client against the Amalgamated Widget Company. Rule 12(a) requires the defendant to file an answer within 20 days after the service of the complaint. The 20 days have now passed, and no answer has been filed. Amanda knows she could proceed to request a default judgment under Rule 55, but for various reasons she would prefer to seek a summary judgment under Rule 56. So she argues as follows:

> **None of the facts alleged in the complaint have been denied in the responsive pleading. Any fact alleged in the complaint that is not denied in the responsive pleading is admitted. If all the facts alleged in the complaint are admitted, my client is entitled to a summary judgment [under Rule 56]. Therefore, my client is entitled to a summary judgment.**

Explain what is wrong with Amanda's reasoning. Hint: focus on the first premise.

6

Intensional Contexts

The case of *Commonwealth* v. *Duchnicz*, 42 Pa. Co. 651 (1914), concerns a man (Duchnicz) who had sexual intercourse with a woman (call her "Jane Doe") who believed he was her husband (call him "John Doe") and who did not discover his identity until he was leaving her bedroom.[1] The court, using the traditional definition of 'rape' as "having intercourse with a woman without her consent," held that no rape had occurred because Jane Doe consented to the intercourse. It made no difference that the "consent" was rooted in mistaken identity.

In times of overcrowded living conditions and bad lighting, cases like *Duchnicz* came up more often than we might expect.[2] In most of them the perpetrator got off because the court could not find a way around the woman's consent. In one Irish case, *R.* v. *Dee*, 15 Cox C.C. 579 (1884), the court found the defendant guilty by holding that the act of marital intercourse is an entirely different act from that of nonmarital intercourse.[3] No other court

---

[1] For a review of rape cases involving impersonation see Anno., 91 A.L.R.2d 591, §§5,8.

[2] Even now, such a case may occasionally happen. See "37-year-old Mistakes Attacker for Boyfriend," *South Bend Tribune* (September 18, 1994), p. A18.

[3] This rationale would not have worked in the case referred to in note 2, where the victim mistook the intruder for her boyfriend.

seems to have taken that approach. It seems obvious that there is a sense in which Jane Doe gave her consent and another sense in which she did not, but how are we to elucidate these senses? This chapter provides the conceptual tools needed to clarify this and related legal conundrums. We will revisit this issue when we have grasped those tools.

§6.1 EXTENSIONAL AND INTENSIONAL CONTEXTS

Consider this argument (which we may call "Smith's Beliefs"):

Smith believes that all Lutherans are Protestants.
Jones is a Lutheran.
So, Smith believes that Jones is a Protestant.

It is clearly possible for the premises of this argument to be true and the conclusion false (Smith may not know Jones or he may mistakenly think that Jones is Catholic); so the argument is invalid. What is puzzling is that we seem to be able to provide a valid symbolization for that invalid argument:

$(x)(Lx \rightarrow Bx)$
Lj
$\vdash Bj$

(Lx = x is a Lutheran, Bx = Smith believes that x is a Protestant, j = Jones) An invalid English argument cannot have a valid symbolization; clearly something has gone wrong.

In order to explain the problem we need to distinguish between the *meaning* (or sense) of an expression and its *reference* (that to which it refers). For example, the meaning of the words 'parking meter' is something like "coin-operated device that registers the purchase of parking time for a motor vehicle" and the reference of the phrase is the class of all such devices. Logicians call the meaning of an expression its *intension* (spelled with an 's') and they call the reference of the expression its *extension*.

Note some ways in which the concepts "intension" and "extension" differ. If you destroy a parking meter you have decreased the size of the class of parking meters and have therefore altered the extension of the phrase 'parking meter', but of course you have not altered its intension (meaning). Adding a further specification to a locution typically augments its intension and decreases its extension; for instance, if you add the term 'broken' to 'parking meter' you inject a new element into the meaning of the phrase but decrease the class which is the phrase's extension. If two locutions have the same intension, then they must also have the same extension. On the other hand, it is quite possible for two locutions with the same extension to have different intensions, for example, 'Hillary Rodham Clinton's husband' and 'the man elected President of the United States in 1992'.

The extension of a general expression (for example, 'parking meter') is a class, and the extension of a singular term ('Bill Clinton', 'the present President of the United States') is the individual to whom the term refers. Let's call general expressions and singular terms "substantive" terms.

A substantive term embedded in a sentence occurs in an *extensional context* when you *cannot* change the truth-value of the sentence by replacing that substantive with another substantive having the same extension. Here is an example:

(S1) 50 is greater than 48.

The terms '50' and 'the number of states in the U.S.A.' have the same extension since they both refer to the same number (the immediate successor of 49). While the terms have the same extension, they have different intensions (meanings). Even though the two terms '50' and 'the number of states in the U.S.A.' differ in intension, you can interchange them in S1 without changing the truth-value of that sentence. In fact, S1 will remain true under all extensionally equivalent substitutions for '50' (for example, 'the number of members of the Indiana state senate'). We summarize this fact by saying that S1 provides an extensional context for '50'.

A substantive embedded in a sentence occurs in a nonextensional or *intensional context* when you *can* change the truth-value of the sentence by replacing that substantive with another substantive having the same extension. Consider this example:

(S2) 50 is necessarily greater than 48.

(S3) The number of states in the U.S.A. is necessarily greater than 48.

While S2 is true, S3 is not. (It is a contingent matter of fact, rather than a matter of necessity, that the number of states in the U.S.A. exceeds 48.) The false statement S3 results from the true S2 by the interchange of extensionally equivalent terms. So S2 provides an intensional context for '50'.

The standard logic presented in this book is an extensional (not an intensional) logic; its rules permit the replacement of one expression by another having the same extension without regard to their intensions.[4] Care must be taken when this logic is applied to sentences involving intensional

---

[4]Some logicians propose to cope with problems such as that presented by "Smith's Beliefs" by developing *intensional logics*, that is, logics that permit the substitution of terms within intensional contexts. See, for example, Edward N. Zalta, *Intensional Logic and the Metaphysics of Intensionality* (Cambridge, MA: MIT Press, 1988). We take a different approach; we offer strategies for treating these problems within standard extensional logic.

contexts. In particular, since the truth-value of a statement can be changed by replacing a term that appears in an intensional context with another term having the same extension, we must be careful not to symbolize any such statement by a formula in which the rules of predicate logic (UO, EO, =O, and EI) will permit such a substitution. How can we avoid doing this? The best way is to watch how we assign meanings to our predicate letters: these are the building blocks of our formulas. We should not define a predicate in such a way that an individual constant or variable attached to it occurs in an intensional context. Otherwise, we will be in danger of creating formulas that can be falsified by legitimate substitutions.

That is what went wrong with our symbolization of the argument we have called "Smith's Beliefs." We assigned a meaning to the predicate 'B' with this dictionary entry: Bx = Smith believes that x is a Protestant. We can show that that interpretation creates a problem by considering the conclusion of "Smith's Beliefs":

(S4) Smith believes that Jones is a Protestant.

Suppose we are sure that S4 is true. Smith knows that Jones is a Protestant because he sees him in church every Sunday. But what Smith doesn't know (nor does anybody else except Jones himself) is that Jones is the thief who stole the Ruritanian Crown Jewels. Then 'Jones' and 'the thief who stole the Ruritanian Crown Jewels' have the same extension. But it is not the case that Smith believes that the thief who stole the Ruritanian Crown Jewels is a Protestant; he has no idea who the thief is. It follows that replacing 'Jones' in S4 with another term having the same extension can change the truth-value of S4. Therefore, 'Jones' is in an intensional context in S4, and 'x' is in an intensional context in the meaning assigned 'Bx' in our dictionary. It is because we used this impermissible assignment of meaning in our symbolization of the first premise and the conclusion of "Smith's Beliefs" that we were led to provide a valid symbolization for that invalid argument.

Can the first premise of that argument be symbolized at all in standard logic? Yes, it can be symbolized in any of the following ways:

Propositional symbolization:
(F5A) **B**   (B = Smith believes that all Lutherans are Protestants)

Property-predicate symbolization:
(F5B) **Bs**   (Bx = x believes that all Lutherans are Protestants, s = Smith)

Relational-predicate symbolization:
(F5C) **Bsp**   (Bxy = x believes y, p = the proposition that all Lutherans are Protestants)

Symbolization F5A presents no problem because it involves no variable. (More generally, since no symbolizations within propositional logic involve variables, intensional terms may be used freely in that branch of logic.) F5B is also not problematic, since the variable in the interpretation given to the predicate 'B' lies outside of the intensional context created by 'believes that'. In short, in the incomplete sentence 'x believes that y', 'x' occurs within an *ex*tensional context while 'y' occurs inside an *in*tensional one. Consider again the first premise of "Smith's Beliefs":

(S5) Smith believes that all Lutherans are Protestants.

The extension of the name 'Smith' is Smith, so if you substitute an extensionally equivalent term for 'Smith' (say, 'Mrs. Smith's husband' or 'the owner of Smith's garage'), the new term will also denote that same individual. S5 and the statement resulting from the substitution will have the same truth-value. F5C presents a third acceptable symbolization of S5. If we view "belief" as a relation that holds between a person and a "proposition,"[5] we can shift the intensional element in S5 into the singular term 'the proposition that all Lutherans are Protestants'. These three acceptable symbolizations have this feature in common: they encapsulate the intensional element present in S5 in some expression represented by a single symbol (a statement letter in F5A, a predicate letter in F5B, and an individual constant in F5C). This encapsulation will prevent the illegitimate substitution of substantives occurring in an intensional context.

We can think of S5 as a *pseudo-universal*. Although it contains the word 'all' as universal statements often do, we have just shown that it cannot be symbolized as a universal formula. There are also *pseudo-existentials*—statements that look as if they could be symbolized by formulas using existential quantifiers, but which, in fact, cannot be. Here is one:

(S6) Harvey needs a pair of shoes.

'Needs', at least in this case, creates an intensional context. Suppose, for instance, that all shoes have wingtips. The following statement could still be false:

(S6A) Harvey needs a pair of shoes with wingtips.

---

[5]In earlier chapters we treated the terms 'proposition' and 'statement' as interchangeable. A proposition so defined is (roughly) a declarative *sentence*. But a Russian and a Brazilian can have the same belief or make the same claim (that smoking causes cancer, for example) and express that belief or claim with different sentences (in different languages). In this chapter we will use 'proposition' to refer to *that which is asserted* when a statement is uttered. The Brazilian and the Russian use two statements to express one proposition.

Accordingly, we cannot symbolize S6 with F6X.

(F6X) $(\exists x)(Sx \,\&\, Nx)$

(Sx = x is a pair of shoes, Nx = Harvey needs x) If that were a proper symbolization of S6, we could combine it with a formula symbolizing the statement that all shoes are wingtips

$(x)(Sx \rightarrow Wx)$

(Wx = x has wingtips) and derive a formula purportedly representing S6A, which we have said is not true:

$(\exists x)(Sx \,\&\, Nx)$
$(x)(Sx \rightarrow Wx)$
$\vdash (\exists x)(Sx \,\&\, Wx \,\&\, Nx)$

The reason we could come up with this valid symbolization for an invalid English argument is that we have assigned a meaning to the predicate letter 'N' that involves placing a variable in the intensional context created by 'Harvey needs'.

Another way to recognize a pseudo-existential is that it will be found on analysis not to have the same existential entailments as its proposed symbolization. Even if Harvey needs a pair of shoes, as S6 asserts, it is not the case that there is *a* pair of shoes that he needs—which is what F6X asserts. There may be a dozen shoe stores in town, and each one of them may have three or four pairs that will serve his purposes equally well. Furthermore, S6 could still be true if there were no shoes at all in the world, whereas F6X could not.

§6.2 RECOGNIZING INTENSIONAL CONTEXTS

We have shown by now that we must use a good deal of care in dealing with expressions that create intensional contexts. But how are we to recognize these troublesome expressions? Most of them contain terms that belong to one or another of the four families shown in the box on page 302, and may be recognized by their resemblance to the examples provided. For two reasons, more must be said. First, the terms in these lists do not always create intensional contexts, and second, there are terms (for example, 'needs' as just discussed) that create intensional contexts but do not belong to any of these four families.

MODAL TERMS	DEONTIC TERMS	PSYCHOLOGICAL AND EPISTEMIC TERMS	TERMS OF INDIRECT DISCOURSE
must	obligatory	believes	says
necessary	required	knows	alleges
possible	permitted	hopes	provides
impossible	forbidden	fears	testifies

We may shed more light on the matter by introducing the concept of a "statement operator." By a *statement operator* we mean an expression containing gaps which becomes a declarative sentence when those gaps are filled with other declarative sentences. We concentrate on the simplest case: statement operators with *one* gap. Here are some examples:

It is necessarily the case that . . .
The law requires that . . .
Jones believes that . . .
The D.A. alleges that . . .

Filling the gap in each of these operators with a sentence (such as 'Smith buys a Cadillac') produces a sentence. For instance:

It is necessarily the case that Smith buys a Cadillac.
The law requires that Smith buys (buy) a Cadillac.
Jones believes that Smith buys (is buying) a Cadillac.
The D.A. alleges that Smith buys (has bought) a Cadillac.

Statement operators generally create intensional contexts.[6] Therefore, when we symbolize a sentence containing a statement operator we should be on guard for intensional contexts.

The concept of "statement operator" helps to unite the four families of terms discussed above, for terms in all four groups may occur in statement operators. Often, where no statement operator is apparent, one seems to lie beneath the surface. For example, the first of the following pair of sentences does not overtly involve a statement operator, yet it is logically equivalent to the second sentence, which does.

---

[6]The statement operators (or connectives) studied in Chapter Two ('not', 'and', etc.) are exceptions; they do not create intensional contexts. That is why it was possible to create an extensional logic for them.

50 is necessarily greater than 48.
It is necessarily the case that 50 is greater than 48.

Both sentences create intensional contexts.

Some of the terms listed above (especially the psychological terms) may also occur in sentences that do not contain statement operators either explicitly or implicitly. Consider these sentences:

A	B
Smith knows that Brown is guilty.	Smith knows Brown.
Smith fears that Brown is guilty.	Smith fears Brown.
Smith hates (the fact) that Brown is guilty.	Smith hates Brown.

The sentences in column "A" contain statement operators and involve intensional contexts; the sentences in column "B" lack statement operators and involve extensional contexts only. Let us examine this phenomenon more closely by focussing on the term 'knows'. Consider the following two incomplete expressions containing this term:

Fred knows that . . .
Fred knows . . .

The first expression is a statement operator. We create sentences by filling its gap with other sentences, as the following examples illustrate:

Fred knows that Brown is a pharmacist.
Fred knows that the man who killed Fred's wife is a pharmacist.

Even if the two expressions 'Brown' and 'the man who killed Fred's wife' denote the same individual, the two sentences may have different truth-values. So the first expression (Fred knows that . . .) creates an intensional context.
 The gap in the second expression (Fred knows . . .) can be filled with a term that denotes an individual; when used in this sense the expression is a predicate rather than a statement operator. When the gap is filled with a singular term, a sentence results, for example:

Fred knows Brown.
Fred knows the man who killed Fred's wife.

If the two expressions 'Brown' and 'the man who killed Fred's wife' denote the same individual, then these two sentences must have the same truth-value. (Note that the second sentence does not claim that Fred knows the identity of his wife's murderer; it says only that her murderer is someone with whom he is acquainted.) The second expression creates an extensional context.

To reiterate, some expressions containing 'knows' create extensional contexts and others create intensional contexts. The latter involve statement operators. It should now be clear why S1 may be symbolized by F1, while S2 is not correctly symbolized by F2X.

(S1) Fred knows Brown.
(F1) **Kfb**

(Kxy = x knows y, f = Fred, b = Brown)

(S2) Fred knows that Brown is a pharmacist.
(F2X) **Kfb**

(Kxy = x knows that y is a pharmacist, f = Fred, b = Brown)

F2X is unacceptable because the meaning assigned to the predicate 'K' in the dictionary places the variable 'y' in the intensional context created by 'knows that'. This in turn allows the individual constant 'b' to occur in an intensional context in F2X. On the other hand, F2 does not involve that error and is an acceptable symbolization:

(F2) **Kf**

(Kx = x knows that Brown is a pharmacist, f = Fred)

Sometimes an English sentence is worded in such a way that you can't tell whether or not it creates an intensional context because you can't tell exactly what it is supposed to mean. Consider this statement:

(S3) The District Attorney intends to prosecute the person who robbed the National Bank.

Does S3 assert that a relationship ("intending to prosecute") holds between two people (the D.A. and the robber), or does it claim that a different relationship ("intending to make true") holds between a person (the D.A.) and a proposition (the proposition that *the D.A. prosecutes the person who robbed the National Bank*)? Is S3 about two people (connected by the relationship of intending to prosecute) or about one person and a proposition (connected by the relationship of intending to make true)? Is the verb 'intends' a fragment of a relational predicate (intends to prosecute), or is it part of a statement operator (intends to make it true that) like those discussed above? These two ways of viewing S3 may be phrased as follows:

> (S3A) The District Attorney intends-to-prosecute the person who robbed the National Bank.

> (S3B) The District Attorney intends-to-make-it-true-that the D.A. prosecutes the person who robbed the National Bank.

Now let us suppose that the bank robber happens to be the governor's only illegitimate son, a fact unknown to the D.A. Therefore, substituting the expression 'the governor's illegitimate son' for 'the person who robbed the National Bank' cannot change the truth-value of S3A, but making the same substitution can change the truth-value of S3B. (Prosecuting the governor's son may be no part of the D.A.'s intentions.) Of course, we summarize this by-now familiar situation by saying that the expression 'the person who robbed the National Bank' occurs in an extensional context in S3A and in an intensional context in S3B.

It will be useful to distinguish between two kinds of statements: *de re* (about the thing) and *de dicto* (about the statement).[7] A *de re* statement ascribes a property to an individual (other than another statement) or asserts that a relation holds between two or more individuals. Examples:

> Hillary Rodham Clinton is an attorney.
> Hillary Rodham Clinton is the spouse of Bill Clinton.

A *de dicto* statement ascribes a property to a statement (or proposition) or asserts that a relation holds between two or more statements (or propositions), for instance:

> The statement that Hillary Rodham Clinton is an attorney is true.
> It is true that Hillary Rodham Clinton is an attorney.

While only the first of these sentences makes explicit reference to a statement, it makes sense to view both as being about statements (since *truth* is a property of statements). Let's include hybrid statements that link an individual with a statement in the category of *de dicto* statements. With this terminology in hand, we can say that S3A is a *de re* statement (and gives a *de re* interpretation to S3), and that S3B is a *de dicto* statement (providing a *de dicto* reading of S3).

---

[7] This distinction derives from one drawn by medieval logicians between ascribing necessity (or possibility) to a *thing* (or to the way a thing possesses a property) and to a *proposition*.

§6.2 Recognizing Intensional Contexts

If we give the *de re* reading to S3 we can symbolize it with F3A; if we give it the *de dicto* reading we will use instead F3B.

(F3A) **Pdr** (Pxy = x intends-to-prosecute y, d = the D.A., r = the robber)

(F3B) **Idp** (Ixy = x intends-to-make-true y, d = the D.A., p = the proposition that the D.A. prosecutes the person who robbed the National Bank)

Of course, in some argumentative contexts nonrelational symbolizations of S3A and S3B will suffice:

(F3A') **Pd** (Px = x intends-to-prosecute-the-robber, d = the D.A.) [*de re*]

(F3A'') **Pr** (Px = the-D.A.-intends-to-prosecute x, r = the robber) [*de re*]

(F3B') **Pd** (Px = x intends-to-make-true the proposition that the D.A. prosecutes the person who robbed the National Bank, d = the D.A.) [*de dicto*]

Note that F3B' is *de dicto* because it links a person to a *proposition* (even though the expression designating the proposition is buried within a property predicate).

When considering *de dicto* interpretations, it is essential to bear in mind that a proposition always has a truth-value and so must be expressed by a complete self-contained locution. Only the first of these two bits of language succeeds in expressing a proposition:

The D.A. prosecutes the person who robbed the National Bank

x prosecutes the person who robbed the National Bank

The second phrase fails to express a proposition because it is incomplete. Some instantiations will be true and others false.[8] Accordingly, the phrase has no truth-value of its own. It follows that it would be wrong to interpret the predicate 'P' in F3B' above as:

Px = x intends-to-make-true the proposition that x prosecutes the person who robbed the National Bank

---

[8] It will be apparent that if a phrase is objectionable on this ground, an expression containing it will be objectionable because it places a variable in an intensional context.

When we give a *de dicto* interpretation of a legal rule we often have to state it in case-specific form, rather than as a general principle, in order to ensure that propositions are properly expressed by self-contained locutions.

Before we attempt to symbolize a sentence that can be understood either *de re* or *de dicto*, we must decide which interpretation best expresses the claim in question.

Statement S4 provides another example.

(S4) Green believes all tenured members of the Philosophy Department to have Ph.D.'s.

Do we understand S4 to assert that Green stands in a certain relation ("believing of an individual that he or she has a Ph.D.") to each member of a given class (tenured philosophers) or does it claim that Green stands in the relationship of belief to a certain proposition (that all tenured members of the Philosophy Department have Ph.D.'s)? Restated, do we give S4 a *de re* (S4A) or a *de dicto* (S4B) reading?

(S4A) Green believes of each and every individual who is a tenured member of the Philosophy Department that he or she has a Ph.D.

(S4B) Green believes (the proposition) that all tenured members of the Philosophy Department have Ph.D.'s.

Note that S4A and S4B are logically distinct statements. For instance, S4A—but not S4B—entails S5.

(S5) If Gray is a tenured member of the Philosophy Department, then Green believes that Gray has a Ph.D.

As we would expect, S4A and S4B receive distinct symbolizations:

(F4A) (x)(Px → Hgx) (Px = x is a member of the Philosophy Department, Hxy = x believes y to have a Ph.D., g = Green)

(F4B) Bga (Bxy = x believes y, g = Green, a = the proposition that all the tenured members of the Philosophy Department have Ph.D.'s)

How do we decide which reading (and therefore which symbolization) of S4 is more satisfactory? Naturally, the best course would be to ask the person who advanced S4 to clarify the claim. If that is impossible, one has to make a judgment about the nature of the claim being made. The context in which S4 was advanced, as well as the basis of Green's belief (if known to the person advancing S4), would be relevant information. For instance, if all the members of the department were friends of Green's and Green had heard each one refer to his or her Ph.D., then S4A would be the right reading. On the other hand, if Green's belief came from learning of the department's policy against tenuring anyone without the doctorate, then S4B would be right.

In the absence of any evidence, the conservative choice would be the *de dicto* reading, because it will block potentially invalid inferences such as the inference to S5, or the one drawn in the "Smith's Beliefs" argument.[9]

The key to solving the puzzle raised by the impersonation rape case discussed at the outset of the chapter is to view the consent (or lack thereof) involved as consent to make true a *proposition*:

(S6) Jane Doe consents to make true the proposition that Jane Doe has intercourse with John Doe.

(S7) Jane Doe consents to make true the proposition that Jane Doe has intercourse with Duchnicz.

S6 is true and S7 false. The definition of 'rape' employed in this case should also be interpreted *de dicto*. For the reason we set forth in the above discussion of *de dicto* statements, we consider not the general definition, but the result of applying that definition to the principals involved:

Duchnicz rapes Jane Doe iff he has intercourse with her, and she does not consent to make true the proposition that Jane Doe has intercourse with Duchnicz.

This statement taken together with the negation of S7 and the truth that Duchnicz and Jane Doe had intercourse entail that rape occurred.

Note that the following argument for S7, which might be advanced on Duchnicz's behalf, is defective.

---

[9]Some logicians and philosophers (one of us included) are suspicious of *de re* interpretations of sentences containing notions like "intends to prosecute" (or "believes to have a Ph.D."). They hold that in claiming that x intends to prosecute y we presuppose that y is described in a certain way (as "the person who robbed the National Bank" or as "the governor's only illegitimate son," for example). On this view there is no such relationship as "intending to prosecute" that holds between two people independently of language.

Perhaps the reason lawyers are more willing than philosophers to give *de re* interpretations to statements involving intentions or other mental states is that they are accustomed to treating such states as matters of "fact" for legal purposes. As Bowen, L.J., put it in a famous dictum: "The state of a man's mind is as much fact as the state of his digestion." *Edgerton* v. *Fitzmaurice*, 29 L.R.Ch.Div. 459, 483 (1884). Many philosophers would find this claim controversial, to say the least.

Jane Doe consented to make true the proposition that Jane Doe has intercourse with the man in her bed.
The man in Jane Doe's bed was Duchnicz.
Therefore, Jane Doe consented to make true the proposition that Jane Doe has intercourse with Duchnicz.

The argument appears to have a valid symbolization only by employing the predicate 'Jane Doe consents to make true the proposition that Jane Doe has intercourse with x'. The predicate is illicit because it attempts to express a proposition with an incomplete phrase ('the proposition that Jane Doe has intercourse with x').

EXERCISES

Each of the following arguments appears to be defective; the defect is rooted in intensional language. Edit the argument so that this problem is eliminated. In some cases, a substantial change in the content of the argument will be required.

1. You can learn from the dictionary that a person who collects butterflies is a lepidopterist. George collects butterflies. So, you can learn from the dictionary that George is a lepidopterist.

*2. The Marines are looking for a few good men. Sam, Charlie, and Willie are a few good men. So, the Marines are looking for Sam, Charlie, and Willie.

3. Samantha is afraid of mice. This animal is a mouse. So, Samantha is afraid of this animal.

4. Pat Buchanan hates Communists. Natasha is a Communist. So, Pat Buchanan hates Natasha.

5. Smith knows that Brown is the mailman. Brown is the lover of Smith's wife. So, Smith knows that the lover of Smith's wife is the mailman.

*6. O'Brien believes that all Irish children should learn Gaelic. Maureen is an Irish child. So, O'Brien believes that Maureen should learn Gaelic.

7. Brown plans to invest in the Ponzi Corporation. The Ponzi Corporation is a fraudulent business. So, Brown plans to invest in a fraudulent business.

8. Lois Lane knows Clark Kent to be a mild-mannered reporter. Clark Kent is Superman. So, Lois Lane knows Superman to be a mild-mannered reporter.

§6.3 EXAMPLES AND STRATEGIES

So far in this chapter, we have tried to show you how to recognize intensional contexts, and how to keep from getting in trouble with them when you do recognize them. We turn now to some examples of arguments that present these difficulties.

THE SHOES

Harvey needs a new pair of shoes.
Harvey cannot afford a new pair of shoes.
Therefore, Harvey needs something he cannot afford.

The first premise and the conclusion of this argument are pseudo-existentials. Since there is no particular pair of shoes uniquely capable of filling Harvey's need, it is not the case that there is a new pair of shoes such that Harvey needs just that pair. The second premise is a genuine universal. It is in fact the case that if anything is a new pair of shoes, Harvey can't afford it. To symbolize this apparently valid argument, we will have to restate both pseudo-existentials, translating them into the nearest genuine existentials we can find. Here is a restatement that will work:

Harvey has a need that will be filled only if Harvey has a new pair of shoes.
Every new pair of shoes is one that Harvey cannot afford.
Therefore, Harvey has a need that will be filled only if he has something that he cannot afford.

This can be given a valid symbolization:

(∃x){Nxh & [Fx → (∃y)(Sy & Hhy)]}
(x)(Sx → −Ahx)
⊢ (∃x){Nxh & [Fx → (∃y)(Hhy & −Ahy)]}

(Nxy = x is a need of y, Fx = x is filled, Sx = x is a new pair of shoes, Axy = x can afford y, Hxy = x has y, h = Harvey)

Notice that our strategy for turning the pseudo-existentials into genuine existentials has involved two steps. First, we recognized the statement operator underlying the term 'needs'. We changed 'Harvey needs a new pair of shoes' to 'Harvey needs for it to be the case that he has a new pair of shoes'. Then, we changed the statement operator to the antecedent of a conditional statement: ". . . needs for it to be the case that . . ." to ". . . has a need that will be filled only if . . . ."

THE PRISONER

The Geneva Convention forbids shooting prisoners.
Sam is a prisoner.
So, the Geneva Convention forbids shooting Sam.

Any student of international law knows that the first premise of this argument is true. Assume that the second premise is true. But how about the conclusion? If it concerns the legal effect of the Geneva Convention, then it is true and the argument is valid. But if the conclusion claims that the Convention contains language specifying that Sam not be shot, then the conclusion is obviously false. Under that interpretation, the argument would be invalid, since both premises would be true but the conclusion would be false.

The second, the invalidating, interpretation makes 'forbids' a term of indirect discourse rather than a description of legal effect. As such, the term creates an intensional context for the word 'prisoners' in the first premise because replacing that word with a list containing the names of all the prisoners currently in custody would change the truth-value of the premise. As a result, it makes that premise a pseudo-universal. It cannot be symbolized:

$(x)(Px \rightarrow Gx)$

(Px = x is a prisoner, Gx = the Geneva Convention contains language specifying that the shooting of x is forbidden) The meaning assigned 'G' in the dictionary creates an impermissible intensional context.

Often, this ambiguity between the "legal effect" and the "indirect discourse" interpretations will cause no problem; it will be obvious which is intended. If there is a problem, though, we should have a way of resolving it. There are a number of possibilities. Here is the one we like best:

Whoever shoots a prisoner violates the Geneva Convention.
Sam is a prisoner.
So, whoever shoots Sam violates the Geneva Convention.

$(x)(y)[(Px \ \& \ Syx) \rightarrow Vy], Ps \vdash (x)(Sxs \rightarrow Vx)$

(Universe: people, Px = x is a prisoner, Sxy = x shoots y, Vx = x violates the Geneva Convention, s = Sam)

In devising this reformulation, we have changed the possibly pseudo-universal first premise into a necessarily genuine universal. The key change was the replacement of 'forbids' by 'violates'. While 'the Geneva Convention forbids shooting x' can be interpreted as creating an intensional context, 'x violates the Geneva Convention' cannot.

THE NOTICE

This notice must be published in a newspaper of general circulation in the county.
The *Globe* and the *Argus* are the only papers of general circulation in the county.
Therefore, this notice must be published in the *Globe* or the *Argus*.

This argument appears to be valid; but before we pronounce on the matter we should compare it with another similar argument (which we can call "Crooks One"):

This notice must be published in a newspaper of general circulation in the county.
The only papers of general circulation in the county are owned by crooks.
Therefore, this notice must be published in a newspaper owned by crooks.

"Crooks One" seems dubious because (on a natural reading) it purports to show that there is some legal necessity about publishing the notice in a newspaper with crooked owners. By contrast, the following argument ("Crooks Two") seems straightforward:

This notice will be published in a newspaper of general circulation in the county.
The only papers of general circulation in the county are owned by crooks.
Therefore, this notice will be published in a newspaper owned by crooks.

The only difference between these arguments is that "Crooks Two" contains the term 'will' where "Crooks One" has 'must'. 'Must' creates an intensional context; 'will' does not.

One may be tempted to symbolize "The Notice" as follows:

($\exists$x)(Nx & Px)
(x)[Nx $\leftrightarrow$ (x = g v x = a)]
⊢ Pg v Pa

(Nx = x is a newspaper of general circulation in the county, Px = this notice must be published in x, g = the *Globe*, a = the *Argus*) The symbolized argument is definitely valid, but the interpretation assigned to the predicate 'P' locates a variable within the intensional context created by the words 'must be published in'. In addition to this difficulty, there is a second problem: the symbolized conclusion corresponds to S2, not S1.

(S1) This notice must be published in the *Globe* or the *Argus*.
(S2) Either this notice must be published in the *Globe* or it must be published in the *Argus*.

The conclusion of the English argument is S1 rather than S2. Furthermore, S2 is false (even if the premises of the argument are true). It is not the case that the notice must be published in the *Globe*, since it may be published in the *Argus* instead; hence, the left disjunct is false. Parallel reasoning shows that the right disjunct is also false. Since both disjuncts are false, S2 itself is false.

We can skirt both of these problems by dropping the concept of necessity from the meaning assigned the predicate 'P'; we can give instead this extensional interpretation to 'P':

Px = this notice will be published in x

This "patch" creates another problem: the symbolized conclusion now corresponds to S4 rather than S3.

(S3) This notice must be published in the *Globe* or the *Argus*.
(S4) This notice will be published in the *Globe* or the *Argus*.

S4 avers that the notice will appear. That seems to distort the purpose of the argument—which is not to bolster a prediction about publication, but to make clear a practical consequence of a legal requirement. Let us, therefore, amend our argument by referring to the rule (call it "R") that we have in mind. If we do that, we can change the deontic statements in our argument into conditional statements. Adding the phrase 'There is compliance with Rule R only if' to the first premise and the conclusion of "The Notice" (and replacing 'must be' by 'is') produces an argument that can be handled in our extensional logic.

There is compliance with Rule R only if this notice is published in a newspaper of general circulation in the county.
The _Globe_ and the _Argus_ are the only papers of general circulation in the county.
Therefore, there is compliance with Rule R only if this notice is published in the _Globe_ or the _Argus_.

$C \to (\exists x)(Nx \,\&\, Px)$
$(x)[Nx \leftrightarrow (x = g \lor x = a)]$
$\vdash C \to (Pg \lor Pa)$

(C = There is compliance with Rule R, Nx = x is a newspaper of general circulation in the county, Px = this notice is published in x) In this formulation no predicate is defined in such a way as to place a singular term in an intensional context. The validity of the symbolized argument assures us of the validity of the reformulated English argument.

What shall we say about the original English formulation of "The Notice"?—mainly that it is not stated clearly enough to be expressed in standard logic. When we say that the notice "must" be published, we are taking a lot of things for granted. We expect our clients to understand that we don't mean they will be shot if they don't publish it; nor do we mean that they will be acting immorally if they don't publish it. We mean that there is a rule they will not be complying with if they don't publish it, and that not complying with that rule will have consequences that they won't like. Before we can put our argument into logical form, we must make this meaning explicit. One of the reasons why it is useful for lawyers to study logic is that it makes them more careful to say exactly what they mean.

COUNTY SURVEYOR
Argument A

(S5) **Section 8501 of the Revised Statutes makes only licensed surveyors resident in the county eligible for the post of County Surveyor.**

(S6) **All the licensed surveyors resident in the county are white males.**

(S7) **Therefore, Section 8501 of the Revised Statutes makes only white males eligible for the post of County Surveyor.**

Argument B

(S7) **Section 8501 of the Revised Statutes makes only white males eligible for the post of County Surveyor.**

(S8) **A statute that makes only white males eligible for a post is unconstitutional.**

(S9) **Therefore, Section 8501 of the Revised Statutes is unconstitutional.**

The conclusion of Argument A recurs as the first premise of Argument B.

On a quick reading both arguments may *seem* valid. But a statute that imposes a legitimate nonracial qualification for a position is not rendered unconstitutional by the fact that there happen to be no members of a particular minority race who qualify. So we have to suspect that something has gone wrong in at least one of the two arguments. The suspicion is well-founded; the first problem with "County Surveyor" is one of equivocation. Sentences S5, S7, and S8 are all ambiguous; the source of the ambiguity is the term 'makes'. We can illustrate the ambiguity latent in that term by giving these two reformulations of S7:

Section 8501 of the Revised Statutes specifies that only white males are eligible for the post of County Surveyor.

All those who satisfy the eligibility criteria for the post of County Surveyor established in Section 8501 of the Revised Statutes are (as a matter of fact) white males.

Let's reword both arguments substituting clarified (and true) replacements for the ambiguous components. (Statement numbers of replacements are tagged with 'A's.)

Argument A'

(S5A) **Section 8501 of the Revised Statutes specifies that only licensed surveyors resident in the county are eligible for the post of County Surveyor.**
(S6) **All the licensed surveyors resident in the county are white males.**
(S7A) **Therefore, all those who satisfy the eligibility criteria for the post of County Surveyor established in Section 8501 of the Revised Statutes are (as a matter of fact) white males.**

Argument B'

(S7A) **All those who satisfy the eligibility criteria for the post of County Surveyor established in Section 8501 of the Revised Statutes are (as a matter of fact) white males.**
(S8A) **A statute that specifies that only white males are eligible for a post is unconstitutional.**
(S9) **Therefore, Section 8501 of the Revised Statutes is unconstitutional.**

Argument A' is obviously valid; it may be treated symbolically in this way:

$(x)(Cx \rightarrow Lx)$
$(x)(Lx \rightarrow Wx)$
$\vdash (x)(Cx \rightarrow Wx)$

(Cx = x satisfies the eligibility criteria for the post of County Surveyor established in Section 8501 of the Revised Statutes, Lx = x is a licensed surveyor resident in the county, Wx = x is a white male)

Argument B' is just as obviously invalid. There is no connection at all between premises S7A and S8A that will support the inference of S9 in argument B'. It would be an error to attempt to symbolize this argument using a predicate like 'specifies that x is eligible for post y' because 'specifies that' produces an intensional context. The following sentences could (and presumably would) differ in truth-value even if the classes "licensed surveyors" and "surveyors who are white males" are coextensive.

> Section 8501 specifies that only licensed surveyors are eligible for post P.
>
> Section 8501 specifies that only surveyors who are white males are eligible for post P.

We have to settle for the following symbolization of argument B':

> (x)(Cx → Wx)
> (x)(Sx → Ux)
> ⊢ Ua

(Cx = x satisfies the eligibility criteria for the post of County Surveyor established in Section 8501 of the Revised Statutes, Wx = x is a white male, Sx = x is a statute that specifies that only white males are eligible for a post, Ux = x is unconstitutional, a = Section 8501 of the Revised Statutes) Symbolized argument B' is, as we would expect, invalid.

EXERCISES

1. Consider this argument for mind-body dualism:

   > **I am absolutely certain that my <u>mind</u> exists.**
   > **I am not absolutely certain that my <u>body</u> exists.**
   > **So, my mind is not identical to my body.**

   Can this argument be properly symbolized as follows? Explain.

   > Cim, −Cib ⊢ m ≠ b

   (i = I, m = my mind, Cxy = x is absolutely certain that y exists, b = my body)

Follow the instructions in each of the following exercises in such a way as to avoid problems of intensionality in your symbolization.

*2. A simplified version of the mathematics graduation requirement for Arts and Sciences students at the University of Miami:

> Unless exempted by placement, all students must take MTH 101 and either MTH 102 or MTH 103.

Reformulate this academic regulation so that its symbolization avoids problems of intensionality. Symbolize the reformulation.

3. Grade school rule:

> Any student who has been absent must bring a note from a parent upon returning.

Reformulate this rule and symbolize the reformulation.

4. Symbolize the following argument:

> **The General Services Administration will furnish Judge Carberry with whatever she needs for her chambers. Judge Carberry needs a desk for her chambers. Therefore, the General Services Administration will furnish Judge Carberry with a desk.**

5. The National Labor Relations Act forbids an employer to fire an employee for union activities.[10] It is decided at a union meeting that Harris will express the employees' dissatisfaction with working conditions by throwing a custard pie at the boss. He does so and is fired. Symbolize his argument that the firing violates the Act, and the employer's argument that it does not.

6. (CHALLENGING) The famous *M'Naghten* rule is that a defendant is not guilty by reason of insanity if his mental condition keeps him from knowing that what he is doing is wrong. Lorenzo is on trial for strangling Gwendolyn. He knew it would be wrong to strangle Gwendolyn, but because of his mental condition he thought she was a chicken. Symbolize the arguments for and against finding him not guilty by reason of insanity. Hint: Treat each argument as containing *de dicto* statements.

---

[10]This rule is stated as a quantified conditional. But for reasons explained in Section 5.1 above, when it is the only ground offered in a claim for relief against the employer, the employer is entitled to treat it as a quantified biconditional.

§6.4 INTENT, RISK, AND INCHOATE CRIMES[11]

Several classic legal puzzles can be clarified by seeing which of the proposed solutions involves *de re* statements with terms in an extensional context, and which involves *de dicto* statements with terms in an intensional context. This kind of examination won't tell us which side is right, but it may help to clarify the arguments on both sides.

Let's begin with this dialogue, which is of course not legal, but which raises the same questions as the legal problems we have in mind:

>   Wife: *Your necktie has a spot on it.*
>   Husband: *Thanks, I will go and change it.*
>   Wife: *Why did you put on a necktie with a spot on it?*
>   Husband: *I didn't mean to.*
>   Wife: *How can you put on a necktie without meaning to?*

The wife here is accusing her husband of sartorial carelessness by suggesting that he intentionally put on a necktie with a spot. He is defending himself by denying that he did so. She is affirming and he is denying the same English sentence:

(S1) Husband intentionally put on a necktie with a spot.

But the meanings that they assign to the sentence are quite different. Hers can be expressed by a *de re* statement:

(S1A) Husband intended-the-putting-on-of a certain necktie, which necktie had a spot.

His calls for a *de dicto* rendition:

(S1B) Husband intended-it-to-be-the-case-that he put on a necktie with a spot.

She is, of course, right to affirm S1A, and he is right to deny S1B.

---

[11] Inchoate crimes are those such as attempt and conspiracy in which the criminal purpose of the actor does not have to be finally accomplished for the actor to be guilty. See Article 5 of the Model Penal Code.

The difference becomes apparent when we symbolize the two versions. Her version, S1A, comes out as:

(F1A) (∃x)(Nx & Phx & Sx)

(Nx = x is a necktie, Pxy = x intended the putting on of y, h = the husband, Sx = x has a spot) His version, S1B, is:

(F1B) Ihp

(Ixy = x intended to make true y, h = the husband, p = the proposition that husband put on a necktie with a spot) Note that F1X would not be a permissible symbolization of S1B:

(F1X) Phs

(Pxy = x intended it to be the case that husband put on y, h = the husband, s = a certain necktie with a spot) That interpretation puts the variable 'y' in an intensional context. Suppose that Husband has only one necktie with a spot, and that is his one paisley necktie. Then the terms 'the necktie with a spot' and 'the paisley necktie' will have the same extension. But substituting one for the other in the interpretation of F1X will change the truth-value, because Husband did intend it to be the case that he put on the paisley necktie, although he did not intend it to be the case that he put on the necktie with a spot.

Now, let's apply this kind of analysis to a famous case, *People* v. *Jaffe*, 185 N.Y. 497, 78 N.E. 169 (1906). In that case, a certain thief had been caught with stolen goods, and was enlisted by the police to help them catch the defendant, whom they believed to be a fence. Therefore, with the owner's permission, the thief retained the goods and offered them to the defendant as goods he had stolen. When the defendant attempted to buy them, he was charged with attempting to receive stolen goods. The court held that since the goods were being used with the permission of the owner, they were no longer stolen, so the defendant could not be found guilty. Had he succeeded in doing what he set out to do, he would not have committed any crime. Therefore, setting out to do it could not be attempting to commit a crime.

The completed crime of receiving stolen goods would have involved three elements: that the property was stolen, that the defendant was aware it was stolen, and that the defendant bought it. In symbols:

(∃x)(Sx & Adx & Bdx) → Rd

(Sx = x is stolen, Axy = x is aware of the stolen status of y, d = defendant, Bxy = x buys y, Rx = x is guilty of receiving stolen goods) The interpretation ascribed to the predicate 'A' must be *de re* to be legitimate; it cannot mean 'x is aware that y is stolen'. The attempt to receive stolen goods involved all the same elements except that it was enough for the defendant to *attempt to buy* the property instead of actually buying it. Note that this understanding of the law, as put forward by the defense and accepted by the court, also requires a *de re* interpretation. It can be symbolized as follows:

(F2) $(\exists x)(Sx \,\&\, Adx \,\&\, Pdx) \to Gd$

(Pxy = x attempts to purchase y (*or* x attempts-the-purchase-of y), Gx = x is guilty of attempting to receive stolen goods)

The prosecution's version of the law, rejected by the court, involves a *de dicto* interpretation:

(S3) Defendant is guilty of attempting to receive stolen goods if defendant attempts to make it be the case that there exists something stolen that defendant knows is stolen and that defendant attempts to purchase.

(F3) $Hdp \to Gd$[12]

(Hxy = x attempts to make true y, p = the proposition that there exists something stolen and defendant knows that it is stolen and defendant purchases it)

Of course, it takes more than logic to determine which of these interpretations of the law is the right one, but we suspect that the court was influenced by the fact that the prosecution's case could not be analyzed any further than F3. Courts are rightly reluctant to establish rules of criminal liability that cannot be broken down into elements capable of being stated in an extensional context.

The drafters of the Model Penal Code (§ 5.01) attempted to occupy a middle ground between F2 and F3. They provided that a person is guilty of an attempt if he

purposely engages in conduct which would constitute the crime if the attendant circumstances were as he believed them to be.

---

[12]The proposition involved in a *de dicto* statement can often be buried in a predicate instead of symbolized by a constant. For instance, S3 could be symbolized more simply as '$Hd \to Gd$'. (Hx = x attempts to make it be the case that there exists something stolen and defendant knows that it is stolen and defendant purchases it)

Under this language, the crime of attempting to receive stolen goods would have two elements, both capable of *de re* statement.[13] First, one must believe a certain object to be stolen property. Second, one must engage in a transaction which one believes to be the purchase of that object (in Jaffe's case, it wouldn't be a real purchase, because the person posing as the seller had no authority to sell). The rule of liability established by these two elements can be symbolized as follows:

(F4) $(\exists x)(Bdx \ \& \ Pdx) \rightarrow Gd$

(Bxy = x believes y to be stolen property, Pxy = x engages in a transaction which x believes to be the purchase of y)[14] Under this rule, Jaffe would have been guilty.

Another classic case of criminal liability for attempting the impossible is that of the pickpocket who is arrested putting a hand in an empty pocket. Such pickpockets have often been held guilty of attempted theft even though there was nothing that they attempted to steal. It is impossible in such a case to make a *de re* statement of the offense. There is no specific thing the pickpocket was trying to steal. Indeed, even if the pocket was not empty, we could not point to a specific thing the pickpocket was trying to steal unless he or she had an advance inventory of the contents of the pocket. On the other hand, a *de dicto* statement of the offense is perfectly possible. The pickpocket did, in fact, attempt to make it be the case that he or she stole something out of the victim's pocket. The rule could be symbolized as:

(F5) $Adp \rightarrow Gd$

(Axy = x attempts to make true y, p = the proposition that defendant steals the contents of the victim's pocket, Gx = x is guilty of attempted theft) Note that the following alternative symbolization

(F5X) $Adc \rightarrow Gd$

(Axy = x attempts to steal y, c = the contents of the victim's pocket) is impermissible for the same reason F1X is. Replacing 'the contents of the victim's pocket' by a list of the contents would change the truth-value of the statement.[15]

---

[13]This discussion disregards § 223.6 of the Code, which provides that a person who receives property "believing that it has probably been stolen" is guilty not of the attempt but of the offense itself.

[14]These two interpretations are typical of those that would be questioned by the skeptical logicians described in footnote 9 on page 309.

[15]Also, if the pocket was empty, the formula would be prohibited by the existential presuppositions of standard logic. See page 291.

F5, of course, is in the same form as F3, the rule rejected by the court in *Jaffe*. Evidently, the reason it is more persuasive in this case than in *Jaffe* is that there is no plausible alternative. Unless the rule is accepted in this form, unsuccessful pickpockets, or even unsuccessful burglars, will not be guilty of anything unless they know exactly what they want to steal before they set out to steal it. See *Clarke v. State*, 86 Tenn. 511, 8 S.W. 145 (1888).

One area of the law of torts presents problems with some similarity to these criminal law problems. That is the area of "result within the risk," part of the area of proximate cause. Here is a hypothetical case raising the problem in its classic form: Waldo gives Calvin, a six-year-old child, a live hand grenade to play with. Calvin accidentally drops it on his foot, breaking a toe. Under familiar rules of tort law, Waldo would clearly have been liable if the grenade had exploded and injured Calvin. His conduct created an unreasonable risk of harm to Calvin, and the harm he was risking would have been the one that occurred. But the risk of a broken toe from the hand grenade was no greater than the risk from a croquet ball, a toy truck, a large picture book, or a bottle of ginger ale.

The applicable rule is that if the defendant creates an unreasonable risk of harm to the plaintiff, and the harm occurs, the defendant is liable. The question is whether the harm that occurs must be the one risked. Defendant in the Hand Grenade case would argue for a *de re* interpretation of the rule:

> (S6) If defendant's conduct creates an unreasonable risk of the occurrence of a certain harm to plaintiff, and also causes the occurrence of that harm, defendant is liable.

That could be symbolized as follows:

> (F6) $(x)[(\exists y)(Hy \mathbin{\&} Rxy \mathbin{\&} Cxy) \rightarrow Lx]$

(Hx = x is a harm to plaintiff, Rxy = the conduct of x creates an unreasonable risk of the occurrence of y, Cx = the conduct of x causes the occurrence of y, Lx = x is liable)[16] Under the interpretation provided by S6 (and F6), defendant is not liable because there is no single harm that both befell the plaintiff and was unreasonably risked by defendant.

Meanwhile, the plaintiff in the Hand Grenade case would have to argue for a *de dicto* interpretation of the rule:

> (S7) If defendant's conduct creates-an-unreasonable-risk-of-its-being-the-case-that plaintiff is harmed, and also causes-it-to-be-the-case-that plaintiff is harmed, defendant is liable.

---

[16] We could have adopted a more elaborate symbolization in which the plaintiff is represented by a variable, but this simplified version will do for illustrative purposes.

The symbolization would be:

(F7) (x)[(Rxp & Cxp) → Lx]

(Rxy = the conduct of x creates an unreasonable risk of making y true, p = the proposition that plaintiff is harmed, Cx = the conduct of x makes y true, Lx = x is liable) As we did with F1B, F3, and F5, we used a constant to designate the proposition governed by the statement operator in S7. As in the other cases, we cannot unpack the proposition in question.

In the famous *Polemis* case, [1921] 3 K.B. 560 (C.A.), Lord Scrutton adopted a rule corresponding to S7, whereas the Privy Council in the *Wagon Mound* case, [1961] A.C. 388, adopted S6. The later authorities, both British and American, have tended to follow *Wagon Mound* rather than *Polemis*. But for our present purposes, what we need to note is not whether S6 is a better legal rule than S7, but the differences between them in logical form. As in the criminal cases, noting the difference won't show which is the better rule, but it may help us see what the people who argue for one rule or the other are arguing about.

EXERCISES

The following exercises concern "intention" and related concepts like "belief," "knowledge," and "declaration." You will be asked to elaborate alternative interpretations of statutes, definitions, and principles involving these concepts. The solutions to these exercises involve the distinction between de re *and* de dicto *interpretations.*

1. Indiana Code 35-44-3-4(a)(3) makes anyone guilty of a class D felony who

   > alters, damages, or removes any record, document, or thing, with intent to prevent it from being produced or used in any official proceeding or investigation.

   The Consolidated Kiddiecar Company has been involved in a series of transactions that might be regarded as violating state and federal antitrust laws, although the authorities responsible for enforcing these laws have not taken any notice. The CEO of Consolidated has a letter in her files that might be very embarrassing if it were to come to light in the course of an investigation. She therefore proposes to destroy it. The statute displayed above can be understood in two distinct ways: one *de re* and the other *de dicto*. (a) Illustrate this ambiguity by twice symbolizing the application of the law to the CEO's contemplated act. For the *de re* interpretation use these symbols:

> Ox = x is an official proceeding or investigation, Axyz = x destroys y with intent to prevent y from being produced or used in z, c = the CEO, l = the letter, Dx = x is guilty of a Class D felony

For the *de dicto* interpretation use these symbols:

> As above: c, l, Dx
> Bxy = x destroys y, Ixy = x intends to make true proposition y, a = the proposition that the letter will not be produced or used in any official proceeding or investigation

(b) Under which interpretation would the CEO's action be a felony?
(c) Why would it not be a felony under the other interpretation?

*2. The offense of larceny was defined at common law as the taking of another person's property with the intent permanently to deprive the owner of it. This definition can be understood in two ways (*de re* and *de dicto*).

Sam takes a car out of Joe's driveway that he erroneously believes to be his own. It really belongs to Joe. Joe would like to have Sam prosecuted, but the prosecutor explains that the courts don't impose liability in such cases as this. (a) Symbolize the two alternative versions of the common law definition as applied to this case: the one Joe would like the prosecutor to use, and the one the prosecutor believes (rightly) that the courts will apply. (b) Explain why Sam is not guilty under the latter version. Use these dictionaries:

> (*Joe's version*) Lx = x is guilty of larceny in the case under consideration, s = Sam, c = the car involved in this case, Bxy = x belongs to y, j = Joe, Txy = x takes y, Dxyz = x intends permanently to deprive y of z

> (*Prosecutor's version*) As above: Lx, s, Txy, Bxy, c
> Ixy = x intends to make true proposition y, a = the proposition that the owner of the car involved in this case will be permanently deprived of it

3. The case of *Regina v. Chapman* is described in exercise 6 of Section 4.2. You will recall that Chapman was found guilty of taking a girl under eighteen from her home with the intention that she should have unlawful sexual intercourse. He argued that the sexual intercourse he intended for her to have was not unlawful since there was no statute making it a crime.

The court rejected this argument. Chapman could have advanced a different argument:

Chapman violates the statute iff he takes the girl from her home and intends to have unlawful sexual intercourse with her. However, Chapman has no such intent. Therefore, he does not violate the statute.

The first premise of this argument can be given both *de dicto* and *de re* interpretations. (Chapman's argument requires the former.) Symbolize both interpretations of the first premise using these dictionaries:

(de dicto *interpretation*) Vx = x violates the statute, c = Chapman, Txy = x takes y from y's home, g = the girl, Ixy = x intends to make true proposition y, a = the proposition that Chapman has unlawful sexual intercourse with the girl

(de re *interpretation*) As above: Vx, c, Txy, g
Sxy = x intends to engage in sexual intercourse with y, Uxy = sexual intercourse between x and y is unlawful

4. In *Screws* v. *United States*, 321 U.S. 91 (1945), defendant police officers were convicted of wilfully violating civil rights protected by the Constitution. They argued that the statute under which they were convicted, 18 U.S.C. § 242, was unconstitutional, because it was too vague. No one but a scholar of Constitutional Law could know what rights are protected by the Constitution, and not even such a scholar could know for sure in the absence of a recent Supreme Court decision directly in point. But the court held that the statute should be interpreted to require a specific intent to deprive someone of a constitutional right, and that as so interpreted it would be constitutional. Note that the defendants' invalidating version of the statute is *de re* and the court's validating version is *de dicto*. Symbolize both.

Instructions: In exercise 5, symbolize the interpretation of the applicable law put forward by the prosecution ("Crown") and accepted by the trial court, and the one adopted by the Court of Appeal. In exercise 6, symbolize the interpretation adopted by the majority and that adopted by the dissent. In each case one interpretation will be de re *and the other* de dicto. *Put your answers in the form of conditional statements each having the consequent 'Defendant is guilty'. Disregard any language in any statute or rule of law that is not relevant to the particular case.*

5. REGINA v. SMITH (DAVID)
[1974] 2 Q.B. 354, 58 Cr. App. 320 (C.A.)

JAMES, L.J. . . . [T]he appellant, David Raymond Smith, was convicted of an offence of causing criminal damage contrary to section 1(1) of the Criminal Damage Act 1971. He appeals against that conviction on a question of law. . . .

The question of law in this appeal arises in this way. In 1970 the appellant became the tenant of a ground-floor flat at 209, Freemasons' Road, E.16. The letting included a conservatory. In the conservatory the appellant and his brother, who lived with him, installed some electric wiring for use with stereo equipment. Also, with the landlord's permission, they put up roofing material and asbestos wall panels and laid floor boards. There is no dispute that the roofing, wall panels and floor boards became part of the house and, in law, the property of the landlord. Then in 1972 the appellant gave notice to quit and asked the landlord to allow the appellant's brother to remain as tenant of the flat. On September 18, 1972, the landlord informed the appellant that his brother could not remain. On the next day the appellant damaged the roofing, wall panels and floorboards he had installed in order—according to the appellant and his brother—to gain access to and remove the wiring. The extent of the damage was £130. When interviewed by the police, the appellant said: "Look, how can I be done for smashing my own property. I put the flooring and that in, so if I want to pull it down it's a matter for me." . . .

The appellant's defence was that he honestly believed that the damage he did was to his own property, that he believed that he was entitled to damage his own property and therefore he had a lawful excuse for his actions causing the damage. In the course of his summing up the deputy judge directed the jury in these terms:

> Now, in order to make the offence complete, the person who is charged with it must destroy or damage that property belonging to another, "without lawful excuse," and that is something that one has got to look at a little more, members of the jury, because you have heard here that, so far as . . . defendant was concerned, it never occurred to [him] and, you may think, quite naturally never occurred to [him] that these various additions to the house were anything but [his] own property. . . . It is said that he had a lawful excuse by reason of his belief, his honest and genuinely held belief that he was destroying property which he had a right to destroy if he wanted to. But, members of the jury, I must direct you as a matter of law, and you must, therefore, accept it from me, that belief by the defendant David Smith that he had the right to do what he did is not lawful excuse within the meaning of the Act. Members of the jury, it is an excuse, it may even be a reasonable excuse, but it is not, members of the jury, a lawful excuse, because, in law, he had no right to do what he did. . . .

It is contended for the appellant that that is a misdirection in law, and that, as a result of the misdirection, the entire defence of the appellant was wrongly withdrawn from the jury.

Section 1 of the Criminal Damage Act 1971 reads:

> (1) A person who without lawful excuse destroys or damages any property belonging to another intending to destroy or damage any such property or being reckless as to whether any such property would be destroyed or damaged, shall be guilty of an offence.

The offence created includes the elements of intention or recklessness and the absence of lawful excuse....

It is argued for the appellant that an honest, albeit erroneous, belief that the act causing damage or destruction was done to his own property provides a defence to a charge brought under section 1(1). The argument [is] that the offence charged includes the act causing the damage or destruction and the element of mens rea [criminal intent]. The element of mens rea relates to all the circumstances of the criminal act. The criminal act in the offence is causing damage to or destruction of "property belonging to another" and the element of mens rea, therefore, must relate to "property belonging to another." Honest belief, whether justifiable or not, that the property is the defendant's own negatives the element of mens rea....

It is conceded by Mr. Gerber [counsel for the Crown] that there is force in the argument that the element of mens rea extends to "property belonging to another." But, it is argued, the section creates a statutory offence and that it is open to the construction that the mental element in the offence relates only to causing damage to or destroying property. That if in fact the property damaged or destroyed is shown to be another's property the offence is committed although the defendant did not intend or foresee damage to another person's property....

If the direction given by the deputy judge in the present case is correct, then the offence created by section 1(1) of the Act of 1971 involves a considerable extension of the law in a surprising direction. Whether or not this is so depends upon the construction of the section. Construing the language of section 1(1) we have no doubt that the actus reus [criminal act] is "destroying or damaging any property belonging to another." It is not possible to exclude the words "belonging to another" which describes the "property." Applying the ordinary principles of mens rea, the intention and recklessness and the absence of lawful excuse required to constitute the offence have reference to property belonging to another. It follows that in our judgment no offence is committed under this section if a person destroys or causes damage to property belonging to another if he does so in the honest though mistaken belief that the property is his own, and provided that the belief is honestly held it is irrelevant to consider whether or not it is a justifiable belief.

In our judgment, the direction given to the jury was a fundamental misdirection in law. The consequence was that the jury were precluded from considering facts capable of being a defence to the charge and were directed to convict.

For these reasons on November 5 at the conclusion of argument we allowed the appeal and ordered that the conviction be quashed.

*6. UNITED STATES v. INTERNATIONAL MINERALS & CHEM. CORP.
402 U.S. 558 (1971)

MR. JUSTICE DOUGLAS delivered the opinion of the Court.
The information charged that appellee shipped sulfuric acid and hydrofluosilicic acid in interstate commerce and "did knowingly fail to show on the shipping papers the required classification of said property, to wit, Corrosive Liquid, in violation of 49 C.F.R. 173.427."

Title 18 U.S.C. § 834(a) gives the Interstate Commerce Commission power to "formulate regulations for the safe transportation" of "corrosive liquids" and 18 U.S.C. § 834(f) states that whoever "knowingly violates any such regulation" shall be fined or imprisoned.

Pursuant to the power granted by § 834(a) the regulatory agency promulgated the regulation already cited which reads in part:

> *Each shipper offering for transportation any hazardous material subject to the regulations in this chapter, shall describe that article on the shipping paper by the shipping name prescribed in § 172.5 of this chapter and by the classification prescribed in § 172.4 of this chapter....*

The District Court, ... ruled that the information did not charge a "knowing violation" of the regulation and accordingly dismissed the information....

Here ... strict or absolute liability is not imposed; knowledge of the shipment of the dangerous materials is required. The sole and narrow question is whether "knowledge" of the regulation is also required. It is in that narrow zone that the issue of "mens rea" is raised; and appellee bears down hard on the provision in 18 U.S.C. § 834(f) that whoever "knowingly violates any such regulation" shall be fined, etc....

... We ... see no reason why the word "regulations" should not be construed as a shorthand designation for specific acts or omissions which violate the Act. The Act, so viewed, does not signal an exception to the rule that ignorance of the law is no excuse and is wholly consistent with the legislative history. [The Court's discussion of the legislative history is omitted.] ...

The principle that ignorance of the law is no defense applies whether the law be a statute or a duly promulgated and published regulation.... [W]e decline to attribute to Congress the inaccurate view that that Act requires proof of knowledge of the law, as well as the facts, and that it intended to endorse that interpretation by retaining the word "knowingly." We conclude that the meager legislative history of the 1960 amendments makes unwarranted the conclusion that Congress abandoned the general rule and required knowledge of both the facts and the pertinent law before a criminal conviction could be sustained under this Act....

[W]here as here, ... dangerous or deleterious devices or products or obnoxious waste materials are involved, the probability of regulation is so great that anyone who is aware that he is in possession of them or dealing with them must be presumed to be aware of the regulation.

Reversed.

MR. JUSTICE STEWART, with whom MR. JUSTICE HARLAN and MR. JUSTICE BRENNAN join, dissenting.

This case stirs large questions—questions that go to the moral foundations of the criminal law. Whether postulated as a problem of "mens rea," of "willfulness," of "criminal responsibility," or of "scienter," the infliction of criminal punishment upon the unaware has long troubled the fair administration of justice.... But there is no occasion here for involvement with this root problem of criminal jurisprudence, for it is evident to me that Congress made punishable only knowing violations of the regula-

tion in question. That is what the law quite clearly says, what the federal courts have held, and what the legislative history confirms. . . .

. . . Other federal courts, faced with the precise issue here presented, have held that the statute means exactly what it says—that the words "knowingly violates any such regulation" means no more and no less than "knowingly violates any such regulation." St. Johnsbury Trucking Co. v. United States, 220 F.2d 393 (C.A.1 1955), . . . Chief Judge Magruder filed a concurring opinion in the St. Johnsbury case, and he put the matter thus: . . . "If a statute provides that it shall be an offense "knowingly" to sell adulterated milk, the offense is complete if the defendant sells what he knows to be adulterated milk, even though he does not know of the existence of the criminal statute, on the time-honored principle of the criminal law that ignorance of the law is no excuse. But where a statute provides, as does 18 U.S.C. § 835, that whoever knowingly violates a regulation of the Interstate Commerce Commission shall be guilty of an offense, it would seem that a person could not knowingly violate a regulation unless he knows of the terms of the regulation and knows that what he is doing is contrary to the regulation. . . ."

7. (CHALLENGING) In Matthew 15: 3-6 and Mark 7: 9-12, Jesus condemns the teaching of certain rabbis who say that if a man tells his parents that anything available for their support is Corban—*i.e.*, dedicated to the Temple treasury—he is free of his obligation to support them. Furthermore, he doesn't actually have to part with any of his property in order to accomplish this result. He has not dedicated to the Temple anything except what is available to support his parents, and anything so dedicated is not so available. It (supposedly) follows that none of his property is available to support his parents, and this result is obtained without implying that any of his property is Corban. Evidently, the rabbis were not comfortable with this result, but were not able to escape the apparent force of the following reasoning:

> **Everything the owner says is Corban is Corban.**
> **Nothing that is Corban is available for parental support.**
> **Joe says that all of his property that is available for parental support is Corban.**
> **Therefore, none of his property is available for parental support.**

(a) What is wrong with the following attempt to symbolize and validate the rabbis' argument?

$$(x)(Sx \to Cx), (x)(Cx \to -Ax), (x)(Ax \to Sx) \vdash -(\exists x)Ax$$

(Universe: items of Joe's property; Sx = Joe says that x is Corban, Cx = x is Corban, Ax = x is available to support Joe's parents) (b) A second attempt differs from the first only in its interpretation for S: Sx = Joe says of x that it is Corban. What is wrong with this attempt? (c) Is there a valid symbolization of this argument with legitimate interpretations? If so, provide it.

8. (CHALLENGING) Professor Henry Hart of the Harvard Law School, one of the greatest scholars in the field of federal jurisdiction, was sometimes over the heads of his students. Though they were proud to take his Federal Courts course, which met at 12:00 three days a week, they often referred to it as "Darkness at Noon." One of the hypotheticals with which he challenged his students' ingenuity was this:

> Jones is about to be tried for a murder that occurred at 3:00 P.M. on a certain day. His defense is an alibi: at the time in question, he was in church praying. He therefore claims that his trial should be removed to a federal court. The prosecution seeks to punish him for what he was doing at the time of the murder. What he was doing at the time of the murder was praying in church. Therefore, the prosecution seeks to punish him for praying in church. A punishment for praying in church violates the First Amendment. A case in which the prosecution seeks to impose a punishment that violates the First Amendment is removable to the federal court. Therefore, this case is removable.

Symbolize Jones's reasoning and show what is wrong with it.

7

Conclusion

§7.1 ANALOGICAL AND DEDUCTIVE LEGAL REASONING

It is often said that legal reasoning, at least in the Anglo-American common law system, is generally by analogy rather than by deduction from premises. We have sometimes been asked how the system presented in this book relates to analogical reasoning (which, by the way, a number of legal writers refer to as "logic").

We believe that the answer lies in the nature of analogical reasoning as an application of the principle *Treat like cases alike*. If present Case A is analogous to prior Case B, then the principle that was applied in B should be applicable in Case A as well (either exactly as employed in B or with minor modifications). Now the application of a principle to a particular instance is a matter of *deducing* a statement about the instance from the principle and other statements setting out the facts of the case. Here is an example. *Myers v. United States*, 272 U.S. 52 (1926) involved a postmaster who was fired by the President (without the Senate's consent) before his four-year term had expired. The applicable statute said that the President could appoint postmasters for four-year terms with the consent of the Senate, and could dismiss them with the same consent. The ex-postmaster sued for his back salary. The Supreme Court held against him on the ground that the statutory requirement of senatorial consent to the firing was unconstitutional.

A few years later, the President fired a member of the Federal Trade Commission. Like postmasters, these members were appointed to fixed terms with the consent of the Senate. But the applicable statute, unlike that covering postmasters, made no provision for their removal except for misconduct in office. In *Humphrey's Executor* v. *United States*, 295 U.S. 602 (1935), the Supreme Court held that the purported firing was not effective. In arguing this case, the executive branch maintained that under the *Myers* case any official appointed by the President with the consent of the Senate was freely removable. Under the Constitution, Congress could not restrict the power to remove such an official. But the court limited the holding in *Myers* to officials performing purely executive functions. Thus, it did not apply to officials such as Federal Trade Commissioners who also had rulemaking and adjudicative functions to perform. Officials of that kind could not be removed except as provided in the statutes creating their offices.

We can state the "rule" of *Myers* as presented by the executive branch as:

> The President may freely dismiss any official appointed by him with the consent of the Senate.

We can symbolize this proposed rule this way:

$(x)(Ax \rightarrow Dx)$

(Ax = x has been appointed by the President with the consent of the Senate, Dx = x may be freely dismissed by the President) Obviously, the executive branch was urging the acceptance of this deductive argument (in symbols):

$(x)(Ax \rightarrow Dx)$
Ah
$\vdash Dh$

(h = Commissioner Humphrey) They claimed to find the rule in the *Myers* decision. But the rule as found by the Supreme Court was:

> The President may freely dismiss any official appointed by him with the advice of the Senate who performs purely executive functions.

$(x)[(Ax \,\&\, Ex) \rightarrow Dx]$

(Ex = x performs purely executive functions) The Supreme Court blocked the argument advanced by the executive branch by including in the antecedent of their version of the rule a property lacked by Commissioner Humphrey ("performing purely executive functions").

Notice that the version of the rule put forward by the executive branch and the one adopted by the Court are equally in accord with the decision in the *Myers* case. Either of the following arguments would support that decision:

(x)(Ax → Dx)
Am
⊢ Dm

(x)[(Ax & Ex) → Dx]
Am & Em
⊢ Dm

The situation here resembles one that is well known to philosophers of science. It is called "the under-determination of theory." Given any set of data, there will always be several possible theories each of which is compatible with those data.[1] We can call the legal counterpart of this fact about the nature of scientific theory, "the under-determination of rules." For any set of facts in a given case at law, there will be (at least) several rules or principles which, when applied to those facts, will support the decision reached by the Court. Therefore, unless the Court in its decision states the rule it is employing with great specificity, there will be room for disagreement about the exact content of the rule that was applied by the Court.[2]

Let's return once again to the contention considered at the beginning of this section, that legal reasoning is analogical and not deductive. There is some truth to the contention. (1) The intellectual process by which the court identifies the analogous case may well not be deductive. (2) When the precedent case has been identified, the process by which the court determines the rule underlying the case may also not be deductive. (The formulation of the rule will be based in some part on what the court that decided the precedent case actually said, but in larger part on how the court deciding the instant case thinks it should come out.) What the objection (that legal reasoning is analogical and, therefore, nondeductive) overlooks is that when the court has gotten the rule, it either argues that the rule applies to the instant case or that it does not apply to this case. The court is either advancing or blocking a deduction. While deduction does not exhaust legal reasoning, it is an inescapable element in it, for deduction is employed whenever principles are applied to facts. A main aim of this book has been to elucidate the structure of the reasoning by which legal principles are applied to cases.[3]

---

[1] In fact, the number of such possible theories will be infinite.
[2] Many legal theorists would go farther and say that it is impossible to frame a rule so specifically as to leave no room for disagreement as to its content. Interpretation is always required.
[3] For a discussion of legal reasoning that endeavors to explain in detail how deduction, induction, and analogical reasoning interrelate, see Scott Brewer, "Exemplary Reasoning: Semantics, Pragmatics, and the Rational Force of Legal Argument by Analogy," *Harv. L. Rev.*, CIX (1996), 923.

Appendix One
Deontic Logic[1]

Deontic logic is the logic of the expressions 'required' (or 'obligatory'), 'permitted', and 'forbidden'.[2] These expressions occur in moral discourse as well as legal discourse; obviously our focus here is on the latter. We adopt the following abbreviating symbols:

O = It is legally required (obligatory) that[3]
P = It is legally permitted that
F = It is legally forbidden that

(For the sake of brevity we will frequently drop the qualifier 'legally'.)

In some deontic logics, deontic expressions attach only to statements; in others, they attach both to statements and to building blocks that compose statements. We may call the former "deontic *propositional* logics" and the latter "deontic *predicate* logics." The system presented here is propositional.

---

[1] Deontic logic is commonly regarded as one of the tools of the legal logician. We therefore treat the subject for the sake of completeness. But we put our treatment in an appendix because we believe, for reasons given below, that deontic logic—or, at least, standard deontic logic—is of limited utility in the analysis of legal reasoning.

[2] The first viable system of deontic logic was presented by G. H. von Wright in "Deontic Logic," *Mind* LX (1951), 1–15. For a survey of the subject, see Dagfinn Føllesdal and Risto Hilpinen, "Deontic Logic: An Introduction," in Hilpinen (ed.), *Deontic Logic: Introductory and Systematic Readings* (Dordrecht: D. Reidel Publishing Company, 1971, 1981), pp. 1–35.

[3] Proponents of legal deontic logic assume that a single clear meaning can be assigned to the expression 'it is legally required that'. However, many legal theorists deny that the expression has a single meaning, or even that it has any meaning at all unless it is accompanied by further information such as who has a right to enforce the requirement, or what consequences will follow from not doing what is required. Such theorists (among them one of the authors of this book) are naturally skeptical about the possibility of using deontic logic in the analysis of legal argumentation.

The sentence,

> Candidates are required by law to report their campaign contributions.

may be symbolized in deontic logic with this formula:

OC

(C = Candidates report their campaign contributions) A deontic formula represents a *normative* or *regulative* English sentence by affixing a deontic operator to the symbolization of a *descriptive* statement. Deontic operators (**O**, **P**, and **F**) can always be distinguished from statement letters because these operators are affixed directly to statement letters (or larger formulas representing statements), while statement letters are only attached to other statement letters with the help of statement connectives. To further emphasize the distinction between letters representing deontic operators and letters abbreviating statements we print the former in boldface.

Even though the system we are presenting is a deontic *propositional* logic, we will allow predicate formulas to appear in it, provided that deontic operators are attached only to formulas that represent statements. For example, the first formula below is permissible while the second is not.

| **O**(∃x)Ax | ['(∃x)Ax' represents a statement.] |
| (∃x)**O**Ax | ['Ax' does not represent a statement.] |

(The rules that determine whether a formula in this system of logic is well formed are stated in Appendix Two.) The scope of a deontic operator is shown by groupers (where needed)—just as with the dash.

The three deontic operators are fully interdefinable:

DEONTIC DEFINITIONS (DD)

OA	=	$-P-A$	=	$F-A$
$O-A$	=	$-PA$	=	FA
$-OA$	=	$P-A$	=	$-F-A$
$-O-A$	=	PA	=	$-FA$

The second line in this group of definitions corresponds to our intuitions that if it is required that something not be done, then doing it is not permitted (and vice versa), and that if something is not permitted it is forbidden (and vice versa). The other lines have similar intuitive support.

There are several nonequivalent systems of deontic propositional logic, but one system has come to be regarded as the standard system; we'll call it "SDL" (short for "Standard Deontic Logic"). SDL embodies the three principles displayed below. We may add SDL to our existing system of logic by adopting the deontic definitions (call them "DD") given above and these four rules of inference:

DEONTIC PRINCIPLES	ASSOCIATED RULES OF INFERENCE	
1. Whatever is entailed by what is required is also required.	Obligation Out (OO)	From **O**A (in a proof) derive A (in a deontic sub-proof).
	Obligation In (OI)	From A (in a deontic sub-proof) derive **O**A (in a proof) *provided that A does not depend on a provisional assumption made within that sub-proof.*
2. Whatever is required is permitted.	Obligation/ Permission (OP)	From **O**A derive **P**A.
3. Whatever is logically necessary is required.	Logical Truth (LT)	If A is free of assumptions, then derive **O**A from it.

The first two principles underlying SDL seem obvious;[4] the third does not. The third principle follows from the first deontic principle, the logical principle that any statement entails a logical truth, and the undeniable claim that at least something is required by law.

We'll show how the definitions and the rules of inference are used by proving the validity of several arguments. We start by showing that in SDL the state of affairs expressed by any given statement, A, is either permitted or forbidden:

(1) −(**PA** v **FA**)   **PA**
(2) −**PA** & −**FA**   1 DM
(3) **FA** & −**FA**   2 DD
(4) **PA** v **FA**   1-3 −O

The standard assumption-dependence principle applies to DD and all the other deontic rules of inference. As the proof above illustrates, DD may be applied to line parts as well as to whole lines. By contrast the other four rules are correctly applied to whole lines only.

---

[4]While the second principle *seems* obvious, we question it below.

Appendix One: Deontic Logic

The deontic rules *Obligation Out* (**OO**) and *Obligation In* (**OI**) involve the concept of a "deontic sub-proof." As the expression suggests, a deontic sub-proof is a portion of a larger proof. We identify the boundaries of the sub-proof by drawing a box or rectangle around it. You may begin a deontic sub-proof at any point in a proof after the assumption lines; you would normally begin one with a line derived by **OO**. You may end a deontic sub-proof at any point above the last line of the proof; typically a deontic sub-proof ends when you are ready to derive a line by **OI**.[5]

A line in a deontic sub-proof may be justified only in one of these ways:

(1) it is derived from a line above the sub-proof by **OO**;
(2) it is introduced as a provisional assumption or by the Identity In Rule; or
(3) it is derived from other lines in the sub-proof.

Only one rule sanctions a move *from a line above* a deontic sub-proof *to a line within* the sub-proof; that rule is **OO**. And only one rule allows a move *from a line within* a deontic sub-proof *to a line below* the sub-proof; that rule is **OI**. A proof for the following argument ("Auto Insurance") illustrates the use of **OO** and **OI**:

It is required that all automobiles be REGISTERED. It is required that all registered automobiles be INSURED. It follows that it is required that all automobiles be insured.

O(x)Rx, **O**(x)(Rx → Ix) ⊢ **O**(x)Ix

(Universe: automobiles)

1	(1)	**O**(x)Rx	A
2	(2)	**O**(x)(Rx → Ix)	A
1	(3)	(x)Rx	1 **OO**
2	(4)	(x)(Rx → Ix)	2 **OO**
5	(5)	−(x)Ix	PA
5	(6)	(∃x)−Ix	5 QE
5	(7)	−Ia	6 EO
1	(8)	Ra	3 UO
2	(9)	Ra → Ia	4 UO
1,2	(10)	Ia	9,8 →O
1,2,5	(11)	Ia & −Ia	10,7 &I
1,2	(12)	(x)Ix	5-11 −O
1,2	(13)	**O**(x)Ix	12 **OI**

---

[5] A proof may contain two (or more) deontic sub-proofs. They may be consecutive or nested one inside another; their boxes may not intersect. A single application of **OO** or **OI** cannot be used to cross *two* nested sub-proofs.

You can see how the rules **OO** and **OI** and the apparatus of deontic sub-proofs embody the principle that whatever is entailed by what is required is required. The deontic sub-proof above shows that '(x)Rx' (line 3) and '(x)(Rx → Ix)' (line 4) entail '(x)Ix' (line 12). Since the first two of these formulas are alleged (in lines 1 and 2) to be legal requirements, we are entitled to conclude (on line 13) that the third formula also expresses what is legally required.

The next two proofs show the logical equivalence of S1 and S2:

(S1) It is required that therapists have a GRADUATE degree and that they be CERTIFIED by the state. (F1) **O**(G & C)

(S2) It is required that therapists have a GRADUATE degree and it is required that they be CERTIFIED by the state. (F2) **OG & OC**

(1)	**O**(G & C)	A
(2)	G & C	1 **OO**
(3)	G	2 &O
(4)	C	2 &O
(5)	**OG**	3 **OI**
(6)	**OC**	4 **OI**
(7)	**OG & OC**	5,6 &I

(1)	**OG & OC**	A
(2)	**OG**	1 &O
(3)	**OC**	1 &O
(4)	G	2 **OO**
(5)	C	3 **OO**
(6)	G & C	4,5 &I
(7)	**O**(G & C)	6 **OI**

The equivalence of F1 and F2 means that the '**O**' operator "distributes across the ampersand in both directions." Note that the '**O**' distributes across the wedge in only one direction. F3 entails F4, but not vice versa.

(F3) **OG v OC**
(F4) **O**(G v C)

Here is an example showing that F4 does not entail F3:

(S4) Ralph is required to file an income tax return or request an extension.

(S3) Either Ralph is required to file an income tax return or he is required to request an extension.

While S4 is true, S3 is false. Ralph is not required to file a return (he could request an extension instead), and he is not required to request an extension (he could file a return).

Similarly, the '**P**' operator distributes across the wedge in both directions, but in only one direction across the ampersand. That is, F5 and F6 are equivalent, but while F7 entails F8[6] the reverse does not hold.

(F5) **P**(A v B)
(F6) **PA** v **PB**

(F7) **P**(A & B)
(F8) **PA** & **PB**

This interpretation shows that F8 does not entail F7:

(S8) Carla is permitted to marry Al, and she is permitted to marry Bob.

(S7) Carla is permitted to marry both Al and Bob.

We utilize the next two rules, *Obligation/Permission* (**OP**) (which sanctions the move from a formula beginning with '**O**' to one beginning with '**P**') and *Logical Truth* (**LT**) (which permits the attachment of an '**O**' to a logical truth), in the proof of this formula:

P(A → A)

1	(1)	A	PA
1	(2)	− −A	1 DN
1	(3)	A	2 DN
	(4)	A → A	1-3 →I
	(5)	**O**(A → A)	4 LT
	(6)	**P**(A → A)	5 **OP**

Line 5 may be derived from line 4 because 4 is free of assumptions. (Any statement derived free of assumptions is a logical truth.)

---

[6]It seems that according to Genesis 30:26 it was permitted that Jacob marry both Rachel and Leah, but not that he marry Rachel only. One might think that this is a case where '**P**(A & B)' is true but '**PA**' (and therefore also '**PA** & **PB**') is false. (A = Jacob marries Rachel, B = Jacob marries Leah). But it is actually not such a case. The statement that Jacob is not permitted to marry Rachel unless he also marries Leah is not expressed by '−**PA**', but by '**F**(A & −B)', which is logically consistent with '**PA**'.

We establish the validity of a deontic argument by constructing a proof of validity; how do we show invalidity? A method related to the technique of propositional analogues may be employed; we call it the method of "non-deontic analogues." We test a deontic argument by evaluating its non-deontic analogue. Within its range of applicability, this technique demonstrates both validity and invalidity. We shall need this definition: A 'canonical deontic formula' is a formula containing only one deontic operator that begins with either an '**O**' or a '**P**', where that operator has the remainder of the formula within its scope. The method proceeds in five steps.

THE METHOD OF NON-DEONTIC ANALOGUES

1. Is the premise set contradictory or the conclusion logically true?
   YES: The argument is valid.[7]
   NO: Go to step 2.

2. Delete every non-deontic premise. Using the rules of SDL transform each deontic formula into one or more equivalent canonical deontic formulas. Was this accomplished?
   YES: Go to step 3.
   NO: Stop the test. Inconclusive result.

3. Does each formula begin with **O**?
   YES: Go to step 5.
   NO: Go to step 4.

4. Does exactly one premise and the conclusion begin with **P**?
   YES: Go to step 5.
   NO: Stop the test. Inconclusive result.

5. Delete the deontic operators (creating the non-deontic analogue). Test the analogue. Is it valid?
   YES: The deontic argument is valid.
   NO: The deontic argument is invalid.

We illustrate the technique by testing two arguments, beginning with "Auto Insurance":

O(x)Rx, **O**(x)(Rx → Ix) ⊢ **O**(x)Ix

The instructions carry us quickly to step 5 where we produce this non-deontic analogue:

---

[7]Any deductive argument—not just a deontic argument—is valid if its premise set is contradictory or its conclusion is logically true.

$(x)Rx, (x)(Rx \rightarrow Ix) \vdash (x)Ix$

The analogue is valid (a result that can be shown by formal proof or propositional analogue); therefore, so is the deontic argument, "Auto Insurance."

For a second example we evaluate "Handguns":

It is required that all police OFFICERS carry HANDGUNS. There are police officers. Hence, it is required that there are people who carry handguns.

$O(x)(Ox \rightarrow Hx), (\exists x)Ox \vdash O(\exists x)Hx$

Notice that in step 2 the second (non-deontic) premise is eliminated. The non-deontic analogue of the argument produced in step 5 $((x)(Ox \rightarrow Hx) \vdash (\exists x)Hx)$ is invalid (a result that can be shown by interpretation); so SDL assesses "Handguns" as invalid. This assessment seems intuitively correct. The premises do not include a requirement that there be policemen, so the conclusion goes beyond the premises.[8]

SDL has certain shortcomings; we will examine three. (1) The paradox of material implication (discussed in Section 5.3) occurs in SDL in particularly severe forms. Consider, for example, this argument ("Rape"):

It is legally required that RAPES not occur. Therefore, it is required by law that if rapes occur, then MURDERS also occur.

$O-R \vdash O(R \rightarrow M)$

(R = Rapes occur, M = Murders occur) The validity of the symbolization is easily demonstrated in SDL; most people will think the English argument invalid. The predicate analogue of "Rape" seems even more outrageous:

It is legally required that no one RAPE$_R$ anyone. Therefore, it is required by law that whoever rapes someone, MURDERS$_R$ that person.

$O(x)(y)-Rxy \vdash O(x)(y)(Rxy \rightarrow Mxy)$

(Universe: people) Since SDL is built upon standard propositional logic, it should come as no surprise that SDL hosts this paradox.

(2) Consider this argument ("Hot Dogs"):

---

[8] A conclusion that can be validly drawn from these premises, "There are people who are required to carry handguns," cannot be expressed in SDL because its symbolization places a deontic operator within the scope of a quantifier, a construction that SDL does not permit.

It is required that all hot DOG vendors have vendor's LICENSES. <u>Sam</u> is a hot dog vendor. Therefore, it is required that Sam have a vendor's license.

O(x)(Dx → Lx), Ds ⊢ OLs

SDL judges the symbolization of this argument to be invalid, yet the English argument seems plainly valid. Indeed, the argument appears to typify a very common pattern of legal reasoning—the application of a law to an individual case. If Sam were your client (in a community having the legal requirement expressed in the first premise), you would undoubtedly advise him that (as a hot dog vendor) he is required to have a vendor's license.

You have probably noticed that non-deontic statements of fact (like premise two of "Hot Dogs") play no role in SDL. This is evident in the SDL method of proof from the fact that non-deontic statements cannot be imported into deontic sub-proofs; and it is evident in the technique of non-deontic analogues from the fact that (in step 2) non-deontic premises are eliminated. SDL judges "Hot Dogs" to be invalid because SDL ignores the second premise and finds that the conclusion is not entailed by the first premise alone.

(3) According to SDL, inconsistent legal obligations are impossible. Formula F9, for instance, is a logical contradiction in that system, as can be shown by deriving a standard contradiction from it.

(F9) O**S** & O2**S**

(1)	O**S** & O−**S**	A
(2)	O**S**	1 &O
(3)	O−**S**	1 &O
(4)	P**S**	2 **OP**
(5)	−O−**S**	4 DD
(6)	O−**S** & −O−**S**	3,5 &I

We contend that statements having the form of F9 are sometimes true; and if they are true, they cannot be contradictory. Consider the case of *Helzberg's Diamond Shops, Inc. v. Valley West Des Moines Shopping Center, Inc.*, 564 F.2d 816 (8th Cir. 1977). Valley West, the operator of a shopping center, leased space to Helzberg's for a jewelry store. The lease provided that Valley West would not permit more than two other full-line jewelry stores in the shopping center. After there were three jewelers in the mall, Valley West entered into a lease with Lord's Jewelers to provide store space for a fourth store. Given the lease with Lord's, the left conjunct of S9 (below) is true. Given the lease with Helzberg's and the fact that the mall already contained three jewelry stores, the right conjunct is also true. Because each conjunct of S9 is true, S9 itself is true.

(S9) Valley West is required to provide SPACE to Lord's Jewelers and Valley West is required not to provide space to Lord's Jewelers.

(F9) O**S** & O−**S**

(S = Valley West provides space to Lord's Jewelers) Through carelessness (or perhaps worse) Valley West got into a situation in which it incurred inconsistent legal obligations. It was in legal difficulty whether or not it honored its contract with Lord's. This type of occurrence is not uncommon. Note that if S9 (F9) can be true, then either the rule **OP** or DD (which are employed in the proof that F9 is inconsistent) must be rejected. The culprit is obviously **OP**, which sanctions the move from the true S10 to the false S11.

(S10) Valley West is required to provide space to Lord's Jewelers.

(S11) Valley West is permitted to provide space to Lord's Jewelers.

What can be said on behalf of SDL in the light of these difficulties? Let's draw a distinction—admittedly not too precise—between *pure* and *applied* legal deontic statements. By a "pure" legal deontic statement we mean a statement (ordinarily general in scope) that expresses either a statute or an entailment of a set of statutes. By an "applied" legal deontic statement we mean a statement that expresses the result of applying a statute (or several statutes) to a specific case, person, etc. An applied legal deontic statement will typically include a reference to some individual. Consider these examples:

PURE DEONTIC STATEMENTS	APPLIED DEONTIC STATEMENTS
It is legally permitted that tobacco be purchased by adults.	Jones is legally permitted to purchase tobacco.
It is legally required that Presidents be natural-born citizens.	Henry Kissinger is forbidden by law to be President.

We may regard SDL as expressing *the logic of pure deontic statements*, and then resist using it when any applied deontic statements are involved. By this maneuver we can avoid the second and third difficulties noted above.

In the area of pure legal deontic logic the assumption that obligations cannot be inconsistent is defensible.[9] If it is discovered that two putative laws contradict one another, one of them will be struck down in order to preserve consistency. Consistency among laws is recognized as a requirement for a satisfactory legal system. But it is not a requirement for a system of laws that it be impossible under them for an individual to incur inconsistent legal obligations.

---

[9]This assumption is also defensible (but not incontrovertible) in moral deontic logic. Ethicists generally maintain that since people cannot be morally required to do the impossible, they can't be subject to inconsistent moral obligations. But it has also been argued that in a polytheistic world "morality" can be defined as "conformity to the will of the gods," and different gods can will inconsistent things. See Anton-Hermann Chroust, Book Review, *Natural Law Forum*, i (1956), 135.

The assumption that obligations cannot be inconsistent (in pure deontic logic) breaks down where there are two systems of law in force in the same territory with no effective way of resolving inconsistencies between them. For instance, from 1297 until 1306, the law of England required all English bishops to pay taxes to the king, while the law of the church forbade them to do so.[10] To avoid this type of case we can confine SDL to statements that make up a *single* system of laws.

By restricting SDL to pure deontic statements (in one legal system) we can avoid two of the three shortcomings pointed out above. The price paid, of course, is a severe limitation in the scope of SDL. The utility of a system of deontic logic so narrowed is greatly diminished.

---

[10]See Robert E. Rodes, Jr., *Lay Authority and Reformation in the English Church* (Notre Dame, IN: University of Notre Dame Press, 1982), p. 10.

Appendix Two
Formation Rules

Some formulas of logic (F1, for instance) are properly constructed, while others are ungrammatical (F2X and F3X, for example). Note that F2X is quite nonsensical, while F3X is merely amphibolous.

(F1) −(A & B)
(F2X) → C &
(F3X) D & E → F

It is possible to lay down precise rules (called "formation rules") for distinguishing between formulas that are "well formed" and those that are not. In this appendix we provide such rules for the systems of propositional, predicate, and deontic logic presented in this book.

We begin by specifying the vocabulary for propositional logic:

sentence letters:	capital letters with or without prime marks
connectives:	&, →, v, ↔, and −
groupers:	(,)

The addition of prime marks gives us an inexhaustible supply of sentence letters. For the sake of simplicity we use only parentheses as groupers. Now we can state the formation rules for propositional logic:

1. **A sentence letter is well formed.**

2. **A formula $-P$ is well formed if P is well formed.**

3. **The formulas $(P \mathbin{\&} Q)$, $(P \rightarrow Q)$, $(P \vee Q)$, and $(P \leftrightarrow Q)$ are well formed if P and Q are well formed.**

4. **No formula is well formed unless its being so follows from the rules above.**

These rules may be applied "recursively." For example, formulas F4 and F5 are categorized as well formed by rule 1; F6 is then covered by rule 3; and, finally, F1 is brought into the fold with the help of rule 2.

(F4) A
(F5) B
(F6) (A & B)
(F1) −(A & B)

None of the first three rules applies to formulas F2X or F3X, and, therefore, rule 4 classifies them as not well formed.

With two exceptions, this set of formation rules matches our past practice in writing propositional formulas: (1) we haven't placed groupers at the beginning and end of formulas built around '&', '→', 'v', and '↔', and (2) we haven't required internal groupers in multiple conjunctions and disjunctions. Let's regard these two practices as producing convenient abbreviations of well-formed formulas (for in each of these cases the additional groupers do no useful work).

The vocabulary for predicate logic consists of the vocabulary of propositional logic plus the following:

individual constants:	lower-case letters from a to v with or without prime marks
variables:	lower-case letters from w to z with or without prime marks
quantifiers:	(x), $(\exists x)$ (where 'x' is any variable)
predicate letters:	capital letters with or without prime marks (when they are followed by one or more individual constants or variables)
the identity sign:	=

The formation rules for predicate logic:

1. **A sentence letter, or a predicate followed by one or more individual constants, is well formed.**

2. **An individual constant followed by the identity sign followed by an individual constant is well formed.**

3. **A formula −P is well formed if P is well formed.**

4. **The formulas (P & Q), (P → Q), (P v Q), and (P ↔ Q) are well formed if P and Q are well formed.**

5. **A formula is well formed if it can be generated from a second well-formed formula by prefixing a quantifier (whose variable, v, does not occur in the**

second formula) and replacing at least one occurrence of an individual constant by *v*.

6. **No formula is well formed unless its being so follows from the rules above.**

Let's see how this set of rules catalogues formula F10 as well formed.

(F7) Gi
(F8) Hi
(F9) (Gi & Hi)
(F10) (∃x)(Gx & Hx)

Rule 1 labels F7 and F8 "well formed," and rule 4 then shows that F9 is constructed correctly. Since F10 can be generated from F9 in the way described by rule 5, F10 achieves the status of being well formed.

Formula F11X, on the other hand, cannot be reached by applying rules 1 through 5. Therefore, by rule 6, it is not well formed.

(F11X) (∃x)(Gx & Hy)

The vocabulary for the version of deontic logic presented in Appendix One consists of the vocabulary of predicate logic plus the three deontic operators: **O**, **P**, and **F**. Here are the formation rules:

1. **Any well-formed formula of propositional or predicate logic is well formed.**

2. **The formulas O*P*, P*P*, and F*P* are well formed if *P* lacks deontic operators and is well formed.**

3. **A formula −*P* is well formed if *P* is well formed.**

4. **The formulas (*P* & *Q*), (*P* → *Q*), (*P* v *Q*), and (*P* ↔ *Q*) are well formed if *P* and *Q* are well formed.**

5. **No formula is well formed unless its being so follows from the rules above.**

These rules do not allow the nesting of deontic operators (as in formulas F12X and F13X); nor do they permit deontic operators to fall within the scope of quantifiers (as in F14X). Some deontic logics accept these formulas as well formed.

(F12X) **OPJ**
(F13X) **O(K → OL)**
(F14X) (∃x)**OMx**

Appendix Three
Alternative Symbols

The symbols employed in this book are not the only ones in use. Here is a list of other symbols:

OUR SYMBOLS	ALTERNATIVE SYMBOLS		POLISH NOTATION
⊢	∴		
&	•	∧	K
→	⊃		C
–	~	¬	N
∨			A
↔	≡		E
(x)	∀x	(∀x)	Πx
(∃x)	∃x	(Σx)	Σx

Polish notation (developed by the Warsaw school of logicians) does not employ grouping symbols. Instead, it places connective symbols before the statement letters they connect; for example, 'If P, then Q' is symbolized as 'Cpq'. Note that groupers are required to distinguish O1 from O2 and O3 from O4, but not to distinguish P1 from P2 or P3 from P4:

our notation   *Polish notation*

(O1) (P & Q) → R   =   (P1) CKpqr
(O2) P & (Q → R)   =   (P2) KpCqr

(O3) (x)Fx → (∃y)Gy   =   (P3) CΠxFxΣyGy
(O4) (x)[Fx → (∃y)Gy]   =   (P4) ΠxCFxΣyGy

Appendix Four
Solutions to Starred Exercises

SECTION 1.3

1. False statements: (b) (no argument with true premises and a false conclusion is valid) and (g).

3(a) Each sentence entails the other. There are three possibilities: Sue takes both exams, neither exam, or exactly one exam. In first two cases both statements are false; in third case both are true. So, if either statement is true, the other must be true as well.

(b) Each sentence entails the other. There are three possibilities: both are permissible, neither is permissible, or exactly one is permissible. In first two cases both statements are true; in third case both are false. So, if either statement is true, the other must be true as well.

(c) Neither sentence entails the other. First sentence is true and second false when applied to the intersection depicted by sign A, and vice versa for sign B.

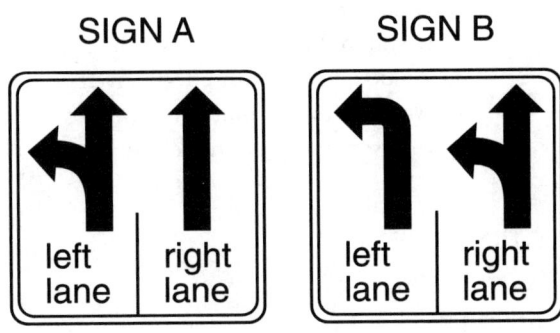

(d) First sentence does not entail second. If Powell and another are both found guilty, the first will be true and the second false.

Second sentence entails first. If first sentence is false then Powell is not found guilty but someone else is. This situation falsifies second sentence. So first sentence cannot be false and second true.

(e) Each sentence entails the other. If one of these sentences is false, then there is a person in fetters and a person not in fetters. This situation falsifies the other sentence. So neither sentence can be false while the other is true.

(f) First sentence does not entail second. In this logically possible scenario first sentence would be true and second false: Juror A will vote acquittal if any juror is bribed, every juror has been bribed, but only juror A votes acquittal.

Second sentence does not entail first. In this logically possible scenario second sentence would be true and first false: Each juror will vote acquittal if and only if every juror has been bribed, but only one has been bribed, and none vote acquittal.

(g) First sentence does not entail second. In this logically possible scenario first sentence would be true and second false: Each substance has some attribute and some attribute inheres in no substances.

Second sentence does not entail first. In this logically possible scenario second sentence would be true and first false: Each attribute inheres in some substance and some substance has no attributes.

(h) First sentence entails second.

    1. Everybody loves my baby.
    2. So, my baby loves my baby. (from 1)
    3. My baby don't love nobody but me.
    4. I am my baby. (from 2 and 3)

Second sentence does not entail first.

(i) First sentence does not entail second. In this logically possible scenario first sentence would be true and second false: Each person who degrades another degrades Whitman; Alice degrades only herself.

Second sentence entails first. If first sentence is false, then someone degrades another without degrading Whitman. The "other" couldn't be Whitman. That situation would falsify second sentence. So first sentence cannot be false and second true.

(j) First sentence entails second.

    1. Waldo admires all who do not admire themselves.
    2. If Waldo does not admire himself, then he admires himself. (from 1)
    3. Waldo admires himself. (from 2)
    4. Waldo admires some self-admirer. (from 3)

Second sentence does not entail first.

SECTION 2.1

1(b) F & W

(f) K & D & L

2(b) simple

(h) The sentence is equivalent to the conjunction of these two statements:

> All persons born or naturalized in the United States and subject to the jurisdiction thereof are citizens of the United States.
>
> All persons born or naturalized in the United States and subject to the jurisdiction thereof are citizens of the State wherein they reside.

A deeper analysis reveals that the amendment is also equivalent to the conjunction of these four statements:

> All persons born in the United States and subject to the jurisdiction thereof are citizens of the United States.
>
> All persons naturalized in the United States and subject to the jurisdiction thereof are citizens of the United States.
>
> All persons born in the United States and subject to the jurisdiction thereof are citizens of the State wherein they reside.
>
> All persons naturalized in the United States and subject to the jurisdiction thereof are citizens of the State wherein they reside.

3(b) (1) C & D & E   A
    (2) D   1 &O

4(b) There is no genuine issue as to any material fact and the moving party is entitled to a judgment as a matter of law.

F & E

SECTION 2.2

1(b) T → B
(f) (P & M) → O
(j) P → [(N → M) & (O → F)]

2(b) Miami wins its last regular-season game, and if New York loses its last regular-season game, then Miami wins the division championship.

3(b) Q → P
(f) C → S
(j) D → S

4(b) (1)   D → E   A
    (2)   F → G   A
    (3)   D & F   A
    (4)   D   3 & O
    (5)   E   1,4 →O
    (6)   F   3 & O
    (7)   G   2,6 →O
    (8)   E & G   5,7 &I

6. (1) A   A
    (2) B   A
    (3) (A & B) → C   A
    (4) A & B   1,2 &I
    (5) C   3,4 →O

10. 1   (1) U → C   A
    2   (2) F   A
    3   (3) (C & F) → R   A
    4   (4) U   PA
    1,4   (5) C   1,4 →O
    1,2,4   (6) C & F   5,2 &I
    1,2,3,4   (7) R   3,6 →O
    1,2,3   (8) U → R   4-7 →I

14(a) **If Haiti achieves political stability, it will enjoy a climate of security. When Haiti has such a climate, it will attract investments. Once Haiti attracts investments, Haitians will have jobs. When Haitians have jobs, they will have money. And when Haitians have money, they will have food. Therefore, our achieving political stability will lead to food for the Haitian people.**

(b) **P = Haiti achieves political stability, C = Haiti has a climate of security, I = Haiti attracts investments, J = Haitians have jobs, M = Haitians have money, F = Haitians have food**

P → C, C → I, I → J, J → M, M → F ⊢ P → F

(c) 1   (1)   P → C   A
    2   (2)   C → I   A
    3   (3)   I → J   A
    4   (4)   J → M   A
    5   (5)   M → F   A
    6   (6)   P   PA
    1,6   (7)   C   1,6 →O
    1,2,6   (8)   I   2,7 →O
    1,2,3,6   (9)   J   3,8 →O
    1,2,3,4,6   (10) M   4,9 →O
    1,2,3,4,5,6   (11) F   5,10 →O
    1,2,3,4,5   (12) P → F   6-11 →I

SECTION 2.3

1(b) W & −V
 (f) −(B & S)
 (k) −(M & G) → E
      This is preferable to the nonequivalent '(−M & −G) → E'.
 (n) A → {−J → [E → (−I & −D)]}

Appendix Four: Solutions to Starred Exercises   **355**

2(b) The defense witness was not granted immunity and the prosecution witness was not granted immunity.

4(b)
1	(1)	−B → C	A	
2	(2)	−B → −C	A	
3	(3)	−B	PA	
1,3	(4)	C	1,3 →O	
2,3	(5)	−C	2,3 →O	
1,2,3	(6)	C & −C	4,5 &I	
1,2	(7)	B	3-6 −O	

6.
1	(1)	S & −H	A	
2	(2)	S → H	PA	
1	(3)	S	1 &O	
1,2	(4)	H	2,3 →O	
1	(5)	−H	1 &O	
1,2	(6)	H & −H	4,5 &I	
1	(7)	−(S → H)	2-6 −I	

10.
1	(1)	S → −A	A	
2	(2)	B → A	A	
3	(3)	S & B	PA	
3	(4)	B	3 &O	
2,3	(5)	A	2,4 → O	
3	(6)	S	3 &O	
1,3	(7)	−A	1,6 → O	
1,2,3	(8)	A & −A	5,7 &I	
1,2	(9)	−(S & B)	3-8 −I	

SECTION 2.4

1(b) E v −E
 (f) −L → (R v F)
 (j) (P v A) → (R v B)

2(b) Cruelty to animals is neither a misdemeanor nor a felony in Florida.

4(b)
1	(1)	(P → O) & (M → O)	A	
2	(2)	P v M	PA	
1	(3)	P → O	1 &O	
1	(4)	M → O	1 &O	
1,2	(5)	O	2,3,4 vO	
1	(6)	(P v M) → O	2-5 → I	

6.
1	(1)	A v M	A	
2	(2)	A → S	A	

3	(3)	S → B	A	
4	(4)	M → B	A	
5	(5)	A	PA	
2,5	(6)	S	2,5 →O	
2,3,5	(7)	B	3,6 →O	
2,3	(8)	A → B	5-7 →I	
1,2,3,4	(9)	B	1,8,4 vO	

10.
1	(1)	L v S	A
2	(2)	(L → K) & (S → C)	A
3	(3)	(K → G) & (C → G)	A
4	(4)	L	PA
2	(5)	L → K	2 &O
2,4	(6)	K	5,4 →O
3	(7)	K → G	3 &O
2,3,4	(8)	G	7,6 →O
2,3	(9)	L → G	4-8 →I
10	(10)	S	PA
2	(11)	S → C	2 &O
2,10	(12)	C	11,10 →O
3	(13)	C → G	3 &O
2,3,10	(14)	G	13,12 →O
2,3	(15)	S → G	10-14 →I
1,2,3	(16)	G	1,9,15 vO

SECTION 2.5

1(b) P ↔ A

(f) C ↔ S

2(b) It is false that Notre Dame wins the national championship iff it wins its bowl game.

4(b)
1	(1)	D → (E & F)	A
2	(2)	E → D	A
3	(3)	D	PA
1,3	(4)	E & F	1,3 →O
1,3	(5)	E	4 &O
1	(6)	D → E	3-5 →I
1,2	(7)	D ↔ E	6,2 ↔I

6.
1	(1)	W ↔ P	A
2	(2)	−P	A
3	(3)	W	PA
1	(4)	W → P	1 ↔ O

Appendix Four: Solutions to Starred Exercises   357

	1,3	(5)	P	4,3 →O
	1,2,3	(6)	P & −P	5,2 &I
	1,2	(7)	−W	3-6 −I
10.	1	(1)	A ↔ B	A
	2	(2)	−(A & B)	A
	3	(3)	A	PA
	1	(4)	A → B	1 ↔O
	1,3	(5)	B	4,3 →O
	1,2,3	(6)	(A & B) & −(A & B)	3,5,2 &I
	1,2	(7)	−A	3-6 −I
	8	(8)	B	PA
	1	(9)	B → A	1 ↔O
	1,8	(10)	A	9,8 →O
	1,2,8	(11)	(A & B) & −(A & B)	10,8,2 &I
	1,2	(12)	−B	8-11 −I
	1,2	(13)	−A & −B	7,12 &I

SECTION 2.6

There are several correct proofs for each argument. We present only one.

1(b)	(1)	C v D	A	
	(2)	−C	A	
	(3)	D	1,2 DA	
	(4)	−−D	3 DN	
3.	(1)	B	A	
	(2)	(B & S) → M	A	
	(3)	−M	A	
	(4)	−(B & S)	2,3 MT	
	(5)	−S	4,1 CA	
7.	1	(1)	P → −M	A
	2	(2)	−M → −A	A
	3	(3)	−P → −M	A
	4	(4)	−−A	PA
	2,4	(5)	−−M	2,4 MT
	1,2,4	(6)	−P	1,5 MT
	2,3,4	(7)	−−P	3,5 MT
	1,2,3,4	(8)	−P & −−P	6,7 &I
	1,2,3	(9)	−A	4-8−O
11.	1	(1)	−(P & C)	A
	2	(2)	−C → −L	A
	3	(3)	L	PA

3	(4)	− −L	3 DN
2,3	(5)	− −C	2,4 MT
2,3	(6)	C	5 DN
1,2,3	(7)	−P	1,6 CA
1,2	(8)	L → −P	3-7 →I
1,2	(9)	−L v −P	8 AR

SECTION 2.7

1(b)

N	L	‖	−N	→	−L	⊢	N	→	L	invalid
T	T		F	T	F			T		
F	T		T	F	F			T		
✓T	F		F	T	T			F		
F	F		T	T	T			T		
				*						

2(b)

A	B	‖	−(A	v	B)		−A	v	−B	not logically equivalent
T	T		F	T			F	F	F	
✓F	T		F	T			T	T	F	
✓T	F		F	T			F	T	T	
F	F		T	F			T	T	T	
				*				*		

4(b)

U	O	‖	−(U	&	O)		−U	⊢	O
✓F	F		T F	F	F		T F		F
				*			*		

6.

P	R	‖	P		P	→	R	⊢	R	valid
T	T		T			T			T	
F	T		F			T			T	
T	F		T			F			F	
F	F		F			T			F	
						*				

10.

P	C	E	‖	−P		C	→	P		−E	→	−P	⊢	−E	invalid
✓F	F	T		T F		F	T	F		F T	T TF			F T	
				*			*				*			*	

14.

T	P	S	‖	T	→	(P	&	−S)		−T	→	(S	&	−P)	⊢ S	↔	−P	valid
T	T	T		F		F		F		F	T		F	F		F	F	
F	T	T		T		F		F		T	F		F	F		F	F	
T	F	T		F		F		F		F	T		T	T		T	T	
F	F	T		T		F		F		T	T		T	T		T	T	
T	T	F		T		T		T		F	T		F	F		T	F	
F	T	F		T		T		T		T	F		F	F		T	F	
T	F	F		F		F		T		F	T		F	T		F	T	
F	F	F		T		F		T		T	F		F	T		F	T	
				*							*					*		

SECTION 3.1

1(b) (x)Wx
  (f) (x)(Ox → Hx)
  (j) −(x)(Lx → Ox) or (∃x)(Lx & −Ox)
  (n) (x)[Bx → (−Dx & Ux)]

2(b) Some people accept bribes.
  (f) Some judges are corrupt.
  (j) All judges are lawyers and government employees.

3(b) (x)(Bx → −Wx)
  (f) (x)−(Bx v Lx) or (x)(−Bx & −Lx)
  (j) −(∃x)(Ax & −Dx)
  (n) (x)Gx & −(∃x)−Gx

SECTION 3.2

1(b)	(4)	(∃x)−(Dx → Fx)	3 QE
	(5)	−(Da → Fa)	4 EO
	(6)	Da → Ea	1 UO
	(7)	Ea → Fa	2 UO
	(8)	Da → Fa	6,7 CH
	(9)	(Da → Fa) & −(Da → Fa)	8,5 &I
	(10)	(x)(Dx → Fx)	3-9 −O
3.	(1)	(x)(Bx → Vx)	A
	(2)	(x)Bx	A
	(3)	−(x)Vx	PA
	(4)	(∃x)−Vx	3 QE
	(5)	−Va	4 EO
	(6)	Ba → Va	1 UO
	(7)	Ba	2 UO
	(8)	Va	6,7 →O
	(9)	Va & −Va	8,5 &I
	(10)	(x)Vx	3-9 −O
7.	(1)	(∃x)(Kx & Rx & −Px)	A
	(2)	(x)(Kx ↔ Px)	PA
	(3)	Ka & Ra & −Pa	1 EO
	(4)	Ka ↔ Pa	2 UO
	(5)	Ka	3 &O
	(6)	Ka → Pa	4 ↔O
	(7)	Pa	6,5 →O
	(8)	−Pa	3 &O

	(9)	Pa & −Pa	7,8 &I
	(10)	−(x)(Kx ↔ Px)	2-9 −I

11. 1	(1)	(x)Cx v (x)−Cx	A
2	(2)	(∃x)−Cx	PA
3	(3)	−(x)−Cx	PA
1,3	(4)	(x)Cx	1,3 DA
2	(5)	−Ca	2 EO
1,3	(6)	Ca	4 UO
1,2,3	(7)	Ca & −Ca	6,5 &I
1,2	(8)	(x)−Cx	3-7 −O
1	(9)	(∃x)−Cx → (x)−Cx	2-8 →I

SECTION 3.3

1(b) The first formula is not a quantification. The argument can be rewritten as:
(x)Cx, (x)Dx ⊢ (x)(Cx & Dx)

(1)	C	A	valid
(2)	D	A	
(3)	C & D	1,2 &I	

(f) Some of the formulas are not universal and the conclusion is not existential. The argument cannot be rewritten in an equivalent form that reaches stage four.

2(b) (x)Ax, (∃x)Bx ⊢ (∃x)(Bx & Ax)

A	B	A	B	⊢ B & A
T	T	T	T	T
F	T	F	T	F
T	F	T	F	F
F	F	F	F	F

3(b) (x)(Hx → Ax), (x)(Rx → −Hx) ⊢ (x)(Rx → −Ax)   invalid

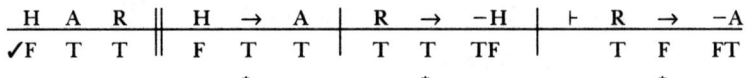

5. (x)(Px ↔ Wx), (x)(−Cx → −Wx) ⊢ (x)(Px ↔ Cx)   invalid

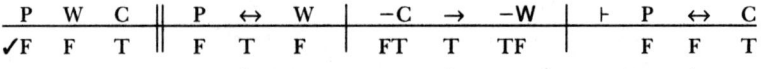

Appendix Four: Solutions to Starred Exercises   361

9.   (x)−Gx, (x)(−Wx → Fx), (x)(Gx → Wx) ⊢ (x)Fx   invalid

G	W	F	‖	−G		−W	→	F		G	→	W	⊢	F
✓F	T	F	‖	TF		FT	T	F		F	T	T		F
				*			*				*			

11(b) (x)(Ax → −Bx), −(∃x)(Ax & Bx) ≡ (x)−(Ax & Bx)

A	B	‖	A	→	−B		−(A	&	B)
T	T	‖	F		F		F		T
F	T	‖	T		F		T		F
T	F	‖	T		T		T		F
F	F	‖	T		T		T		F
				*				*	

12. (∃x)(Ax & Bx), (∃x)(Bx & Ax)   equivalent

A	B	‖	A	&	B		B	&	A
T	T	‖		T				T	
F	T	‖		F				F	
T	F	‖		F				F	
F	F	‖		F				F	

13. (x)(Ax → −Bx), (x)(−Bx → −−Ax)   not equivalent

	A	B	‖	A	→	−B		−B	→	−−A
	✓F	F	‖	F	T	TF		TF	F	FTF
or	✓T	T	‖	T	F	FT		FT	T	TFT
					*				*	

SECTION 3.4

3. (x)[(Ax & Bx) → Sx] ⊢ (x)[(Bx & Sx) → Ax]

   Universe: animals; Ax = x is a dog, Bx = x is male, Sx = x is a mammal

   All dogs that are male are mammals.   **(T)**
   So, all male mammals are dogs.   **(F)**

7. (x)(Bx → Lx) & (x)(Ux → Dx)
    ⊢ (x)(Lx → Bx) & (x)(Dx → Ux)

   Universe: animals; Bx = x is a dog, Lx = x is a canine, Ux = x is a cat, Dx = x is a feline

   All dogs are canines, all cats are felines.   **(T)**
   So, all canines are dogs, and all felines are cats.   **(F)**

9(b) (∃x)(Ax & −Bx), (∃x)(Bx & −Ax)

Universe: people; Ax = x is a mother, Bx = x is female

Some mothers are not females. (F)
Some females are not mothers. (T)

SECTION 3.5

1(b) −(∃x)−Lxg or (x)−−Lxg or (x)Lxg
 (f) (x)(∃y)(Ty & Dxy)
 (j) (∃x)[Jx & (y)(Ay → Rxy)]
 (n) (x)(y)[(Dx & Mx & Sy) → −Txy] or (x)[(Dx & Mx) → (y)(Sy → −Txy)]

2(b) Everyone loves everyone.
 (f) There is a person who loves, and is loved by, some person.
  or There are people who love each other.
  The first sentence is preferable because it, like the formula, is compatible with the lover and the beloved being the same person.
 (j) Each person is loved by someone whom he or she loves.

3(b) (∃x)[Px & (y)(Ey → Dxy)]
 (f) (x)(y)[(Fxy & Dy) → −Lxy] or
  (x)(y)[(Fxy & Dx) → −Lyx]
 (k) *This sentence is amphibolous. The two meanings and their symbolizations:*

  Anyone whom a male jockey marries is a tall woman.

  (x){(Mx & Jx) → (y)[Axy → (Ty & Wy)]} or
  (x)(y)[(Mx & Jx & Axy) → (Ty & Wy)]

  Each male jockey marries some tall woman.

  (x)[(Mx & Jx) → (∃y)(Axy & Ty & Wy)]

 (n) (x)(y)(z)[(Mx & Wyxz) → Eyxz]

5(b) (3) (∃y)−(x)Lxy   2 QE
   (4) −(x)Lxa   3 EO
   (5) (∃x)−Lxa   4 QE
   (6) −Lba   5 EO
   (7) Lba   1 UO
   (8) Lba & −Lba   7,6 &I
   (9) (y)(x)Lxy   2-8 −O

7. (1) Bsa   A
   (2) Psag   A
   (3) (x)(y)(z)[(Bxy & Pxyz) → Bxz]   A
   (4) (Bsa & Psag) → Bsg   3 UO

	(5)	Bsa & Psag	1,2 &I
	(6)	Bsg	4,5 →O

11.	(1)	(x)(y)[Dxy → (Ax & Ay)]	A
	(2)	−Am	A
	(3)	Dsm	PA
	(4)	Dsm → (As & Am)	1 UO
	(5)	As & Am	4,3 →O
	(6)	Am	5 &O
	(7)	Am & −Am	6,2 &I
	(8)	−Dsm	3-7 −I

A more difficult six-line proof is possible.

15.	(1)	Da	A
	(2)	−(x)[(Px & Hxa) → (∃y)(Dy & Hxy)]	PA
	(3)	(∃x)−[(Px & Hxa) → (∃y)(Dy & Hxy)]	2 QE
	(4)	−[(Pb & Hba) → (∃y)(Dy & Hby)]	3 EO
	(5)	Pb & Hba & −(∃y)(Dy & Hby)	4 AR
	(6)	−(∃y)(Dy & Hby)	5 &O
	(7)	(y)−(Dy & Hby)	6 QE
	(8)	−(Da & Hba)	7 UO
	(9)	−Hba	8,1 CA
	(10)	Hba	5 &O
	(11)	Hba & −Hba	10,9 &I
	(12)	(x)[(Px & Hxa) → (∃y)(Dy & Hxy)]	2-11 −O

17. (x)(Kxx → Bx) ⊢ (x)[(∃y)Kxy → Bx]

The conclusion may also be symbolized '(x)(y)(Kxy → Bx)'.

Universe: people; Kxy = x hates y, Bx = x is a self-hater

All who hate themselves are self-haters. (T)
So, whoever hates anyone is a self-hater. (F)

21. (x)[Ax → (∃y)Cyx], (x)(Sx v Ix), (x)[Ix → −(∃y)Cxy]
    ⊢ (∃x)[Sx & (y)(Ay → Cxy)]
    The third premise may also be symbolized '(x)[Ix → (y)−Cxy]'.

Universe: positive integers; Ax = x is odd, Cxy = x is greater than y, Sx = x is greater than 1, Ix = x is less than 2

For any odd positive integer there is a greater. (T)
**Every positive integer is either greater
   than 1 or less than 2.** (T)
**Any positive integer less than 2 is greater
   than no positive integer.** (T)
**So, there is a positive integer greater than 1
   that is greater than all odd integers.** (F)

SECTION 3.6

1(b) $(x)(Rx \to g \neq x)$
 (f) $(x)(y)(x \neq y \to Bxy)$
 (j) $(x)(y)[(Cx \& Txy) \to y = g]$

2(b) Someone loves another.
 (f) Mother Teresa loves everyone else.
 (j) Exactly one person loves himself or herself.

3(b)
(2)	$(\exists x)(x \neq x)$	1 QE
(3)	$a \neq a$	2 EO
(4)	$a = a$	=I
(5)	$a = a \ \& \ a \neq a$	4,3 &I
(6)	$(x)(x = x)$	1-5 $-$O

5.
(1)	Wd	A
(2)	$(x)(Wx \to x = c) \ \& \ Wc$	A
(3)	$(x)(Wx \to x = c)$	2 &O
(4)	$Wd \to d = c$	3 UO
(5)	$d = c$	4,1 $\to$O

9.
(1)	$Td \ \& \ (x)(x \neq d \to -Tx)$	A
(2)	$(\exists x)(Px \ \& \ Tx)$	A
(3)	$(x)(x \neq d \to -Tx)$	1 &O
(4)	$Pa \ \& \ Ta$	2 EO
(5)	$a \neq d \to -Ta$	3 UO
(6)	Ta	4 &O
(7)	$--Ta$	6 DN
(8)	$-(a \neq d)$	5,7 MT
(9)	$a = d$	8 DN
(10)	Pa	4 &O
(11)	Pd	9,10 =O

14. $(x)(y)(x \neq y \to Dxy) \vdash (x)(y)Dxy$

Universe: integers; Dxy = x is either larger than or smaller than y

Each integer is either larger or smaller than every other integer. (T)
So, each integer is either larger or smaller than every integer. (F)

The conclusion is false because some (in fact, every) integer is neither larger nor smaller than itself.

SECTION 3.7

2. The Louisiana Creationism Act is consistent with the Establishment Clause of the First Amendment only if (1) the Louisiana legislature's preeminent purpose in adopting the law was secular, (2) the Act's principal effect neither advances

nor inhibits religion, and (3) the Act does not result in an excessive entanglement of government with religion. The Louisiana legislature's preeminent purpose in adopting the law was not secular. Therefore, the Louisiana Creationism Act is inconsistent with the Establishment Clause of the First Amendment.

(1)		C → [S & −(A v I) & −E]	A
(2)		−S	A
(3)		C	PA
(4)		S & −(A v I) & −E	1,3 →O
(5)		S	4 &O
(6)		S & −S	5,2 &I
(7)		−C	3-6 −I

A more difficult five-line proof is possible.

6 (a)   (x)(y)(z)((Sxy & Kxz & Hz) → {(Myx ↔ Fx) & [Ayx ↔ (Ux & −Fx)]})

(c)			
	(1)	(S & K) → [A ↔ (U & −F)]	A
	(2)	S & K	A
	(3)	U & −F	A
	(4)	A ↔ (U & −F)	1,2 →O
	(5)	(U & −F) → A	4 ↔O
	(6)	A	5,3 →O

SECTION 4.2

2 (a)   Plaintiff
         B → D, B ⊢ D

   (b)   Defendant (and Lower Court)   N: B → D
         D → N, −N ⊢ −D

   (c)   Court of Appeals   N: D → N
         B → D, B ⊢ D

6 (a)   Prosecution
         (x)(y){[Gx & Tyx & (∃z)(Azx & Uz & Cyz)] → Ly},
         Gg & Tcg & (∃x)(Axg & Ux & Ccx) ⊢ Lc

   (b)   Chapman   N: Gg & Tcg & (∃x)(Axg & Ux & Ccx)
         (x)(y)(z)[Bxyz → (−Sx → −Ux)],
         (x)(y)(z)[(Bxyz & My & Fz & Oz) → −Sx],
         Bacg & Mc & Fg & Og ⊢ −Ua

   (c)   Court   N: (x)(y)(z)[Bxyz → (−Sx → −Ux)]
         (x)(y)(z)[(Bxyz & −Wyz) → Ux], Bacg & −Wcg ⊢ Ua

SECTION 4.3

1(b) Allegation: Nd & Dds & Lt
     Nx = x drove negligently, Dxy = x was driving at speed y, s = 65 mph, Lx = the speed limit was x, t = 30 mph

Denial: Defendant denies driving negligently. Defendant also denies driving at sixty-five miles per hour or at any other speed in excess of the speed limit.

N: Nd
N: Dds v (∃x)(∃y)(Ddx & Ly & Mxy)
Mxy = x is more than y

(f) Allegation:

(∃x)(Sx & Mdx & Ndx & Ppx & Ipx)

Sx = x is a slide, Mxy = x is the manufacturer of y, Nxy = x was negligent in the manufacture of y, Pxy = x was playing on y, Ixy = x was injured while playing on y

Denial: Defendant is without information as to whether Plaintiff sustained injuries while playing on a slide, denies that it was the manufacturer of any slide on which Plaintiff was playing, and denies that it was negligent in the manufacture of that or any other slide.

N: (∃x)(Sx & Ppx & Ipx)
N: (∃x)(Sx & Mdx & Ppx)
N: (∃x)(Sx & Mdx & Ndx & Ppx) v (∃x)(Sx & Mdx & Ndx)

3. Plaintiff: (x)(Rx → Odxp), Rt ⊢ Odtp
(Rx = x is a reasonable attorney's fee, Oxyz = x is obliged by contract to pay y to z, t = the sum of $2,000)

Defendant: N: [Rt v (∃x)(Gx & Rx)]
(Gx = x is a sum greater than $500)
Defendant denies that $2,000 or any sum greater than $500 is a reasonable attorney's fee in the case.

7. Plaintiff: (x)(y)(z)(w)[(Nx & Eyxzw & Dyxzw) → Oyzx], Ni & Edipf & Ddipf ⊢ Odpi

Defendant: N: [Edipf v Ddipf v (∃x)(Edipx v Ddipx)]
Defendant denies that he either executed or delivered the note at Fairfield, Iowa or anywhere else.

SECTION 4.4

2(a) Government: (x)(Mx → Jx), Mc ⊢ Jc
(Universe: cases; Mx = x involves military personnel, Jx = x is subject to court martial jurisdiction, c = this case)

(b) Court:

D: (x)(Mx → Jx)
C: (x)[(Mx & Sx) → Jx]
N: (x)[(Mx & −Sx) → Jx]
N: −Sc
(Sx = x is service connected)

6(a) Angélique: (x)(y)(z)(Lxy → Sxyz) ⊢ Lta → Star
(Sxyz = x submits to y's wish in the matter of z, r = the timing of Angélique's response to Thomas's proposal)

(b) Thomas:

D: (x)(y)(z)(Lxy → Sxyz)

C: (x)(y)(z)[(Lxy & −Pzxy) → Sxyz]
N: (x)(y)(z)[(Lxy & Pzxy) → Sxyz]
(Pxyz = x pertains to y's possession of z)

SECTION 4.5

2. Defendant's argument: (x)(y)(z)(w)[(Exyz & Swyx) → −Lyzw], (∃x)(Exdp & Sadx) ⊢ −Ldpa
(Exyz = x is an act of entry onto y's property by z, Sxyz = x is an act by y to stop act z, Lxyz = x is liable to y on account of z, a = defendant's act of cutting loose plaintiff's sailboat)

Plaintiff's legal premise distinguished:
D: (x)(y)(z)[(Ixy & Hxz) → Lyzx]
C: (x)(y)(z){[Ixy & Hxz & −(∃w)(Ewyz & Sxyw)] → Lyzx}
N: (x)(y)(z){[Ixy & Hxz & (∃w)(Ewyz & Sxyw)] → Lyzx}
(Ixy = x is an intentional act of y, Hxy = x harms y)

Plaintiff's avoidance:
(x)(y)(z)(w)[(Exyz & Nx) → (Swyx → Lyzw)], (∃x)(Exdp & Nx & Sadx) ⊢ Ldpa
(Nx = x is necessary to avoid danger)

Defendant's legal premise distinguished:
D: (x)(y)(z)(w)[(Exyz & Swyx) → −Lyzw]
C: (x)(y)(z)(w)[(Exyz & −Nx & Swyx) → −Lyzw]
N: (x)(y)(z)(w)[(Exyz & Nx & Swyx) → −Lyzw]

6. Defendant's argument: (x)(y)(z)[(∃w)Rwxyz → −Lyxz], (∃x)Rxpda ⊢ −Ldpa
(Rxyzw = x is a release executed by y releasing z for liability on account of w, Lxyz = x is liable to y on account of z, a = the act complained of)

Plaintiff's legal premise distinguished:
D: (x)(y)(z)(Nxyz → Lyzx)
C: (x)(y)(z){[Nxyz & −(∃w)Rwzyx] → Lyzx}
N: (x)(y)(z){[Nxyz & (∃w)Rwzyx] → Lyzx}
(Nxyz = x is y's act of negligently running over z)

Plaintiff's avoidance:
(x)(y)(z){[Nxyz & (w)(Rwzyx → Fzw)] → Lyzx}, (x)(Rxpda → Fpx), Nadp ⊢ Ldpa
(Fxy = x was induced by fraud to execute y)

Defendant's legal premise distinguished:
D: (x)(y)(z)[(∃w)Rwxyz → −Lyxz]
C: (x)(y)(z)[(∃w)(Rwxyz & −Fxw) → −Lyxz]
N: (x)(y)(z)[(∃w)(Rwxyz & Fxw) → −Lyxz]

SECTION 5.1

1(b) (A & B) → C
 (f) (A → C) & (B → C)

2(b) False. Invalid: A → C ⊢ (A v B) → C

4(a) (x)[Ex ↔ (Ax & Cx & Ix)] or
 (x)[−Ex ↔ −(Ax & Cx & Ix)]

 (b) *(literal)* (x){[−(Ax & Cx) → −Ex] & (−Ix → −Ex)} or
 (x)[−(Ax & Cx & Ix) → −Ex] or
 (x)[Ex → (Ax & Cx & Ix)]

 (c) majority view: each of the listed qualifications is a necessary condition for eligibility; jointly they are a sufficient condition for eligibility.
 dissenting view: each of the listed qualifications is a necessary condition for eligibility; no sufficient condition for eligibility is provided.

8. (x)(−Ex → Sx) ⊢ (x)(Ex → −Sx)

(Universe: claimants who die; Ex = x's claim is extinguished by x's death, Sx = substitution of the proper parties may be ordered by the court in x's case)

Propositional analogue: −E → S ⊢ E → −S

E	S		−E	→	S	⊢	E	→	−S
✓T	T		FT	T	T		T	F	FT
				*				*	

Augmented argument (P1): (x)(−Ex ↔ Sx) ⊢ (x)(Ex → −Sx)
Propositional analogue: −E ↔ S ⊢ E → −S

E	S		−E	↔	S	⊢	E	→	−S
T	T		F	F				F	F
F	T		T	T				T	F
T	F		F	T				T	T
F	F		T	F				T	T
				*					*

12. (x)[Rx → (Ax & Bx)] ⊢ (x)(−Rx → −Ax)

(Universe: rulemaking proceedings; Rx = rules are required by statute to be made on the record in proceeding x after opportunity for an agency hearing, Ax = § 556 applies to x, Bx = § 557 applies to x)

Propositional analogue: R → (A & B) ⊢ −R → −A

R	A	B		R	→	(A	&	B)	⊢	−R	→	−A
✓F	T			F	T		T			TF	F	FT
					*						*	

(It is immaterial what value is assigned to 'B'.)

Augmented argument (P1):
(x)[Rx → (Ax & Bx)] & (x)[(Ax v Bx) → Rx]
⊢ (x)(−Rx → −Ax)

Note that this tempting symbolization of the augmented premise is incorrect:
'(x) [Rx ↔ (Ax & Bx)]'.

Propositional analogue: [R → (A & B)] & [(A v B) → R] ⊢ −R → −A

1	(1)	[R → (A & B)] & [(A v B) → R]	A
2	(2)	−R	PA
1	(3)	(A v B) → R	1 &O
1,2	(4)	−(A v B)	3,2 MT
1,2	(5)	−A & −B	4 DM
1,2	(6)	−A	5 &O
1	(7)	−R → −A	2-6 →I

SECTION 5.2

1(b) (x)(Tx → Mx)
 (f) (x)(Wx → Px)
 (j) (x)[(Dx & Hx) → Cx]
 (n) (∃x)(Jx & Ax)
 (r) (x)[Tx → (y)(Bxy → Hxy)]
 (v) (x)[(Ax & −Sx) → −Bx]

2(b) (x)[(Sx & −Ex) → Ix]
 (f) (x)(y)[(Jx & Pyx) → (Oxy & Dxy)]
 (j) (x)(y)(z){[(Pxyz v Mxyz) & (Szx v Szy)] → −Vxy}

3. (F1) (x)[Ax → (Fx & Rx)]
 (F2) (x)[Fx → (Ax → Rx)]
 (F3) (x)[Rx → (Ax → Fx)]

7. (F1) (x)[(Ox v Ex v Lx) → Px]
 (F2) (x)(Ox → Px) & (x)(Ex → Px) & (x)(Lx → Px) *or*
  (x)[(Ox → Px) & (Ex → Px) & (Lx → Px)]

SECTION 5.3

2. −(x)(Sx → −Ux)
 (Sx = x is an act of sexual intercourse with a girl over the age of 16, Ux = x is unlawful)
 Note that this symbolization would be incorrect:
 '(x)−(Sx → −Ux)'.

6. (x)(y)(z)[Ixy → (Vzyx → Hxz)]
 (Ixy = x is the innocent party to a putative marriage to y, Vxyz = x is a right in y's property that z would have if y and z were validly married, Hxy = x has y)
 Note that this symbolization would be incorrect:
 '(x)(y)(z)[(Vxy → Hzy) → (Ixy → Hzy)]'.

 (Vxy = x is validly married to y, Hxyz = x has right y in z's property)

SECTION 5.4

1(b) ($\exists$x)Ax → (y)Ry
(x)[Ax → (y)Ry]
(x)[($\exists$y)Ay → Rx]
(x)(y)(Ax → Ry)

2(b) (x)[Vx → ($\exists$y)(Qy & Sxy)]
(x)($\exists$y)[Vx → (Qy & Sxy)]

(f) (x)[(Px & Hxa) → ($\exists$y)(Dy & Hxy)] or
(x)($\exists$y) [(Px & Hxa) → (Dy & Hxy)]

4(b)
	(1)	P → (x)Fx	A
	(2)	−(x)(P → Fx)	PA
	(3)	($\exists$x)−(P → Fx)	2 QE
	(4)	−(P → Fa)	3 EO
	(5)	P & −Fa	4 AR
	(6)	P	5 &O
	(7)	(x)Fx	1,6 →O
	(8)	Fa	7 UO
	(9)	−Fa	5 &O
	(10)	Fa & −Fa	8,9 &I
	(11)	(x)(P → Fx)	2-10 −O

1	(1)	(x)(P → Fx)	A
2	(2)	P	PA
3	(3)	−(x)Fx	PA
3	(4)	($\exists$x)−Fx	3 QE
3	(5)	−Fa	4 EO
1	(6)	P → Fa	1 UO
1,2	(7)	Fa	6,2 →O
1,2,3	(8)	Fa & −Fa	7,5 &I
1,2	(9)	(x)Fx	3-8 −O
1	(10)	P → (x)Fx	2-9 →I

5.
	(1)	($\exists$x)Ix → (x)Ix	A	(F1 entails F2)
	(2)	−(x)[($\exists$y)Iy → Ix]	PA	
	(3)	($\exists$x)−[($\exists$y)Iy → Ix]	2 QE	
	(4)	−[($\exists$y)Iy → Ia]	3 EO	
	(5)	($\exists$y)Iy & −Ia	4 AR	
	(6)	($\exists$y)Iy	5 &O	
	(7)	Ib	6 EO	
	(8)	($\exists$x)Ix	7 EI	
	(9)	(x)Ix	1,8 →O	
	(10)	Ia	9 UO	
	(11)	−Ia	5 &O	

(12)	Ia & −Ia	10,11 &I	
(13)	(x)[(∃y)Iy → Ix]	2-12 −O	

(1)	(x)[(∃y)Iy → Ix]	A	(F2 entails F3)
(2)	−(x)[Ix → (y)Iy]	PA	
(3)	(∃x)−[Ix → (y)Iy]	2 QE	
(4)	−[Ia → (y)Iy]	3 EO	
(5)	Ia & −(y)Iy	4 AR	
(6)	−(y)Iy	5 &O	
(7)	(∃y)−Iy	6 QE	
(8)	−Ib	7 EO	
(9)	(∃y)Iy → Ib	1 UO	
(10)	Ia	5 &O	
(11)	(∃y)Iy	10 EI	
(12)	Ib	9,11 →O	
(13)	Ib & −Ib	12,8 &I	
(14)	(x)[Ix → (y)Iy]	2-13 −O	

(1)	(x)[Ix → (y)Iy]	A	(F3 entails F4)
(2)	−(x)(y)(Ix → Iy)	PA	
(3)	(∃x)−(y)(Ix → Iy)	2 QE	
(4)	−(y)(Ia → Iy)	3 EO	
(5)	(∃y)−(Ia → Iy)	4 QE	
(6)	−(Ia → Ib)	5 EO	
(7)	Ia & −Ib	6 AR	
(8)	Ia → (y)Iy	1 UO	
(9)	Ia	7 &O	
(10)	(y)Iy	8,9 →O	
(11)	Ib	10 UO	
(12)	−Ib	7 &O	
(13)	Ib & −Ib	11,12 &I	
(14)	(x)(y)(Ix → Iy)	2-13 −O	

1	(1)	(x)(y)(Ix → Iy)	A	(F4 entails F1)
2	(2)	(∃x)Ix	PA	
3	(3)	−(x)Ix	PA	
3	(4)	(∃x)−Ix	3 QE	
2	(5)	Ia	2 EO	
3	(6)	−Ib	4 EO	
1	(7)	Ia → Ib	1 UO	
1,2	(8)	Ib	7,5 →O	
1,2,3	(9)	Ib & −Ib	8,6 &I	
1,2	(10)	(x)Ix	3-9 −O	
1	(11)	(∃x)Ix → (x)Ix	2-10 →I	

SECTION 5.5

2. (x)(Bx → −Px), (x)(Bx → Fx) ⊢ −(x)(Fx → Px)
   (1) Universe: Catholics; Bx = x is a Lutheran, Px = x is a Catholic, Fx = x is a Protestant

   No Lutherans are Catholics. (T)
   All Lutherans are Protestants. (T)
   So, it is false that all Protestants (in the universe of Catholics) are Catholics. [Restated: There are Protestants (in the universe of Catholics) who are not Catholics.] (F)

   (2) Bats exist.
   (x)(Bx → −Px), (x)(Bx → Fx), (∃x)Bx ⊢ −(x)(Fx → Px)
   (3) Conclusion transformed into a quantification: (∃x)−(Fx → Px)
   Proof of propositional analogue:

(1)	B → −P	A
(2)	B → F	A
(3)	B	A
(4)	F	2,3 →O
(5)	−P	1,3 →O
(6)	F & −P	4,5 &I
(7)	−(F → P)	6 AR

6 (a) (x)(Tx → Ux), (x)(Px → −Ux)
   (b) Universe: people; Tx = x is a Lutheran, Ux = x is a Protestant, Px = x is a Moslem

   All Lutherans are Protestants. (T)
   No Moslems are Protestants. (T)

   (c) Some coin combinations totalling 50¢ contain pennies.
   or
   Some coin combinations containing pennies total 50¢.

   (d)
(1)	(x)(Tx → Ux)	A
(2)	(x)(Px → −Ux)	A
(3)	(∃x)(Tx & Px)	A
(4)	Ta & Pa	3 EO
(5)	Ta → Ua	1 UO
(6)	Ta	4 &O
(7)	Ua	5,6 →O
(8)	Pa → −Ua	2 UO
(9)	Pa	4 &O
(10)	−Ua	8,9 →O
(11)	Ua & −Ua	7,10 &I

Appendix Four: Solutions to Starred Exercises

SECTION 6.2

2. The Marines wish to recruit a few good men. Sam, Charlie, and Willie are good men. Therefore, they belong to a class of people from which the Marines wish to recruit a few members.

6. O'Brien believes it to be the case that all Irish children should learn Gaelic. Maureen is an Irish child. So, she belongs to a class of people of which O'Brien believes it to be the case that all members should learn Gaelic.

SECTION 6.3

2. *One possible solution:*
Students SATISFY the mathematics requirement only if they are EXEMPTED by placement or they take MTH 101 and either MTH 102 or MTH 103.

(x)(Sx → {Ex v [Ax & (Bx v Cx)]})

(Universe: students; Ax = x takes MTH 101, Bx = x takes MTH 102, Cx = x takes MTH 103)

SECTION 6.4

2 (a) Joe's version *(de re)*:   Ls ↔ (Tsc & Bcj & Dsjc)
Prosecutor's version *(de dicto)*:   Ls ↔ (Tsc & Bcj & Isa)

(b) Sam is not guilty under the second version because he does not intend that the owner of the car be deprived of it. Therefore the third conjunct is false.

6. Majority *(de re)*: (∃x)(∃y)(Rx & Vyx & Kdy) → Gd
(Rx = x is a regulation for the safe transportation of corrosive liquids, Vxy = act x violates y, Kxy = x knowingly performs y, d, Gx)

Dissent *(de dicto)*: Tda → Gd
(Txy = x knowingly makes true proposition y, d, a = the proposition that defendant violates a regulation for the safe transportation of corrosive liquids, Gx)

Appendix Five
Rules of Inference

PROPOSITIONAL LOGIC

	THE TEN PRIMITIVE INFERENCE RULES	
	IN	**OUT**
&	From A and B derive $A \& B$.	From $A \& B$ derive either A or B.
$\rightarrow$	From the derivation of B from assumption A (and perhaps other assumptions) derive $A \rightarrow B$.	From $A \rightarrow B$ and A derive B.
$-$	From the derivation of $B \& -B$ from assumption A (and perhaps other assumptions) derive $-A$.	From the derivation of $B \& -B$ from assumption $-A$ (and perhaps other assumptions) derive A.
v	From A derive either $A \vee B$ or $B \vee A$.	From $A \vee B$, $A \rightarrow C$, and $B \rightarrow C$ derive C.
$\leftrightarrow$	From $A \rightarrow B$ and $B \rightarrow A$ derive $A \leftrightarrow B$.	From $A \leftrightarrow B$ derive either $A \rightarrow B$ or $B \rightarrow A$.

	ASSUMPTION-DEPENDENCE PRINCIPLES
A	An assumption depends on itself.
$\rightarrow$I	$A \rightarrow B$ depends on all of the assumptions on which B depends—less A.
$-$I	$-A$ depends on all of the assumptions on which $B \& -B$ depends—less A.
$-$O	A depends on all of the assumptions on which $B \& -B$ depends—less $-A$.

Appendix Five: Rules of Inference

THE SEVEN DERIVED INFERENCE RULES		
CHAIN ARGUMENT	**CH**	From $A \to B$ and $B \to C$ derive $A \to C$.
MODUS TOLLENS	**MT**	From $A \to B$ and $-B$ derive $-A$.
DISJUNCTIVE ARGUMENT	**DA**	From $A \lor B$ and $-A$ derive B. From $A \lor B$ and $-B$ derive A.
CONJUNCTIVE ARGUMENT	**CA**	From $-(A \& B)$ and A derive $-B$. From $-(A \& B)$ and B derive $-A$.
DOUBLE NEGATION	**DN**	From A derive $--A$ and vice versa.
DE MORGAN'S LAW	**DM**	From $-(A \& B)$ derive $-A \lor -B$ and vice versa. From $-(A \lor B)$ derive $-A \& -B$ and vice versa.
ARROW	**AR**	From $A \to B$ derive $-A \lor B$ and vice versa. From $A \to B$ derive $-(A \& -B)$ and vice versa. From $-(A \to B)$ derive $A \& -B$ and vice versa.

PREDICATE LOGIC

UO	From a universal quantification derive any instance of it.
EO	From an existential quantification derive any instance of it, *provided that* the individual constant being introduced does not occur in the symbolization of the argument being tested or on any line above the line derived.
EI	Derive an existential quantification from any instance of it.
QE	From $-(x)Ax$ derive $(\exists x)-Ax$ and vice versa. From $-(\exists x)Ax$ derive $(x)-Ax$ and vice versa.
=O	From $a = b$ (or $b = a$) and Fa derive Fb.
=I	$a = a$ may be introduced at any point in a proof.

INDEX

A

Abduction, 8n.
Admission, 210
Affirmative defense, 232
Affirming the consequent, 32
'All', 117
Alternative symbols, 350
'Although', 19
Ampersand (&), 19
Ampersand In Rule, 22
Ampersand Out Rule, 22
Amphiboly, 11–12, 29
Analogical reasoning, 333–35
Analysis by instantiation, method of, 199–201
   contrasted with propositional analogue method, 201
'And', 17–21
'And/or', 60
Antecedent, 27
'Any', 118
Argument:
   defined, 4–5
   evaluating content, 3–4
   evaluating form, 3–4
   logician's and lawyer's concepts distinguished, 209
   overall structure, 21–22
Aristotelian logic, 116, 116n., 120n.
Arrow (→), 27

Arrow In Rule, 35
   assumption-dependence principle, 37
   restrictions, 40
   strategy, 39
Arrow Out Rule, 30
Arrow Rule, 82, 84
Arrow Strategy, 87
Assumption, 23
   original, 36
   provisional, 36
Assumption-dependence column, 36–38
   relaxation of requirement, 131
Assumption-dependence principles:
   Arrow In Rule, 37
   Assumption Rule, 37
   Dash In Rule, 50
   Dash Out Rule, 52
   Identity In Rule, 188
   standard, 37
Assumption Rule, 36
   assumption-dependence principle, 50
Asymmetry, 169–70
Attempt and intensionality, 320–23

B

'Because', 21

Biconditional:
  defined, 73
  symbolizing, 73–74
'But', 19

C

Chain argument pattern, 34–35
Chain Argument Rule, 82
Charts (*see* Tables)
Comics and cartoons, 41, 42, 44, 45, 51, 58, 86, 94, 103, 106, 123, 157, 158, 182, 196, 226
Compound statement, 17–18
Concedo (C:), 210
Conclusion, 4
Conclusion-introducing terms, 21
Conditional:
  with conditional antecedent, 261–62
  counterfactual (or subjunctive), 265–66
  defined, 27
  double and multiple, 29–30
  grouping principle, 29
  negated, 262–64
  symbolizing, 27–30
Condition contrary to fact, 265–66
Conditions (sufficient, necessary):
  Inclusio unius . . . , principle of, 238–39
  and legal principles, 236–39
  necessary, 33–34, 235
    disjunctive and conjunctive, 236
    exhaustive sets, 239
  necessary and sufficient, 73, 236
  sufficient, 33–34, 235
    disjunctive and conjunctive, 236
    exhaustive sets, 239
    sole, 236–38
Conjunct, 18
Conjunction:
  defined, 18
  double and multiple, 24
  symbolizing, 18–21
Conjunctive argument pattern, 82–83
  counterfeit pattern, 82–83
Conjunctive Argument Rule, 82
Connective, 18
  scope, 29–30
Consent and intensionality, 297–98, 309–10
Consequent, 27
'Consequently', 21
Constant, individual, 115
Contradiction, standard, 50

Contradictory, 146
Contrapositive, 52
Conversion, 146
Counterfactual conditional, 265–66

D

Dash (-), 47
  contrasted with minus sign, 48
Dash In Rule, 50
  assumption-dependence principle, 50
  strategy, 52
Dash Out Rule, 52
  assumption-dependence principle, 52
  strategy, 53
De dicto statement, 306–9
Deduction, 5, 7–8, 8n.
  in legal reasoning, 333–35
Definition of key terms, 4–8
Demurrer, 215, 232
De Morgan's Law Rule, 82, 83–84
De Morgan's Law Strategy, 87
Denial, 209–10
  argumentative, 217
  distinguished from refutation, 209
  and factual premises, 215–17
  legal and ethical rules concerning, 217–18
  and legal premises, 214–15
  rules for, 217
Denying the antecedent, 51
Deontic logic, 336–46
  defined, 336
  formation rules, 349
  non-deontic analogues, method of, 342–43
  operators, 336–37
  principles, 338
  proofs, 338–41
  pure vs. applied, 345–46
  rules of inference, 338
  shortcomings, 343–46
De re statement, 306–9
Dilemma, 64
Disjunct, 59
Disjunction:
  defined, 59
  inclusive vs. exclusive, 59–60
  grouping principle, 61
  multiple, 60–61
  symbolizing, 59–61
Disjunctive argument pattern, 82–83
  counterfeit pattern, 82–83

Disjunctive Argument Rule, 82
Disputation, 207–10, 230–33
Distinguishing, technique of, 221–24, 232
   and public policy, 224–25
   rules, 232–33
Distinguo (D:), 221–22
   in Molière play, 224
Double arrow (↔), 73
Double Arrow In Rule, 74
Double Arrow Out Rule, 74
Double Negation Rule, 82, 83

E

'Each', 118
Entailment, 6–7
   contrasted with implication, 28n.
Enthymeme, 2, 214
'Even though', 19
'Every', 118
Exercises, solutions to starred, 351–73
Existential import, 285–92
   defined, 286, 290
   existential presuppositions of standard logic, 291–92
   of general statements, 286–90
   hypothetical vs. categorical interpretations, 288
   hypothetical worlds, 290
   and presupposition, 291n.
   of singular statements, 290–92
   universal vs. particular statements, 286–88
   vacuously true universals, 289
Existential quantifier [(∃x)], 115
Existential Quantifier In Rule, 133
Existential Quantifier Out Rule, 127
   restrictions, 132–33
Extension, 298–99
Extensional context, 299

F

'False', 47
Falsity, 7
'For', 21
Formal proof:
   checking off lines, 87–88
   defined, 23, 38
   in identity logic, 185–88
   introduced, 23–25
   in property logic, 124–34
   in relational logic, 166–69
   strategy, 39–40, 87–91

tests:
   logical truth, 66
   validity, 23
Formation rules, 29, 117, 337, 347–49
   deontic, 349
   predicate, 348–49
   propositional, 347–48

G

General statement, 119
   basic forms, 117
   compound, 120
General term, 116
Ghosts, mathematical ability of, 289
Groupers, 29

H

'Hence', 21
'However', 19

I

Identity:
   'is' of, 185
   negation of (≠), 184
   qualitative vs. numerical, 184–85
   symbol (=), 184
   symbolization, 184–85
Identity In Rule, 188
Identity logic, 185
Identity Out Rule, 186
'If and only if', 73
'Iff', 73–74
'If . . . then', 27–30
   contrasted with arrow, 260–61
Implication, contrasted with entailment, 28n.
Inchoate crimes, 319n., 320–23
Inclusio unius . . . , principle of, 238–39
Indirect discourse and intensionality, 312
Individual, 115–16
Individual constant, 115
Individual variable, 115
Induction, 7–8, 8n.
Inference rules (*see* rules of inference)
Inference to the best explanation, 8n.
Intension, 298–99
Intensional contexts, 299
   families of terms creating, 302–3
   recognizing, 302–4
   statement operators creating, 303–5
   stricture on variables in, 300–302

Intensionality and the law:
  and attempt, 320–23
  and consent, 297–98, 309–10
  and indirect discourse, 312
  and intent, 319–20
  and risk, 323–24
Intent and intensionality, 319–20
Interpretation, method of:
  construction hints, 154
  epistemic requirement, 154
  establishing invalidity, 153–54
  establishing nonequivalence, 155
  in identity logic, 188–89
  logical requirements, 154
  in relational logic, 170–72
Intransitivity, 169–70
Invalidity, 5
  of legal transactions, 7n.

L

Law and logic, 8–14
Legal reasoning, 333–35
  analogical vs. deductive, 333–35
  and analysis by instantiation, 199–201
  under-determination of rules, 335
Legislation, drafting and interpretation of, 11–13, 224–25, 236–39, 252–53
Logic, 3
  abductive, 8n.
  Aristotelian, 116, 116n., 120n.
  deductive, 5, 7–8, 8n.
  deontic, 336–46
  identity, 185
  inductive, 7–8, 8n.
  and the law, 8–14, 333–35
  predicate, 114, 116n.
  property, 163n.
  propositional, 18
  relational, 163n.
  standard (symbolic), 13–14
  syllogistic, 116, 116n.
  symbolic, 13, 14n.
Logical equivalence, 7
Logical truth, 66

M

Material implication, paradox of, 260–61
  avoidance strategies, 266–71
    burying conditional in predicate, 268–70
    replacing conditional by relational predicate, 266–68
    using universal quantification, 270–71
  conditionals with conditional antecedents, 261–62
  conditions contrary to fact, 264–66
  negated conditionals, 262–64
Mental state and intensionality, 309n.
Modus ponens, 31–32
  use of by Supreme Court, 31
Modus tollens pattern, 49, 51
Modus Tollens Rule, 82
'Moreover', 19

N

Natural deduction, 23n.
'Necessary and sufficient condition', 73, 236
'Necessary condition', 33–34, 235
Negation:
  defined, 47
  grouping principle, 47
  symbolizing, 47–48
Negative Pregnant Rule, 215–16
Nego (N:), 210
'Neither . . . nor', 62
'No', 117
'Not', 47–48
'Not all', 118

O

'Only if', 32–34
  symbolization guide, 33
Operator, statement, 303–5
'Or', 59
  inclusive vs. exclusive, 59–60

P

Paradox of material implication (*see* Material implication)
Particular affirmative statement, 117
Particular negative statement, 117
Performative utterance, 5n.
Pleading, 2, 214–18
  rules of disputation concerning, 218
  (*see also* Affirmative defense, Denial)
Polish notation, 350
Predicate, 116, 116n.
  letter, 115–16
  logician's and grammarian's concepts distinguished, 116n.

property, 116, 161
relational, 161
Predicate logic, 114, 116n.
 contrasted with propositional logic, 114
 formation rules, 348–49
 symbols, 115
Premise, 4
Premise-introducing terms, 21
Proof (*see* Formal Proof)
Property logic, 163n.
Proposition, 18, 301n.
 expressed by a complete phrase, 307–8
Propositional analogue, method of, 139–40
 determining equivalence, 144–46
 determining validity, 140–44
 extending the method, 143–44
Propositional logic, 18
 formation rules, 347–48
 limitations, 113–14
'Provided that', 27
Pseudo-existential, 301–2, 311
Pseudo-universal, 301, 312

Q

Quantification, 124–25
 existential, 124
 instance, 124–25, 166
 universal, 124
Quantifier:
 counterpart, 275
 existential [($\exists$x)], 115
 order, 163, 280–82
 scope (*see* Scope)
 shift, illicit, 171
 universal [(x)], 115
Quantifier-connective pairing, 117, 163–64, 280
Quantifier Exchange Rule, 129

R

Reductio ad Absurdum, 54
Refutation by logical analogy, 154–55
Relational logic, 163n.
Relational predicate, 161–62
Relational symbolization, 161–66
 and quantifier order, 163, 280–82
 and quantifier scope, 117, 165–66, 273–82
 in stages, 164
Relations, properties of dyadic, 169–70

Risk and intentionality, 323–24
Rules of inference, 374–75
 Ampersand In, 22
 Ampersand Out, 22
 Arrow, 82, 84
 Arrow In, 35
 Arrow Out, 30
 Assumption, 36
 Chain Argument, 82
 completeness, 76, 133
 Conjunctive Argument, 82
 consistency, 76, 81, 133
 Dash In, 50
 Dash Out, 52
 defined, 22
 De Morgan's Law, 82, 83–84
 derived, 81–91, 134
 Disjunctive Argument, 82
 Double Arrow In, 74
 Double Arrow Out, 74
 Double Negation, 82, 83
 Existential Quantifier In, 133
 Existential Quantifier Out, 127
 Identity In, 188
 Identity Out, 186
 Modus Tollens, 82
 part-line vs. whole-line, 84
 Quantifier Exchange, 129
 reversible, 84
 Universal Quantifier Out, 125
 Wedge In, 61
 Wedge Out, 64

S

Scope:
 connective, 29–30
 quantifier, 117, 165–66, 273–82
  and antecedents, 273–77
  conjunctions and disjunctions, 278–79
  and consequents, 277–78
Simple statement, 17
'Since', 21
Singular statement, 120
Singular term, 115
'So', 21
Solutions to starred exercises, 351–73
'Some', 117
Standard contradiction, 50
Standard symbolic logic, 13–14
 existential presuppositions, 291–92
 and extensionality, 299–300
Statement, 5
 compound, 17–18
 connective, 18

de dicto, 306–9
de re, 306–9
general, 119
operator, 303–5
simple, 17
singular, 120
Subjunctive conditional, 265–66
'Sufficient condition', 33–34, 235
Syllogism, categorical, 116n.
Syllogistic logic, 116, 116n.
Symbolic analysis, levels of, 189–90
Symbolic logic, 13, 14n.
   standard (*see* Standard symbolic logic)
Symmetry, 169–70

T

Tables:
   alternative logic symbols, 350
   basic general statement forms, 117
      alternative expressions for, 118
   deontic definitions, 337
   deontic principles and rules of inference, 338
   instance, 125
   intensional terms, families of, 303
   non-deontic analogues, method of, 342–43
   propositional analogue equivalence test, 145
   propositional analogues of predicate formulas, 140
   propositional analogue validity test, 141
   quantification, 125
   quantifier scope in conjunctions and disjunctions, 279
   rules of inference, 82, 374–75
   stages of a Dash Out predicate proof, 130
   symbolization:
      and antecedent quantifier scope, 277
      and consequent quantifier scope, 279
      identity statements, 186
      property statements, additional, 119
      property statements, basic, 117
      relational statements, common forms, 164
      relational statements, less common forms, 165
      universal statements, problematic, 248
   and universe of discourse, 119
   three methods for evaluating predicate arguments, 156
'Therefore', 21
'Thus', 21
Transitivity, 169–70
Truth, 7
Truth tables:
   asterisk, 98
   basic, 96–97
   brief, 102–4
      invalidity test, 103
      nonequivalence test, 104n.
   check marks, 99
   defined, 98
   and formal proofs, 101
   full, 98–102
      equivalence test, 101
      validity test, 100
   guide columns, 98
   number of rows, 102
   principles, 97
Truth-value, 96
Turnstile (⊢), 23

U

Universal affirmative statement, 117
Universal negative statement, 117
Universal quantifier [(x)], 115
Universal Quantifier Out Rule, 125
   shortcut, 167
Universal statements, symbolizing problematic, 245–53
   'all . . . and', 247
   'all . . . are not', 248
   'all . . . except', 246–47
   definitional truths, 249
   definitions, 248–49
   lacking quantifier terms, 250–51
   legislative, 252–53
   'none but', 245–46
   normative, 252
   'only', 185–86, 245
   'the only', 246
   in singular form, 249
Universe of discourse, 119–20
'Unless', 48–49
Use-mention distinction, 4n.

V

Validity:
   defined, 5

of legal transactions, 7n.
technical and ordinary meanings contrasted, 7
and truth, 5–7
Variable, individual, 115

W

Wedge (v), 59

Wedge In Rule, 61
Wedge Out Rule, 64
  strategy, 66
Well-formed formula (*see* Formation rules)

Y

'Yet', 19

TABLE OF CASES

Adams v. State (1982), 177
Asseltyne v. Fay Hotel (1946), 200
Bain v. Gillispie (1984), 203–4
Barber v. Vincent (1680), 230–32
Beachcomber Coins, Inc. v. Boskett (1979), 166
Bonkowski v. Arlan's Dept. Store (1968), 233
Canfield v. Tobias (1863), 243
Clarke v. State (1888), 323
Commonwealth v. Duchnicz (1914), 297, 309
Curnow v. Phoenix Ins. Co. (1896), 219
Denham v. Cuddeback (1957), 219
Edgerton v. Fitzmaurice (1884), 309n.
Edwards v. Aguillard (1987), 202
Endicott Johnson Corp. v. Perkins (1943), 211
Federal Crop Insurance Corp. v. Merrill (1947), 270–71
Fuller v. Preis (1974), 212
Garratt v. Dailey (1955), 79
Globe Woolen Co. v. Utica Gas & Electric Co. (1918), 263, 269
Grapin v. Grapin (1984), 74
Harisiades v. Shaughnessy (1952), 72–73
Helzberg's Diamond Shops, Inc. v. Valley West Des Moines Shopping Center, Inc. (1977), 344–45

Humphrey's Executor v. United States (1935), 239, 334
Irwin v. Ashurst (1938), 234
Kavafian v. Seattle Baseball Club Ass'n (1919), 234
Leathers v. State (1937), 202–3
Lindberger v. General Motors Corp. (1972), 234
Marshall Mfg. Co. v. Dickerson (1916), 219–20
McDonald v. Santa Fe Trail Transp. Co. (1976), 244
Morris v. Webber (1587), 70–71
Myers v. United States (1926), 333–35
N.L.R.B. v. Hearst Publications, Inc. (1944), 224–25
Nashville & K.R.R. v. Davis (1902), 227–28
Norman v. Emp. Sec. Agency (1960), 272
O'Callahan v. Parker (1969), 11, 226
Palsgraf v. Long Island R.R. (1928), 227
Parker v. Motor, Inc. Boat Sales (1941), 227
People v. Jaffe (1906), 320
People v. Kevorkian (1994), 165
People v. Roberts (1920), 165
Plessy v. Ferguson (1896), 67
Ploof v. Putnam (1908), 233
Polemis, In re (1921), 324

Ragan v. Merchants Transfer & Warehouse Co. (1949), 31
Regina v. Chapman (1958), 213, 324–26
Regina v. Dee (1884), 297
Regina v. Jackson (1891), 243
Regina v. Jordan (1956), 201–2
Regina v. Smith (David) (1974), 326–28
Screws v. United States (1945), 326
711 Kings Highway Corp. v. F.I.M.'s Marine R.S., Inc. (1966), 219
Sheppard v. Maxwell (1966), 106
Sherbert v. Verner (1963), 272
Sommersett's Case (1772), 243
Southern Pac. Co. v. Jensen (1917), 227
Spano v. Perini Corp. (1969), 211–12
Spencer v. Turney & Co. (1897), 220
U.S. Term Limits, Inc. v. Thornton (1995), 240
United States v. Florida East Coast Railway (1973), 213–14, 243–44
United States v. Fordice (1992), 181
United States v. International Minerals & Chem. Corp. (1971), 328–30
United States v. Joe Grasso & Son, Inc. (1967), 212
United States v. Lee (1882), 148
United States v. McLain (1987), 180
Valerie D., In re (1992), 176
Wagon Mound, The (1961), 324
Willinger v. Mercy Catholic Medical Center (1976), 220
Wingfoot California Homes Co. v. Valley Nat. Bank (1956), 219
Wong Yang Sung v. McGrath (1950), 212–13
Yakus v. United States (1944), 135, 199–200
Yniguez v. Arizonans for Official English (1995), 181

TABLE OF STATUTES AND RULES

United States Constitution
  Article I, 240–41
  Article III, 26
  Amendment V, 11–12, 258–59
  Amendment XIV, 26, 252
5 U.S. Code
  §553, 213–14, 243–44
  §554 (Administrative Procedure Act, §5), 212–13
  §556, 255
  §557, 213–14
  §704, 255
18 U.S. Code §242, 326
29 U.S. Code
  §152, 225, 318
42 U.S. Code §402, 236–37
  §1981, 244
Federal Rules of Civil Procedure
  Rule 8, 217, 219, 232, 295
  Rule 11, 217
  Rule 12, 2, 215, 295
  Rule 14, 212
  Rule 15, 272
  Rule 19, 272–73
  Rule 21, 264, 267–68
  Rule 23, 69, 241–42
  Rule 24, 69
  Rule 25, 243
  Rule 26, 234, 243
  Rule 37, 238
  Rule 55, 295
  Rule 56, 26, 218, 295
  Form 9, 2, 217
Federal Rules of Evidence
  Rule 201, 249
  Rule 607, 255
Conn. Gen. Stat. §45a-717(f)(2), 176
Fla. Stat. §794.011, 92
Indiana Code 35-44-3-4(a)(3), 324
Model Rules of Professional Conduct
  Rule 1.8, 239n.
  Rule 3.1, 218
Code of Judicial Conduct Canon 3(B)(3), 255
Model Penal Code
  §1.06, 250
  Article 5, 319n.
  §5.01, 321
Model Business Corporation Act §41, 262–64, 267–69
Geneva Convention, 312
Code of Canon Law, Can. 1090, 256